Vacuum Science and Technology
Pioneers of the 20th Century

Vacuum Science and Technology
Pioneers of the 20th Century

**A Volume Commemorating the
40th Anniversary of the
American Vacuum Society**

Edited by

Paul A. Redhead
*National Research Council
Ottawa, Ontario, Canada*

**Published for the
American Vacuum Society**

Library of Congress Cataloging-in-Publication Data
Vacuum science and technology: pioneers of the 20th century: history of vacuum science and technology 2 / edited by Paul A. Redhead.
   p. cm.
   "A volume commemorating the 40th anniversary of the American Vacuum Society."
   Includes bibliographical references and index.
   ISBN 978-1-56396-248-6
   1. Vacuum technology — History.   2. American Vacuum Society.
I. Redhead, P. A. II. American Vacuum Society.   III. Title: History of vacuum science and technology 2. IV. Title: History of vacuum science and technology two.
TJ940.V258   1993
621.5'5–dc20   93-28714
                                                        CIP

Contents

Contributors ix

Foreword xi

The American Vacuum Society at 40 1
Jack H. Singleton

Section I. Some Pioneers **23**

Cecil Reginald Burch *(1901–1983)* 25
W. Steckelmacher

Pieter Clausing *(1898–Present)* 28
H. Adam and W. Steckelmacher

Saul Dushman *(1883–1954)* 32
J. M. Lafferty

Wolfgang Gaede *(1878–1945)* 43
Günter Reich

Fernand Holweck *(1890–1941)* 59
P. S. Choumoff

Rudolf Jaeckel *(1907–1963)* 68
H. G. Nöller, H. L. Eschbach, and H. Pauly

Martin Knudsen *(1871–1949)* 75
H. Adam and W. Steckelmacher

Irving Langmuir *(1881–1957)* 79
George Wise

Marcello Pirani *(1880–1968)* 83
H. Adam and W. Steckelmacher

John Yarwood *(1913–1987)* 86
K. J. Close

Section II. Some Major Advances **89**

History of Vacuum Science: A Visual Aids Project 91
J. M. Lafferty

History of the Development of Diffusion Pumps 107
B. B. Dayton

Early Development of the Molecular-Drag Pump 114
Günter Reich

Comments on the History of High Vacuum Turbopumps 126
M. H. Hablanian

The Quest for Ultrahigh Vacuum (1910–1950) 133
P. A. Redhead

Ultrahigh Vacuum Technology at Westinghouse Research Laboratories (1947–1957) .. 144
Daniel Alpert

Section III. Reproductions of Historical Papers (1900–1960) **151**

About Slow Cathode Rays
Verhandlungen der Deutschen Physikalischen Gesellschaft—1908
Translated by G. Lewin 153
O. von Baeyer

The Emission of Electrons from Tungsten at High Temperatures: An Experimental
Proof that the Electric Current in Metals is Carried by Electrons
Philosophical Magazine—1913 156
O. W. Richardson

Theory and Use of the Molecular Gauge
Physical Review—1915 162
S. Dushman

A High Vacuum Mercury Vapor Pump of Extreme Speed
Annals of Physics—1915 180
I. Langmuir

Oil, Greases, and High Vacua
Nature—1928 184
C. R. Burch

The First Oil Condensation Pump: An Example of the Impact of One
Technology on Another
Chemistry & Industry—1949 185
C. R. Burch

The Flow of Highly Rarefied Gases through Tubes of Arbitrary Length
Translated from Annalen der Physik—1932 187
P. Clausing

High-Vacuum Gauges
Philips Technical Review—1937 198
F. M. Penning

Surface Phenomena Useful in Vacuum Technique
Industrial and Engineering Chemistry—1948 206
L. R. Apker

Some Recent Developments in Vacuum Techniques
Report on 10th Annual Conference on Physical Electronics—1950 208
D. Alpert and R. T. Bayard

Vacuum Factor of the Oxide-Cathode Valve
British Journal of Applied Physics—1950 214
G. H. Metson

Ultra-High Vacuum Technology
Vacuum Symposium Transactions—1954 219
D. Alpert

Electronic Ultra-High Vacuum Pump
Review of Scientific Instruments—1958 226
L. D. Hall

Contributors

H. **Adam** is retired from Leybold AG and lives at Heinrich-Esser-Str. 33, D-5040 Brühl, Germany.

Daniel Alpert is Director Emeritus at the Center for Advanced Study, University of Illinois, 912 West Illinois Street, Urbana, IL 61801.

P. S. Choumoff is a past-president of the French Vacuum Society and lives at Avenue Daniel Lesueur 9, 75007 Paris, France.

K. J. Close is professor of Microelectronics, University of Westminster, 115 Newcavendish Street, London, W1M 8JS, UK.

B. B. Dayton is retired from the Consolidated Vacuum Corp. and lives at 209 Hillandale Drive, East Flat Rock, NC 28726.

H. L. Eschbach lives at Ispralaan 37, B 2400 Mol, Belgium.

M. H. Hablanian is with Varian Vacuum Products, 121 Hartwell Avenue, Lexington, MA 02173.

J. M. Lafferty is retired from the General Electric Research Laboratories and lives at 1202 Hedgewood Lane, Schenectady, NY 12309.

G. Lewin translated the articles by G. Reich and O. von Baeyer from the German; he is retired from the Princeton University Plasma Physics Laboratory and lives at 21 Yale Terrace, West Orange, NJ 07052.

H. G. Nöller is retired from Leybold AG and lives at Hanrathstrasse 47, D 5303 Bornheim 3, Germany.

H. Pauly works at the Max Planck Institut für Strömungsforschung, Bunsenstrasse 10, D 3400 Göttingen, Germany.

P. A. Redhead is Researcher Emeritus at the National Research Council, Ottawa, ON K1A OR6, Canada.

Günter Reich is retired from Leybold AG and lives at Vochemestrasse 9, D-5000 Köln 51, Germany.

Jack H. Singleton is retired from the Westinghouse Research Laboratories and lives at 1184 St. Vincent Drive, Monroeville, PA 15146.

W. Steckelmacher is a Visiting Senior Research Fellow at the University of Sussex in England and lives at 177 Rusper Rd., Ifield, Crawley, West Sussex, RH11 0HT, UK.

George Wise is a communications specialist with General Electric Corporation, R/D, P.O. Box 8, Schenectady, NY 12301.

Foreword

On the occasion of the 30th Anniversary of the American Vacuum Society in 1983, a volume entitled *History of Vacuum Science and Technology* was published by the AVS to commemorate the anniversary of the Society. As the 40th Anniversary of the Society approached, the AVS history committee decided to prepare another historical volume for publication at the time of the 1993 national symposium to mark the 40th birthday of the AVS. The board of directors of the AVS agreed to finance this project and to present a copy to all those attending the 1993 symposium. It was also decided to continue these historical volumes as an occasional series devoted to the history of those branches of science and engineering of interest to the AVS; thus this volume becomes the second in this series. The first two volumes of the series are concerned with the history of the production, measurement, and use of vacuum. It is intended that succeeding volumes will be concerned with the history of other areas of science and engineering of interest to the AVS, including surface science, thin films, plasma science, vacuum metallurgy, semiconductor processing, and so forth.

This volume opens with a history of the AVS at its 40th birthday. The remainder of the volume is concerned with vacuum science and technology in the first part of the 20th century, i.e., approximately 1900 to 1960. Section I contains biographies, some brief, some more extended, of some of the scientists and engineers who made outstanding contributions to the advancement of the field. Section II contains historical reviews of some of the major developments of the period. Finally, Section III contains reproductions of papers from the period 1910 to 1960, which describe important new developments. Many of these papers are cited in the reviews in Section II.

The advances outlined in this volume form the basis on which modern vacuum technology is built. It is this ubiquitous technology that has made possible many major advances in recent times in science and engineering.

It is a pleasure to acknowledge the assistance of the members of the AVS history committee and the labors of the authors who so kindly agreed to prepare articles. The work of Dr. Gerry Lewin in translating three of the articles from the German is gratefully acknowledged.

P. A. Redhead

The American Vacuum Society at 40

Jack H. Singleton

I. INTRODUCTION

June 18, 1993, marks the 40th anniversary of the establishment of the Committee on Vacuum Techniques (CVT), the forerunner of the American Vacuum Society (AVS), and it is fitting how to review the history of the Society. The intent is to provide an overview of the entire history, with a fairly detailed account of the past 10 years, integrated with a briefer summary of the first 30 years. The summary is largely drawn from three previously published accounts, written by H. W. Schleuning (1953–73) (1), J. L. Vossen and Nancy L. Hammond (1973–83) (2), and J. M. Lafferty (1956–83) (3), which, with a few exceptions, provide greater detail for the years prior to 1983 than is given here. In addition, extensive information about the development of the short course program of the Society has been assembled by V. J. Harwood and A. W. Czanderna (4). The interested reader is urged to consult these earlier articles.

II. ORIGINS

On June 18, 1953, 56 people gathered in New York City to consider the need for a forum to discuss problems and applications of high-vacuum technology. Almost 60% of the attendees were associated with vacuum equipment suppliers; 18% had interest in vacuum tubes (radio tubes and lighting), and 9% in vacuum optical coating. Six days after the first meeting the Committee on Vacuum Techniques was organized, with the mandate to initiate a symposium and develop programs for education and for standards. The group was formally incorporated in Massachusetts on October 19, 1953. The first national symposium was held June 16–18, 1954, in Asbury Park, New Jersey, attracting 295 registrants from several countries for the presentation of 35 papers. The rapidity with which the group became productive is a measure of the dedication they brought to their new Society. The objectives of the CVT, developed in the first two organizing

meetings, remain fully consistent with those of the present AVS. The continued vitality of the AVS is a testament to the dedication of the original and present members.

Until 1961, the Society essentially comprised *only* the people attending the annual symposium, whose vote elected the officers and committee chairs for the succeeding year. At the 1957 symposium, the membership voted to change the name of the group to the American Vacuum Society, Inc., and the articles of incorporation were accordingly amended in January 1958. A revised Constitution and By-Laws were adopted in 1961, establishing the current procedure of a mail ballot for election of officers and directors. The membership of the Society has increased over the years, as shown by the year-end totals in Fig. 1. The AVS is interdisciplinary and serves a stable core group. A number of members have only a peripheral interest in the topics covered by the Society and leave when their interests change; membership figures indicate that the annual turnover rate can be as high as 24% of the membership (data for 1985).

Eligibility for membership in the Society has changed considerably over the years. As noted above, membership in CVT was limited to those people who attended the annual symposia. The constitution adopted in 1961 defined three classes of membership:

Members, defined as any individual expressing an interest in vacuum science or engineering.

Honorary Members, elected by vote of the board of directors, and ratified by the membership at business meeting.

Sustaining Members, limited to institutions and companies providing supplemental income to support the objectives of the Society. This category has apparently never been used.

In 1968, the Society moved its office into the American Institute of Physics (AIP) in New York City. Having been an affiliate member of AIP since 1963, AVS was now considering application for full membership in that organization. In preparation for this, the requirements for AVS membership were modified to be more consistent with those used by AIP member societies. Only the *Member* grade was changed, being divided into four categories, setting specific requirements for each:

Senior Member, requiring evidence of professional maturity, and endorsement by three Honorary or Senior members.

Member, requiring evidence of professional competence, and the signatures of two Honorary, Senior, or regular members.

Associate Member, open to anyone interested in the field of vacuum science and technology, and requiring one signature from an Honorary, Senior, or regular member.

Student Member, open to any full-time college student.

The Membership Committee quickly discovered a problem in fairly administering admission to the Senior Member category, and eventually suggested that it should be eliminated. In 1976, a referendum proposing such a change was approved by 77% of the membership, and the constitutional change was confirmed by a 90% majority in 1978.

In 1978 the *Emeritus Member* category was created, providing substantial benefits for long-term members retired from full employment. The eligibility requirements were reduced in 1983.

In 1992, AVS members approved the creation of the Fellows of the Society membership category: Eligibility requires a high level of professional achievement, and the number accepted each year is limited to 0.5% of the total membership; this grade of membership is administered by the trustees of the Scholarships and Awards Committee (see Section VIII), and requires confirmation by a majority vote of the AVS board.

The objectives of the Society, as adopted in 1961, were concerned with the support of vacuum science and engineering by the dissemination of information through symposia and publication, promotion and coordination of

research and education, development of standards, support of divisions and sections, provision of scholarships for student education, and administration of awards. These objectives stood unchanged until 1990, when the areas of interest and the types of activity were broadened to reflect more accurately the present character of the Society.

III. SYMPOSIA

The first annual symposium of the CVT in 1954 was the major focus of the group's activities, and it has remained a major part of the Society's activities ever since. The growth of the symposium and of the activities that are closely related to it, provides a measure of the development of the Society. In that first symposium, a total of 35 papers were presented over a 3-day period; there were no parallel sessions. The papers were subsequently published as transactions of the meeting, but were not refereed. The first seven symposia followed the pattern of the first, in that they primarily provided a forum for the presentation of papers, and for the annual business meeting, at which the Society leadership was elected.

The eigth annual symposium, held in Washington, DC, in 1961, was a particularly important occasion since it was combined with the Second International Vacuum Congress, organized by the AVS on behalf of the International Organization for Vacuum Science and Technology (IOVST),[1] resulting in a meeting roughly double the size of the preceding symposia. The particular importance of this combination of two nominally separate symposia arises from the fact that, while the annual AVS symposia are designed to be essentially self-supporting, the triennial International Congress must, in addition, provide the major source of income for the IOVST (later, IUVSTA). The added number of papers at such a congress increases the cost of publication of the proceedings very substantially. Further, in 1961 we were still in the era where the IOVST required that the abstracts be accepted in *any* of the three languages of the conference—French, German, or English—while the manuscript could also be submitted in any of these languages. The sum of the several financial requirements placed a severe burden on the operation of the meeting, required financial support from outside agencies and, almost inevitably, an increase in the registration fees. Despite all of these problems, the meeting was an unqualified success, bringing together a great wealth of talent.

The first Vacuum Show to be sponsored by the AVS, held in conjunction with this IOVSTA Congress, provided a showcase for the newest in vacuum technology; it was also a source of additional revenue to the meeting. The fact that this first show had been so

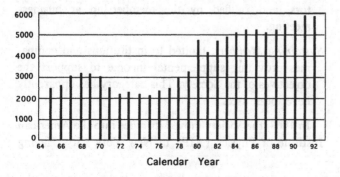

FIG. 1. AVS membership totals (year end).

---

[1]The IOVST was replaced by the International Union of Vacuum Science, Technique and Applications (IUVSTA) in December 1962.

successful encouraged the AVS to repeat the experiment, first in alternate years, and, starting in 1965, every year. This broadening of the scope of the meeting brought with it some problems. At some early shows there were complaints from exhibitors that competitors were exhibiting their wares in hotel suites in order to avoid renting a booth in the formal exhibit area; in truth, most suites housed only very popular hospitality rooms. But of greater consequence was the fact that the sites for the symposium had already been booked several years ahead, and some could not provide adequate exhibit space. This came home to roost in the Pittsburgh Hilton in 1968, where the exhibit had to be squeezed into every available space, sometimes with totally inadequate electrical power. This resulted in serious power outages to some sections of the exhibit, causing substantial disruption. It was fortunate that, starting in 1965, the shows were managed by the American Institute of Physics, and the experience of their representative, Ed. Greeley, was fully tested in blunting the impact of these problems so that they did not fall upon the rather small local arrangements group of that time—a total of eight people. Over the years the vacuum show has continued to grow, and at present it is the largest vacuum equipment show held in the United States. The size of this show, and the specialized facilities it requires, are such that there are limited venues throughout the country that can satisfy the total requirements of the annual symposium. One point of importance is that the vacuum show provides a valuable source of information on state-of-the-art equipment and, as such, is considered by the Internal Revenue Service to provide an educational service consistent with the nonprofit status of the Society. The negotiation of equipment sales is therefore prohibited at the show.

The publication of the Transactions of the Symposia was continued through 1963. These Transactions contain a great wealth of valuable information, but there was considerable pressure to provide for refereed, *archival* publication, both of the symposium papers and of other general papers dealing with topics of interest to the Society.

With the publication of the first volume of the *Journal of Vacuum Science and Technology* in the fall of 1964, it was hoped that most papers given at the symposium would be submitted to the new journal. As a consequence, the Transactions were discontinued and only the abstracts for the symposium papers were included in the *Journal*. This practice continued for one further year, but it became apparent that, while a substantial fraction of the contributers were happy with the submission of a brief abstract, most attendees preferred a more substantial account of the material that had been presented. Consequently for 2 years, 1966–67, each author was asked to provide a two-page extended abstract of the presentation, including figures, and these were reproduced by photo-offset to provide a hardbound copy, substantially representing the material given at each symposium. The compromise suffered from the fact that the abstracts were not published (in order not to compromise publication of the material in an archival journal), and were unavailable to those who did not attend the symposium.

The symposium held in 1968 may perhaps be considered as the prototype of the present format of the meetings. It is the meeting that first combined technical sessions, a vacuum show, and a short course program. This was also the first meeting that benefited from the presence of an AVS staff member in the person of Nancy Hammond, the new executive secretary. In addition, starting with this symposium, a radical new approach was adopted, in an attempt to ensure publication of all papers presented orally at the symposium. Contributions would be selected for inclusion in the symposium based upon the submission of a 150-word abstract. If an abstract was accepted, the second requirement for presentation at the meeting was the submission of a complete manuscript (3000-word limit for contributed papers), with a commitment to publish either in *JVST* or *in any other archival journal*. Manuscripts for *JVST* would be submitted to standard *JVST* reviewing prior to the symposium. For those papers to be included in the special *JVST* issue, final revision and refereeing would be completed at the symposium, permitting publication in the January/February issue of the *Journal*. Note that submission of the paper was sufficient to permit an oral presentation; acceptance of a paper by the reviewers could affect only the publication. The procedure worked remarkably well, providing for timely publication of the proceedings, but requiring a great deal of cooperation from all concerned, including the authors, who had to print a reasonable account of what had been promised in their earlier (occasionally too-optimistic) abstract, and the reviewers and *Journal* editor, who were working against tight deadlines. A major problem with this procedure was the need for early submission of the abstracts; post-deadline sessions at the symposium were conceived later, as a means of accommodating late-breaking developments. With time, there has been a move away from including a record of every symposium paper in the *Journal*, and a relaxation in the word limit for contributed papers. Since 1988 only full papers have been included in the proceedings; the maximum acceptable length has been increased by one page and, at the same time, the practice of including extended abstracts has ended.

The growth of the symposium from 1968 is shown in Table 1, which lists data for the symposia from 1983 through 1992. The attendance figures for the technical sessions, short courses, and exhibits are shown separately, in so far as they are recorded. The measure of the growth in the meeting is seen by comparing the figures for 1954 with those for 1991: Attendance for technical sessions

TABLE 1. *The AVS National Symposia: 1967–1992*

Year	Location	Attendance Symposium	Short courses*	Exhibits†	Number of papers	Number of exhibit booths	Program Chairman	Local Arrangements Chairman
1967	Kansas City, MO	841	0	—	~84	92	Edward E. Donaldson	Paul J. Bryant
1968	Pittsburgh, PA	831	~140	—	84	71	William J. Lange	Jack H. Singleton
1969	Seattle, WA	826	~80	—	~125	82	J. Roger Young	Rolland R. LaPelle
1970	Washington, DC	591	100	—	~131	77	J. Roger Young	Joseph T. Scheurich
1971	Boston, MA	1081	~100	~400	310	94	Jack H. Singleton	Marsbed H. Hablanian
1972	Chicago, IL	711	102	—	~153	75	N. Rey Whetten	John S. Moenich
1973	New York, NY	671	131	—	139	88	Robert B. Marcus	Jerome J. Cuomo
1974	Anaheim, CA	637	120	—	169	83	Leonard C. Beavis	Joseph P. Davis
1975	Philadelphia, PA	740	118	—	158	95	Robert Rosenburg	Thomas A. Jennings
1976	Chicago, IL	777	132	—	179	88	John L. Vossen	Brooke V. Thorley
1977	Boston, MA	1351	283	—	193	107	William R. Bottoms	John J. Sullivan
1978	San Fransisco, CA	1518	386	—	282	104	Edward N. Sickafus	Howard G. Patton
1979	New York, NY	1737	399	—	251	127	Raymond S. Berg	Tony M. Messina
1980	Detroit, MI	1731	502	—	287	128	John A. Thornton	James M. Burkstrand
1981	Anaheim, CA	1202	450	700	395	154	William D. Westwood	Joseph P. Davis
1982	Baltimore, MD	1311	450	745	417	197	Lawrence L. Kazmerski	James S. Murday
1983	Boston, MA	1557	458	964	397	230	Leonard J. Brillson	David J. Diamond
1984	Reno, NV	1442	542	150	471	258	Galen B. Fisher	Howard G. Patton
1985	Houston, TX	1359	675	400	549	265	Jerry M. Woodall	Alex Ignatiev
1986	Baltimore, MD	1950	445	649	858	235	Theodore E. Madey	James S. Murday
1987	Anaheim, CA	1834	465	799	637	297	J. William Rogers, Jr.	George E. Aguilu
1988	Atlanta, GA	1739	407	293	646	272	John R. Noonan	John F. Wendelken
1989	Boston, MA	2091	426	805	662	335	H. F. Dylla	David J. Diamond
1990	Toronto, Ontario	1692	421	440	654	338	Frank Jansen	Frank R. Shepherd
1991	Seattle, WA	1490	304	326	913	321	Richard J. Colton	Eric M. Stuve
1992	Chicago, IL	1466	405	492	925	299	John H. Weaver	William D. Sproul

*Number of people taking short courses. Many take more than one course, so these figures are smaller than those plotted in Fig. 2.

†Number of people registered for the Vacuum Show only. No data available prior to 1981.

was 295 compared to 1692; 35 papers were given in single sessions, as compared to 912 papers given in as many as 9 parallel sessions.

In 1971 the AVS again served as host for an International Union of Vacuum Science, Technique and Applications (IUVSTA) Congress. By this time the Society had a vigorous and demanding Surface Science Division (SSD). The SSD was in competition with a number of other groups in attracting the surface science community, and it was therefore essential to provide the best possible setting for surface science papers. Under the dynamic leadership of C. B. Duke, a separate conference was developed within the Congress program, entitled the 1971 International Congress on Solid Surfaces (ICSS) and, most significantly, sponsorship of the International Union of Pure and Applied Physics was secured for this meeting. This concept was accepted by IUVSTA, as a result of the persuasive gentleness of Luther Preuss, the General Chairman of the Organizing Committee, and was a great success, accounting for about half of all the papers presented at the combined meeting. It is gratifying to note that the ICSS has continued in association with the International Vacuum Congress. From the point of view of logistics, the 1971 conference program turned out to be easier to organize than had been the case in 1961. In the then-current economic climate, it was necessary to minimize all costs, such as printing the final program, while essentially complying with the authors' option of submitting the abstracts in any of the three official congress languages. Fortunately, during the previous year, for the first time, the AVS program had been efficiently assembled using camera-ready copy submitted by the individual authors, thus avoiding any retyping or proofreading. Permission was given by the IUVSTA to use a similar procedure for the assembly of the Congress program, but each contributor was required to submit an abstract in English, since the reviewing committee was entirely from the US, and was given the option of supplying an additional abstract in either French or German, for inclusion in the book of abstracts. The program committee received 380 abstracts, approximately 40% from outside North America, but only four authors included an additional abstract.

By 1976 the annual AVS symposium stretched over a 3 1/2-day period, with three parallel sessions, morning and afternoon, and some evening sessions. In order to accommodate the ever-increasing number of papers that were submitted, the Society's first poster session was held. The call for papers carefully noted that a poster session was an option being considered, and all authors were asked to specify, on their submitted abstract, if they did *not* wish it to appear in such a session. One poster session, "Control of Thin Film Properties," was scheduled in the final program; it included a total of 14 papers and, to the great relief of the Program Committee, the session was generally viewed as satisfactory. Since 1976 poster sessions

have continued annually, providing an essential expansion mechanism to accommodate the steadily increasing numbers of submitted papers. Considerable effort has been expended to increase the acceptance of the poster sessions; in 1982 the authors gave brief (3-minute) oral presentations at the opening of each poster session, and wine and cheese were laid out to sustain the weary. A survey in 1983 confirmed that the poster sessions had achieved a high level of acceptance, with 80% of the participants responding that they should be continued, and that they personally would be willing to repeat the experience. In the 1991 symposium, fully 45% of the contributed papers were presented in poster sessions, which were well attended.

In 1986 the AVS was once again the host for the IUVSTA Congress; the 10th International Congress and the 6th ICSS, were combined with the 33rd National Symposium of the AVS. The meeting was held in Baltimore, where the combination of the unique location on the Inner Harbor and a superbly operated convention center provided an excellent venue. It is interesting to compare some of the statistics for this meeting with those of the one hosted 15 years earlier in Boston. The number of papers accepted for presentation, including invited papers, was 860 as compared to 318, with a rejection rate of about 13% compared to 20%. About 44% of the papers were from outside North America, quite close to the 40% in 1971. Approximately one third of the papers were presented in poster sessions, an innovation not yet in place in 1971. One disturbing feature of the meeting was that 36 authors did not appear to present their papers, including three invited speakers; the authors of about five contributed papers were "no-shows" in 1971.

The very large growth in the annual symposium was already a concern in 1981, but the board specifically disapproved the possibility of holding two national symposia in a given year, out of concern for the loss of synergism between the diverse groups within the Society, and because of logistical difficulties. In an attempt to provide a mechanism for expansion separate from the symposium, an increase in the number of topical conferences was specifically encouraged. Up until this time, many conferences had been sponsored or co-sponsored by the Society, though never under the specific designation of "topical." The first meeting to be scheduled with such a designation, a "Topical Symposium on Sputtering," sponsored by the Thin Film Division, was held in April 1984, immediately preceding the Vacuum Metallurgy Division's 11th International Metallurgical Coatings Conference. At the 1985 AVS town hall meeting, which was organized by the Long Range Planning Committee, the possibility of grouping several topical conferences as a spring national symposium was raised.

In the spring of 1987 two topical conferences were planned to coincide with the annual symposium of the Northern California Chapter, and in 1986 three topical conferences and one workshop were scheduled around the Baltimore annual symposium; yet another was in conjunction with the 1987 annual symposium. Given this rapid expansion of topical conferences, the board established a new committee in November 1986, under the chair of H. F. Dylla, to coordinate their development. The major concerns were to avoid excessive overlap, to match the number of conferences to the perceived needs of the scientific community, and to stay within the available resources of the Society. Topical conferences have clearly become an important mode for expansion beyond the limits imposed by the crowded schedule of the annual national symposium. The Society currently sponsors or co-sponsors an average of 20 meetings each year, but the practice of scheduling such conferences around the national symposium surely exacerbates the feeling of information overload.

Many other symposia and conferences have been held over the years, sponsored in whole or in part by the Society or by its component divisions and chapters. For example, the Vacuum Metallurgy Division (VMD) of the AVS has been active since its inception in 1961 in supporting a variety of conferences, including the Vacuum Metallurgy Conference (1962–67), the Structure-Property Relationships in Thick Films and Bulk Coatings conferences of 1974 and 1975, and, starting in 1976, the International Conference on Metallurgical Coatings (ICMC), which developed into *the* major conference in this field. In 1990 the ICMC was preceded by a thin-film topical conference, and in the following year the scope of ICMC was broadened to include thin-film technology within the overall meeting. The 1967 vacuum metallurgy conference was organized as the International Vacuum Metallurgy Conference and became the forerunner of a separate, and continuing, series of such conferences. It could be argued the VMD has comprised a fairly distinct subdivision of the Society, enjoying close cooperation when dictated by a particular topic.

Many chapters of the Society hold an annual symposium of one or more days, in some cases serving the local community, and in others being truly national in scope. The combination of such symposia, with an exhibit and a group of short courses, provides a complete group of services to the community, often generating surplus funds that can be used to support other chapter activities, such as are discussed in Section VII.

IV. PUBLICATIONS

The first major publishing activity of the CVT and AVS was the transactions of the first ten national symposia. Then, in September 1964, the *Journal of Vacuum Science and Technology* was established (1). Daniel Alpert was the first editor, followed by Franklin M. Propst (1967),

Paul A. Redhead (1970), and Peter Mark (1974). During 1979 Peter Mark became terminally ill and, unknown to the majority of the membership, John L. Vossen assisted in maintaining the editorial work until the editorial duties were assumed, in January of 1980, by Gerald Lucovsky, the current editor. As is described by Vossen (2), the *Journal*'s growth has been primarily driven by the publication of the proceedings of the annual AVS national symposium, and of other conferences. In addition to regular articles, and shop notes, critical reviews have been actively solicited for the *Journal* since 1970. In 1983 the *Journal* was divided into two parts, *JVST-A*, devoted to topics in vacuum, surfaces, and films, and *JVST-B*, devoted to microelectronics. Lucovsky assumed the position of editor-in-chief, while remaining the editor of *JVST-A*. Thomas Mayer became the first editor of *JVST-B*, to be followed in 1990 by Gary McGuire. In July 1990 the scope of *JVST-B* was further expanded to emphasize nanometer structures, their processing, measurement, and phenomena. The continued expansion of the *Journal* imposes a large financial burden upon the Society, and the sources of income are limited. The budget for the *Journal* is currently about 40% of the total Society budget.

The journals are provided to all members of the Society, and a fraction of their membership dues (equal to the cost of printing, binding, and postage) is allocated to the *Journal* account. The other main sources of income allocated to the *Journal* are nonmember subscriptions, page charges, and advertising; only the page-charge income increases when more pages are published, whereas the expense increases almost linearly with those pages. All authors and/or their institutions are requested to honor the page charge for any published article, and the Society is fortunate that the honoring of these charges is high. Consequently it has been possible to maintain a policy that the acceptance of a paper for publication is decided solely on its merits, and publication proceeds as expeditiously as possible, irrespective of payment of page charges, a procedure that recognizes the fact that such payments are not always within the control of the author. The financing of the *Journal* has been controlled by adjusting the member and nonmember dues, the page and advertising charges, and the number of pages printed. To smooth out such adjustments, the budget is averaged over several years, rather than annually, and the Society maintains a reserve fund roughly equal to the annual cash flow to cover against a major miscalculation or an unpredictable external event. From such financial considerations flow the limitations imposed on the number of pages allocated to a manuscript for the annual symposium, or the allocation of the number of pages in any agreement to publish the proceedings of a particular meeting.

The problem of having to limit the size of a meeting because of the associated cost of publishing the transactions came to a sudden climax at the 1984 meeting in

Reno. During the week, the Electronic Materials and Processing Division Executive Committee had decided that their program for the 1985 symposium could best be served by the introduction of up to five half-day topical meetings. However, it was realized that such an expansion would devastate the *JVST* publication budget. In a number of ad hoc discussions, a proposal was developed that the Society embark upon a new monograph series, utilizing publishing from camera-ready copy supplied by the authors, and publishing in the already-established AIP Conference Proceedings series. Such a format would facilitate rapid publication in a well-known venue at a fraction of the cost of the *Journal*. Plans to convene a special board meeting were made late in the afternoon of the last full day of the meeting, and the somewhat-embarrassed secretary went off to find the President, Ed. Sickafus, to invite him to a meeting which he alone could convene! In that 1 hour meeting the *AVS Topical Conference Monograph Series*, a subset of the AIP series, was born. There are at present 13 volumes in publication, and several of these volumes have been very successful, both in terms of sales and in the frequency with which they have been cited in the literature. However, the series has been perceived as less desirable than publication in the *JVST* series. To remedy this perception, starting with the topical conferences at the 1992 Annual Symposium, the articles will be subjected to the same standards of review as are articles in *JVST* (at least two reviewers), and will be electronically typeset. Manuscripts will be accepted primarily on computer disk, and figures and pictures will be input by electronic scanning. Such changes should serve to bridge the gap between the original monograph format and that of the conventional typeset *JVST* volumes, and may well provide the model for future publication procedures.

At the 1987 national meeting, an ad hoc committee focused on the needs of the surface science community for a comprehensive atlas of reference spectra, and in 1988 the Applied Surface Science Division (ASSD) gave formal sponsorship to creation of a surface science spectral database. Input from the surface science community was solicited through the medium of surveys and joint ASSD/ASTM workshops and through ASTM task groups; on the international level, a VAMAS Surface Analysis Task Group effort (Project 24) was established. The resultant plan, for the establishment of an AVS publication, *Surface Science Spectra*, was approved by the AVS board in 1990. Publication was under the direction of a board having Lee as the chair, and Bryson and McGuire as co-editors. The first issue was published in June 1992, with an initial list of 50 subscribers.

In the vacuum field, as in most other technical areas, there are a frustrating number of books, long out-of-print, which remain of enduring value. To alleviate this problem, the Society has embarked upon the production of the *AVS Classic Series in Vacuum Science and Tech-*

nology, the first four volumes of which were scheduled for issue in the spring of 1993.

The Education Committee of the Society has, through the years, been an important sponsor of various publications. Their monograph series provides concise accounts of various topics, some developed as adjuncts to specific short courses. In the past 10 years, six new or revised volumes have been issued, and a bibliography, *Silicides for Microelectronics Application*, distributed on floppy disks, was completed in 1986.

In 1960 the Committee produced two 20-minute films, entitled *Leak Detection* and *Fundamentals of Vacuum Technology*, with considerable financial support from the vacuum industry; these are still in use, having recently been transferred to video format. Two extensive courses, comprising sets of multiple video tapes, *Sputter Deposition* and *Ion Beam Processes*, presented by W.D. Westwood (1986), and *Properties of Vacuum System Materials*, presented by W. F. Brunner (1986), have been produced more recently for the Society at the Lawrence Livermore Laboratories, and have been well received.

In 1988, at the Atlanta national meeting, two topical symposia, *Probing the Nanometer Properties of Surfaces and Interfaces* and *High T_C Superconductivity: Thin Films, Devices, and Characterization*, were videotaped, as was a workshop, *Calibration and Use of Mass Spectrometers for Partial Pressure Analysis*, in 1989, providing a vivid record of the meetings. The taping of a meeting in real time can be done at relatively reasonable cost, since there is no recourse to retaping any sections; it provides an accurate record of the proceedings and, as such, any discontinuities in presentation are as readily accepted as they are in the live presentation. In 1992 the concept of such real-time taping was applied to the production of a video course on capture pumps, markedly reducing production costs, and perhaps retaining some degree of spontaneity, such as is often present in the live presentation. Whether this approach will succeed must be determined by the response of the intended users. The innovative use of videotapes expands the availability of informational/educational products to those who cannot attend an actual presentation, but it will take some time before the full potential is clear.

The AVS newsletter has developed into a timely means of communication with the membership. The first one appeared in 1959, and until 1971 as many as four issues were printed each year, but with some irregularity; unfortunately the AVS records do not contain a complete set, and so the exact pattern is not documented. Publication became very spotty in 1972, and this precipitated the establishment of a regular schedule of four issues, starting in 1973. It is interesting to consider the factors that placed emphasis on a timely publication of the newsletter. In the first few years of publication, *JVST* was used to record many newsworthy AVS events, including announcements of meetings and awards, and the membership roster—items of interest mainly to the Society membership, but having no value in the transfer of technical information. Analysis showed that the size of the *Journal* was expanding, not only because of the technical articles, but also, to a surprising extent, by the publication of such "Society" material. For example, in 1969 the *Journal* contained the proceedings of both the Annual Symposium and the International Conference on Thin Films (651 pages) plus regular articles (217 pages), all of which were supported by page charges honored at a 73% average rate.[2] But there were 214 pages devoted to Society business, with no supporting page charge, an expensive proposition! This analysis motivated the move to more economical modes of publication for much of the Society business, specifically to the newsletter for general information, and to a separate booklet for the roster; these changes permitted more timely publication of both. The newsletter has gradually increased in size and in sophistication, incorporating a series of changes, such as greater use of photographs. During much of this evolution it was edited by J. Lyn Provo as a solo operation; since 1989 Donna Bakale-Sherwin has joined Provo as a co-editor, and there has been expanded support to facilitate publication.

V. SHORT COURSES

Early in 1968 the Instrument Society of America, approached W. J. Lange, the program chairman for the upcoming national symposium, to suggest that ISA and AVS co-sponsor a 2-day workshop on basic vacuum for technicians, to take place at the symposium. The proposal followed a suggestion by John Kurtz of IBM, who approached ISA with the offer to develop such a course. The ISA suggestion was accepted by the board in May, with the proviso that the Education Committee assure themselves that a quality course would result. A 2-day course was announced by AVS and ISA, with Kurtz as the instructor, to be assisted by one or two outside speakers. It was to be self-supporting, with a $30 fee, and an announced limit of 40 people; the arrangements would be handled by the ISA. Applications to attend the course registration reached 140, with more turned away, and the course was run in two sessions. The majority of the attendees were technicians and engineers, mainly drawn from the AVS, with relatively little interest from the ISA membership. The course demonstrated the need for such training and was clearly a success. There are no records to show whether such courses were given earlier, at the local level, but the AVS board minutes report that similar training courses were held during 1968 in Dallas, Cleve-

land, the Delaware Valley, New England, and the Seattle area. Following the Pittsburgh course, ISA cooperation with the AVS continued. In 1969 courses were co-sponsored by ISA with the New England section, and with the AVS at the national symposium in Seattle. In April 1970 a New York Thin Film Chapter/ISA course had an overflow registration, and a further course was proposed to be held in New York in late 1970; the minutes do not record if it was ever held.

The course offered at the Seattle national symposium in 1969 was advertised as a 2-day course to be taught by Gene Culver of the Oregon Technical Institute, with the assistance of William Brunner and Norman Milleron; the course text was *Practical Vacuum Techniques* by Brunner and Batzer. Once again the course was a great success, with an attendance of around 80, but there were some difficulties in coordinating the arrangements made by ISA in Pittsburgh with those of the AVS Local Arrangements Committee in Seattle, and this was the last national course operated in conjunction with ISA.

The Education Committee initiated its short course program in 1970, with Vivienne Harwood serving as the coordinator, a function which she served, with great effectiveness, throughout the time that the Education Committee had jurisdiction over the program. This first short course program totally run by the Education Committee at the 1970 national meeting in Washington, attracted 110 registrations in the middle of a deeply depressed economy. There were two parallel vacuum technology courses, lasting 2 1/2 days, one for technicians, the other an advanced session. Both were limited to 50 people. Different instructors taught specific topics in these courses, and a total of 11 were used for the two courses. Currently, a single instructor would teach such a course.

The steady expansion, both in the number of short courses offered by the Education Committee, and in the attendance at these courses, has been described by Harwood and Czanderna (4), for the period through 1981. By 1983 the annual cash flow of the short course program was approaching $500,000, or over 25% of the total AVS budget, and in this year the responsibility for the program was separated from the Education Committee and entrusted to a Short Course Executive Committee (SCEC), reporting directly to the board. The first committee, chaired by Frank Ura, included Paul Holloway, Don Mattox, Vivienne Harwood, and Marion Churchill.

It was planned that the short course programs offered by this committee should continue to expand at a substantial rate, and in general this has been the trend. But, in order to understand the growth of the program, it must be realized that a substantial number of courses are given at the local level, and only some of these are co-sponsored by the Committee. First, let us consider the short courses presented at the national symposia, which

---

[2]Page charge honoring was 88% in 1968; at international meetings, the number of papers from non–U.S. institutions increases, and these normally have a somewhat lower honoring rate.

since 1970 have been run at the national level. Data for the entire period is plotted in Fig. 2,[3] showing that major growth started in 1977. After the formation of the SCEC, in 1983, the number of registrants for the national courses surged to 804 in 1984, but decreased thereafter, dropping to an average of 570 for the years 1985 through 1991, close to the average attending in the years 1981 to 1983. Clearly, from 1978 to date, the attendance at the symposium courses has levelled out, with an average of around 590.

Beyond the courses at the symposium are those run at the local level. The number of these expanded very rapidly, starting in 1968. Many chapters have always run their own programs, sometimes drawing upon the same pool of instructors as the national organization, and at other times using different instructors, frequently from within their local organization. For a number of years, the National Education Committee was a co-sponsor of some chapter courses, and in 1979 began to organize a larger group of courses each spring in association with a local organization, with the intent of complementing the very large offerings at the fall symposium. The Committee now regularly cooperates with several chapters, divisions, and non–AVS organizations in the operation of their short-course programs; in such cases the national organization provides instructors and other logistical support, insuring the offering against financial loss, and sharing a portion of any gain. The attendance at SCEC co-sponsored courses, excluding the fall symposium, is plotted in Fig. 2, and shows a general upward trend, but with rather wide swings. These swings are partially a result of changes in the economy, but they also reflect the changes in the number of chapters that have chosen to accept co-sponsorship from year to year.

Harwood (4) has assembled data for the years 1979, 1980, and 1981, that show the number of courses in the combined spring and fall Education Committee short course offerings were only 51%, 62%, and 50%, respectively, of the total number of courses offered throughout the Society in each of these years. Unfortunately this study is the only detailed information available for the attendance at such courses; they clearly add up to a very extensive offering.

Local groups can offer courses tailored to meet their specific needs, and such courses frequently provide income for the support of other chapter activities. Chapters may choose to operate courses at minimal cost in order to serve the broadest possible base, or teach courses that are different in content from the established courses at the national level to address a specific local industry. As one example, consider the technician level, the group at which the very first AVS courses were aimed. From the

beginning many engineers have attended such courses, and have seen a great expansion in the courses that serve them. However, technicians, especially those who are just starting to work in the vacuum field, have less access to the national courses, unless they happen to live in the immediate vicinity of a meeting; few employers are prepared to send an inexperienced technician to a 4- or 5-day course, often in an expensive hotel. Local groups have been active in addressing this need. Thus the Northern California Chapter offered annually a popular course entitled "Vacuum Basics for the Novice." In 1985 this comprised two parallel courses, one for the nontechnical novice, and the other for the technically oriented novice. Similar basic courses have been held in many other chapters; Gerry Lewin taught a 2-day nonmathematical course for a number of years, based upon his 1987 Education Committee monograph, at both the chapter and national level. The vitality of the short-course program is a prime example of the energy displayed by the members of the Society. The courses are generally of a high standard, and many of the instructors are at the peak of their expertise in the area they teach. Written appraisals from the course attendees, used to monitor course content and presentation, are generally enthusiastic.

A number of short courses have been held outside the United States. Possibly the first of these courses was held in March 1982 in Mexico City, where the New Mexico Chapter ran a 3-day course in vacuum technology with an attendance of 55 students. In October 1984, the national organization offered two courses, again in Mexico, one on leak detection, and the other on thin-film

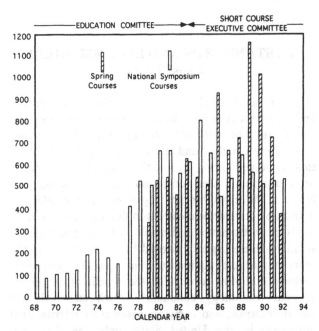

FIG. 2. Attendance at the Spring and Fall (Symposium) short courses sponsored by the Education and Executive Short Course Committees.

---

[3]The attendance plotted in this figure reflects the total number of "seats" occupied at the courses. Since many people take more than one course, this number is greater than the attendance at the short courses which is used in Table 1.

techniques. Courses have since been offered in Brazil and in Canada.

For many years, the national short-course program has provided core courses in basic vacuum technology and in thin-film techniques, repeated several times each year and serving a very wide audience. These courses are taught by a number of instructors, often using more than one instructor for a multiple-day course, and there exist several versions of some core courses. In 1991 the instructors for the vacuum technology courses spent a day and a half in reaching consensus on the detailed content of each day of these courses, to ensure that there would be no gaps in coverage between the versions used by different instructors. The core courses have frequently been over-subscribed, and they usually provide a substantial income to the total program. Beyond these are the shorter, specialized courses, which may not attract a large attendance, but the financial surplus from the core programs permits these courses to operate on quite limited attendance, while still allowing the entire national program to operate with a surplus. At the first presentation in 1968 there was only a single course; at Chicago in 1992 there were 42.

For many years the board maintained a policy that all AVS courses should be fully open to the public. In December 1983 this policy was relaxed, permitting courses to be given to restricted audiences at government laboratories, and in May 1984 approval was extended to permit private courses to be given for any organization, the so-called near site or on-site courses. In 1985 six such courses were successfully held; a seventh was cancelled very close to the scheduled time. Such courses do not at present constitute a large segment of the total short-course program.

VI. STANDARDS AND RECOMMENDED PRACTICES

One early emphasis of the Society was the development of tentative standards for a wide variety of procedures and equipment. The work expended on these standards was exceedingly detailed, and it required considerable intestinal fortitude to bring them to the stage where they could be offered to an often-contentious, and even nitpicking, board of directors for final approval. The satisfaction derived by committee members in developing these standards came from the undeniable fact that many of the standards were urgently needed by industry and by the users of vacuum equipment, and thus the committee performed a vital service for the benefit of the entire vacuum technology community. Such standards are purely informal, and carry no national or international authority; in the United States only the American National Standards Institute (ANSI) can establish such standards. By 1973, 26 tentative standards had been

published, but following publication of two standards concerning helium leak detector calibration and bakable all-metal valves in *JVST* (1973), the committee activity remained very low for a number of years. In fact, at the May meeting of the AVS board in 1984 there was sentiment to abolish the committee altogether, and it was indeed fortunate that this sentiment triggered, during a coffee break, a most persuasive argument from Don Santeler, a past president of the Society and a spokesman for the vacuum technology segment of the Society; by a fortunate coincidence, he was present, not at the board meeting, but at a concurrent chapter meeting. As a consequence of his advocacy, at the next board meeting John J. Sullivan was asked to head up the Standards Committee. He proposed changes in the mode of operation of the committee, which were soon published in a chapters and divisions newsletter, soliciting comment. The procedures proposed were in part a response to the possibility of financial exposure to the Society as a consequence of promulgating "standards." The caution was well founded, since in 1982 the Supreme Court of the United States had confirmed the imposition of a crushing settlement on the nonprofit American Society of Mechanical Engineers, Inc., in favor of the Hydrolevel Corporation, resulting from the promulgation of a standard by ASME that was ruled to have unfairly excluded a particular product.

By December 1984 two study groups, on Pump Speed Measurements and Leak Rate Standards, had been set up, and thus began a revitalization of the committee. In a board meeting in February 1985 the committee asked that its name be changed to the Recommended Practices Committee, and discussed its intended mode of operation. The group would work to establish a practice, which would then be published in the *Journal* as a technical review, for the purpose of gathering comments from all interested parties. After reconciling the published document with any comments, a final version would be published as an AVS recommended practice. Thereafter, it was hoped to develop a standard, in cooperation with ANSI. Sullivan guided the Committee through 1987, by which time one recommended practice had appeared in *JVST-A*, and six working groups were very active; his leadership skills must be credited with the re-emergence of the enterprise. Cathy Stupak replaced Sullivan in 1988, and was in return replaced by Robert E. Ellefson in 1991. At present, six AVS recommended practices have been published in *JVST-A*.

VII. DIVISIONS, SECTIONS, AND CHAPTERS

The first technical subdivision within the Society was the Vacuum Metallurgy Division, formed in May 1961. The previous year, R. F. Bunshah approached the AVS Board with a suggestion that an existing group become a part of the Society. Under the leadership of Bunshah, this group had organized an annual Vacuum Metallurgy Conference from 1957 through 1960. After joining the AVS, the new division maintained this conference as its own annual symposium, since there was perceived to be only limited overlap in the interests of the people who attended this meeting with those attending the AVS national symposium. However, vacuum-metallurgy–oriented sessions have been incorporated into the national symposium from time to time, encouraging the synergism that is a key part of these meetings. Although the major part of the activities of the division remained centered around a distinct meeting, the International Conference on Metallurgical Coatings, considerable interaction developed with the Thin Film Division, which in 1992 became a co-sponsor of the meeting under the name International Conference on Vacuum Metallurgical Coatings and Thin Films.

The Thin Film Division of the Society was officially formed in July 1964, with Klaus Behrndt playing an important role. At this time the division served a major part of the total AVS membership, and through 1969 it ran a separate 1-day symposium, devoted to thin-film topics, on the day preceding every national symposium. Following the tragic death of Behrndt in 1971, the first full day of the thin-film programs at the 1972 and 1973 symposia was named the Klaus Behrndt Memorial Session.

The Surface Science Division was initially proposed by F. M. Propst at an AVS board meeting in February 1968. A surface science symposium, with eight invited speakers, was scheduled at the 1968 AVS symposium to promote this possibility. The session concluded with an open discussion on the formation of the new division, which elicited unanimous support from the approximately 250 people present. The Surface Science Division was approved the next day by the AVS board, and J. P. Hobson was appointed as its first chair. The development into its present position as a leading forum for this technical area has been the result of vigorous cultivation by a number of people, for at the outset there were several alternative forums which seemed attractive to the community. The division was enormously effective in the 1971 IUVSTA meetings, where it initiated the first International Congress on Solid Surfaces, which accounted for about half of the entire papers given at the meeting. This success must be attributed to the groundwork laid by the division leaders, and in particular to the program committee under C. B. Duke. The involvement of a number of Nobel Laureates with that ICSS meeting was no accident!

The vacuum technology community had been the major force during the original formation of the Committee on Vacuum Techniques and, with the formation of the first three technical divisions, a loss of identity began to be felt in this community. This prompted the formation in 1970 of the fourth division, Vacuum Technology, with R. A. Denton and J. M. Lafferty playing important roles.

The Electronics Materials and Processing Division was founded in 1978, with W. E. Spicer and C. B. Duke being the main proponents. The intent was to provide a forum for the rapid developments taking place in this technical area, and to attract audiences not traditionally affiliated with the AVS. Spicer was appointed as the first chairman.

The Fusion Technology Division was founded in 1980, due largely to Manfred Kaminsky. The division serviced a fairly limited, but very active, group of workers. In January 1987 the scope of the division was expanded to include general plasma-materials interactions, plasma diagnostics, plasma-assisted etching and coating, and plasma processing technology, following a proposal initially made by H. F. Dylla and supported by A. Hunt, later approved by the division membership. The name was changed to the Plasma Science and Technology Division.

The Applied Surface Science Division was approved in July 1985 as an outgrowth of the activities of the E-42 Committee on Surface Analysis of the American Society for Testing and Materials, which for 8 years had sponsored joint technical sessions at the AVS national symposia. Cedric Powell played a major role in its formation.

The most recent division to be formed, in February 1992, is the Nanometer Science and Technology Division, largely through the advocacy of Richard Colton, Lawrence Kazmerski, and James Murday, with the objective of providing a forum for the expanding technical interest in this area.

Membership of each division is open to all members of the AVS on payment of a one-time registration fee; an annual fee was originally imposed, but this produced negligible net income and was eliminated in 1972 to reduce paperwork. The divisions sponsor meetings in their spheres of interest, either individually or in co-sponsorship with other AVS divisions, chapters, or outside nonprofit groups.

The program committee for the annual national symposium was originally a small group with members from each division, functioning as a single body in the selection of abstracts. As the number of papers increased, the committee developed into separate groups from each division. These groups, often under the leadership of the chair-elect of the division, have the responsibility of identifying topics and suitable invited speakers and, after abstract selection is complete, they cooperate in the assembly of the final program. In recent years the program committee has held its first meeting one full year ahead of the symposium, permitting a meeting with the members of the previous committee to ensure continuity and providing a more adequate lead time.

The CVT was originally formed in the eastern United

States, but local interest in the Society was such that regional groups were started quite early. The first was the Pacific Northwest Section, founded in 1962. Originally, local sections of the Society levied dues against their members, but in 1968 this mode of operation was changed and the sections were funded from the national organization, using a formula based upon the number of members and the number and type of meetings held each year. Two years later the local organizations, now called chapters, were grouped together into a series of regional administration groups, each containing several chapters, which could be of specific or general interest (affiliated with one or more divisions). All AVS members were assigned to the chapter that served their specific area, with the option of belonging to a second chapter on payment of a one-time fee. The regional groups did not noticeably facilitate governance of the Society, and they were consequently dissolved in 1983, leaving each chapter or division to report directly to the board; a Chapters and Divisions Committee now assists in coordination. There are currently 21 chapters, unchanged from 1983, a new one being formed in western Pennsylvania, in 1988, and the two existing chapters in southern California combining in January 1990. Vossen (2), provides a listing of the chapters and their dates of formation, up to 1983, together with a map of locations. Although several assorted presidents and other nationally active members had lived within the area served by the Western Pennsylvania Chapter, at least two prior organizing attempts had failed. A viable organization was developed only under the guidance of a member who was currently active in a sales organization!

Chapters have been involved in an exceedingly wide range of activities over the years. Typically these include several meetings each year, variously involving individual speakers, panel discussions, and visits to facilities, and an annual 1- or 2-day meeting, which may involve a symposium, short courses, and exhibits. Some of these meetings are small, while others are very sizeable indeed. Education at all levels has been strongly supported by many chapters by granting scholarships for attendance at both local and national meetings and short courses; awarding prizes for the best student or general paper at a meeting and for science fair projects; and allocating funds to support the major awards programs of the AVS and to allow high-school science teachers to attend symposia.

The Northern California Chapter is currently one of the largest and most active within the AVS, and two of their programs are of particular interest. In 1987 the chapter initiated an effort to fund a facility at the Valley Campus of Chabot College (now Las Positas College) in Livermore, to be dedicated to the teaching of vacuum technology. The chapter initially made a substantial cash contribution to the fund. By November of that year the fund exceeded its goal by virtue of a single contribution from the MDC Vacuum Products Corporation. This left the chapter free to concentrate its efforts on equipping the new facility. A second initiative of the chapter has been a long-time commitment to The Exploratorium in San Francisco, for help in the design and procurement of the exhibits in the areas of interest to the Society.

In 1986 the Northern California Chapter assembled a series of experiments in vacuum technology for use at the high-school level, which proved very popular. This theme was followed by the Education Committee, at the 1990 Symposium in Toronto, playing host to around 40 teachers from 24 high schools in the Toronto area, at a Science Educators Day. Included, was the demonstration of a basic pump/bell jar combination kit for use in such experiments. The schools these teachers represented were offered grants to purchase their own demonstration kits through the Society, with the generous cooperation of two manufacturers. In the following 2 years the educators day was repeated at the Seattle and Chicago Symposia, both for teachers from the local area and for others from around the country, who were sponsored by, and supported with, travel grants from local AVS chapters. The courses were very positively received, and hopefully they will contribute, through the enthusiasm of the teachers who attended, to improved programs at the high-school level.

VIII. SCHOLARSHIPS AND AWARDS

The Society, together with its divisions and chapters, sponsor a large number of scholarships and awards at every level, from the science fairs for precollege students, through graduate student participation at technical meetings, to the recognition of major achievements in science and technology. The awards made by the national AVS have since 1975 been administered by a group of six trustees who are elected by the membership and are totally responsible for the selection of the recipients of the awards. These trustees serve as the Scholarships and Awards Committee. Recommendations of the committee are subject to acceptance by the board, but only because of the fiscal responsibility required of that body.

The first major award to be administered by the Society was the Medard W. Welch Award, established in 1969, to encourage outstanding experimental and theoretical research in the technical areas of interest to the Society. The award was funded first by the AIRIES foundation, for ten awards, and then, with an additional gift in 1972. It was further augmented by the AVS board in 1980 following the death of Mr. Welch. It has been given annually since 1970, with the exception of the years 1980 and 1982. Figure 3 shows the presentation of the first Welch Award to Erwin Mueller, in Washington, D.C., 1970. The Welch Award winners for the period 1984–92 are given in Table 2; the earlier winners of this and the other major AVS awards are listed by Vossen (2).

The Gaede-Langmuir Prize was established in 1977 by a grant from an anonymous donor to recognize and encourage outstanding single discoveries and inventions in the areas of interest of the Society. It was to be awarded no more frequently than every 2 years. In 1982, on the 30th anniversary of the AVS, the donor was revealed to be Kenneth C. D. Hickman, winner of the third Welch Award, who had died in 1979. The prize winners for 1984–92 are listed in Table 3.

An account of some of the events leading to the establishment of both the above awards has been given by James M. Lafferty (3), who was intimately involved in their establishment.

The third award to be established, in 1979, was the Peter Mark Memorial Award. Peter Mark was a very active member of the Society who, in the fall of 1974, agreed on very short notice to take over the editorship of JVST from Paul Redhead. He died at a very young age in 1979, and this award is dedicated to the recognition of young scientists for outstanding theoretical or experimental work, at least part of which had been published in the JVST.

The award was endowed by members of the Mark family, private individuals, the Greater New York Chapter, and the Electronic Materials and Processing Division. Recent award winners are given in Table 4.

The Albert Nerken Award was established in 1984 by Veeco Instruments Incorporated in honor of its founder, Albert Nerken, for his role as a founding member of the AVS, his early work in the field of high vacuum and leak detection, and his contributions to the commercial development of that instrumentation. It is awarded to recognize sustained excellence in the solution of technological problems by the use of vacuum and surface science principles. All winners of this award, since its inception, are listed in Table 5.

The John A. Thornton Memorial Award and Lecture was established in 1988, as a memorial to Thornton, who

FIG. 3. Presentation of the first M. W. Welch Award to Erwin W. Mueller, at the 1970 Symposium in Washington, D.C. (*L to R*): Luther E. Preuss, Medard W. Welch, Erwin W. Mueller.

TABLE 2. *M. W. Welch Awards*

Year	Recipient	Abbreviated Citation
1984	William E. Spicer	Photoelectron spectroscopy
1985	Theodore E. Madey	Investigation of surface properties
1986	Harald Ibach	Electron energy-loss spectroscopy
1987	Mark J. Cardillo	Interaction of molecular beams with surfaces
1988	Peter Sigmund	Physical sputtering and related phenomena
1989	Robert Gomer	Pioneering contributions to surface science
1990	Jerry M. Woodall	Compound semiconductor science and technology
1991	Max G. Lagally	Order and growth of crystal structures
1992	Ernst Bauer	Thin Film nucleation and growth

TABLE 4. *Peter Mark Award*

Year	Recipient	Abbreviated Citation
1984	Barbara J. Garrison	Computer modelling of ion-solid interactions
1985	Franz J. Himpsel	Understanding electronic structure of materials
1986	Richard A. Gottscho	Use of spectroscopic techniques
1987	Raymond T. Tung	Growth and properties of epitaxial silicides
1988	Jerry Tersoff	Theory of surfaces and interfaces
1989	Randall W. Feenstra	Applications of scanning tunneling microscopy
1990	Stephen M. Rossnagel	Magnetron and ion beam sputtering
1991	William J. Kaiser	Applications of electron tunneling techniques

died suddenly in November 1987. John Thornton had been a leader in the AVS, serving in many roles, including president (1982). The award recognizes outstanding research or technological innovation, with emphasis on the fields of thin films, plasma processing, and related topics. The award winners are listed in Table 6.

In July 1984 the board instituted a policy of an annual transfer of funds from the AVS general accounts into each of the endowments of the major awards, in order to build up sufficient reserves to sustain the awards indefinitely.

In addition to the major awards, these are those for students. In May 1966 the AVS board approved a trial funding of scholarships in support of vacuum technology at San Jose College, and in 1967 initiated the annual award of Student Scholarships (called *Student Prizes* after 1983), to recognize excellence in graduate studies in science and technology. The selection of the scholarships was delegated to an appointed committee, which was replaced in the election of 1974 by the Scholarships and Awards Committee elected by the membership. Since 1982 the winner of The Russell and Sigurd Varian Fellowship Award has been selected from among the

student prize winners. This award was established to recognize excellence in graduate studies in vacuum science, and consists of a miniature replica of the first Vac-Ion pump, a certificate, and a 1-year fellowship award of $1500. It is funded annually by Varian Associates.

The Nellie Yeoh Whetten Award was established in 1989 as a memorial to a woman who died as the result of an accident at the very beginning of her professional career. The award was endowed by her husband Tim, the family and friends of Nellie, and by the AVS. The award consists of a cash prize, a certificate, and a subsidy of travel expenses to the national symposium, and is open to all women enrolled as graduate students in North American universities. It is gratifying that, following the publicity surrounding the announcement of this award, there was a marked upswing in applications for *all* student awards, and especially by women.

The Society has honored outstanding service, particularly to the Society, through election to the position of Honorary Member. This was the first of all AVS awards, being initiated by the CVT in 1955, with the election of A. S. D. Barrett, in recognition of his outstanding leadership in the field of vacuum technology in Europe. The number of living honorary members is limited to less than 0.5% of the current membership at the time of election; there are now 31. Honorary members elected since 1983 are listed in Table 7.

TABLE 3. *Gaede-Langmuir Prize*

Year	Recipient	Abbreviated Citation
1984	Alfred Benninghoven	Static secondary ion mass spectrometry
1986	Rointan F. Bunshah	Vapor-phase deposition of refractory films
1988	Alfred Y. Cho and John R. Arthur, Jr.	Invention and development of MBE
1990	Francois M. d'Heurle	Electromigration and silicide materials
1992	Russell Young	Invention of the Topografiner

IX. POMP AND CIRCUMSTANCE

Until 1970 an evening banquet, attended by a respectable fraction of the attendees, was part of every annual symposium. This provided a forum for the introduction of honored guests, Society notables, the new board, and the people who were responsible for the symposium arrange-

TABLE 5. *Albert Nerken Award*

Year	Recipient	Abbreviated Citation
1985	John L. Vossen	Control of thin-film deposition/etching techniques
1986	Donald J. Santeler	Contributions to vacuum technology
1987	Marsbed Hablanian	Contributions to vacuum science and technology
1988	Stanley L. Milora	Fueling systems for fusion devices
1989	Martin P. Seah and Charles D. Wagner	Application of electron surface spectroscopies
1990	J. Peter Hobson	Production and measurement of ultrahigh vacuum
1991	Harold R. Kaufman	The electron-bombardment broad-beam ion source
1992	Paolla della Porta	Invention and distribution of getters

ments. There was usually a noted speaker, with talks on topics ranging from science to the ascent of Everest. At the head table were the officers and other luminaries, resplendent in formal dress, in honor of the occasion. Figure 4 shows half of the head table at the 1960 symposium in Cleveland, and Fig. 5 provides a broader view of the banquet in San Francisco in 1966.

At Pittsburgh in 1968 formal dress was replaced by the business suit, but the banquet persisted until 1972, when an informal mixer replaced it. The following year a formal luncheon was inaugurated, motivated, at least in part, by the desire for an appropriate setting for the presentation of the first Medard W. Welch Award (see Fig. 3). Luncheons remained the venue for such awards through 1987. Then, at Atlanta in 1988, the luncheon was replaced by an Awards Assembly, held in the evening, and followed by a reception for the attendees. This format has continued to date.

The informal mixer, first held in 1973, was revived on a much-expanded scale at the 1984 symposium in Reno, replete with an orchestra and a substantial buffet. This affair was announced as the President's Reception, a replacement for a much smaller function which had been limited to those involved in Society management, foreign visitors, and other guests. The function was continued until 1988, when it was combined with the Awards Assembly.

TABLE 6. *John A. Thornton Memorial Award and Lecture*

Year	Recipient	Abbreviated Citation
1989	Eric Kay	Film growth phenomena
1990	Maurice H. Francombe	Thin-film processes and materials
1991	Joseph E. Greene	Effects of ion bombardment on semiconductors
1992	Thomas Anthony	Applications of CVD diamond technology

X. ADMINISTRATION AND OPERATIONS

The administration and operations of the Committee on Vacuum Techniques and during the first years of the AVS were discussed briefly earlier in this article and in more detail by Schleuning (1). Beginning with the adoption of a revised constitution and bylaws in 1961, the governing body has comprised a president, president-elect, past president, secretary,[4] and treasurer, all of whom serve a 1-year term, and six directors, who serve 2-year terms, three being elected each year to ensure continuity. Over the years, most recently in 1988, proposals have been made to add board members elected from each division. The rationale behind the proposals has been to assure better representation of each division's views, particularly in the case of the smaller divisions. The proposals have been rejected in favor of the at-large mode of election, arguing that the directors must represent *all* the members, and that the attendance of a division (or chapter) representative at a board meeting provides an effective way of presenting any case.

Until the election of 1965, the slate of candidates for election presented only one nomination each for all officers and directors, but in that year the incoming president, Steinhertz, argued that at least two candidates should be offered for each opening for director; his suggestion was put into effect with the 1966 election. At this same time, a motion to offer at least two candidates for each of the three officers was defeated. Although the official slate continued to offer only a single candidate for each of the officers, an additional candidate for president-elect was nominated by petition in that same year. The petition candidate was elected, and the message conveyed by this election eventually lead to a change in policy; starting with the election of 1969, two candidates have been nominated *every* year. It is important to note that two *viable* candidates must be nominated, for the difference in votes between candidates can be very close indeed, being a margin of one vote (0.1%) in 1981 (a

---

[4]To conform with the requirements for Incorporation of the Society in the Commonwealth of Massachusetts, the official appelation of the secretary is "Clerk."

TABLE 7. *Honorary Members of the AVS*

Year	Recipient
1983	Daniel G. Bills
	J. Peter Hobson
1984	J. Roger Young
1985	Richard A. Denton and
	Kai Siegbahn
1986	Manfred S. Kaminsky
1988	Jack H. Singleton
1989	Nancy L. Hammond and
	John L. Vossen, Jr.
1991	J. Lyn Provo and
	John Coburn
1992	Marsbed Hablanian
	Donald J. Santeler

single vote has since separated candidates in a divisional election). However, with one exception, only one nomination has been offered in each election for the positions of secretary and treasurer. That exception was in 1969, when Dorothy Hoffman and C. A. Neugebaeur were on the ballot for secretary. Continuity *is* essential in these positions; for example, it requires about 6 months to transfer the office of treasurer, and for this reason, the holders have always been renominated for at least a second year. Singleton held the position of secretary from 1972 through 1984, when he was succeeded by the current holder, W. D. Westwood; Young held the post of treasurer from 1973 until 1983, to be succeeded by the current holder, N. R. Whetten. The slate of candidates

FIG. 4. Head table at the Banquet, Cleveland Symposium, October, 1960. *Back row (L to R)*: Wilfrid G. Matheson, E. Thomas, Medard W. Welch, George W. Carr, Albert Nerken, Richard A. Denton. *Foreground (L to R)*: Hans A. Steinherz, Werner Bachler, Gunther Reich.

FIG. 5. General view of the Banquet at the 1966 Symposium in San Francisco.

for director has most frequently offered a choice of three out of six nominees. Here again it should be noted that the elections are often very closely fought,[5] so that it has always been clear that the nomination of unqualified candidates is fraught with danger. The president of AVS for the years 1984 through 1993 are shown in Fig. 6. Photographs of the presidents from 1953 to 1983 can be found in the article by Lafferty (3).

During the early years of the Society, as many as eight board meetings were held during the year. From 1967 through 1971 there were usually six, and two of these were at the annual fall symposium. Because it was computed to be the most central and economical location to assemble the board, many meetings were held just outside the Chicago airport, the appropriately named Flying Carpet Inn being a favorite location! Beginning in 1973, the number of meetings was reduced to four each year, one meeting at the symposium, and another, the

budget meeting, usually at the AVS office in New York City. It became policy to hold the remaining two meetings in coincidence with chapter or divisional meetings around the country, so as to provide an opportunity for interaction with the local membership. All board meetings are open to any AVS member, but the entertainment value can never be predicted! Figure 7 shows the AVS Board at the Anaheim Symposium in 1974. In February of 1982, in conjunction with a board meeting, the Long Range Planning Committee scheduled an intensive, 1-day study meeting of the board and representatives of all the divisions and several committees, to discuss both the expansion of the national symposia program, and the structure and financing of the *Journal*. This meeting was the forerunner of the "town hall" meetings, which have since been convened to address topics of widespread importance to the Society.

The governance of each of the divisions and chapters is selected in parallel fashion to that of the national board, as mandated by the constitution and bylaws of each group. Every division and chapter is an integral part of

[5]The term "fought" is perhaps injudicious. In 1969 the Board stipulated that the material included in the ballots should be limited to a resume, excluding all other material.

the overall AVS, and all are governed by the overall charter of the Society. Thus, the AVS Board is responsible for *every* action carried out at the local level, a fact that is of particular importance in the control of financial matters and in ensuring conformity with the requirements of the Internal Revenue Service for a nonprofit scientific and educational organization.

The AVS was first granted tax-exempt status by the IRS on December 5, 1967. In 1974, following an audit of the tax years 1970 and 1971, the IRS claimed that the Society was not conforming to the requirements for a tax-exempt organization under its original 501(c)(3) classification. After a lengthy confrontation the matter was resolved satisfactorily. In that investigation, the only Society activity cited as unrelated business, and therefore subject to taxation, was advertising in *JVST*. The short course and scholarship programs of the Society were seen as crucial to the retention of our present nonprofit classification, but, perversely, the publication of *JVST* was not considered of significance, since that particular IRS reviewer believed the *Journal* to be of *more significance in the enhancement of author reputation, than in the transmission of scientific information* (5). The retention of the current IRS classification is not a trivial matter, being, for example, a necessary condition for the Society to retain its membership in the American Institute of Physics.

Until 1967 the Society contracted to have certain office functions performed by a small business organization in Boston. AVS became an Affiliate Member of the American Institute of Physics (AIP) in 1963, and the Institute increasingly provided services to the Society, publishing the new *Journal of Vacuum Science and Technology* in 1964, managing the equipment exhibit at the annual symposium starting in 1965, and collecting the Society and divisional dues from 1966. In January 1968 the AVS office was moved to the AIP building in New York City. This move facilitated access to a wide range of services and an association with many other nonprofit groups having a strong background in the discipline of physics. The AIP manages the pension and health services for AVS employees, has published various monographs, rosters, and smaller items and, since 1984, has published the AVS Topical Conference Monograph series.

In May 1968 a full-time executive secretary, Nancy L. Hammond, was engaged by the AIP for the Society; she attended her first board meeting on May 17, and by the

AMERICAN VACUUM SOCIETY PRESIDENTS, 1984-93

1984
EDWARD N. SICKAFUS

1985
DONALD M. MATTOX

1986
JACK H. SINGLETON

1987
PAUL H. HOLLOWAY

1988
JOHN W. COLBURN

1989
JOSEPH E. GREENE

1990
DAVID W. HOFFMAN

1991
LAWRENCE L. KAZMERSKI

1992
JAMES S. MURDAY

1993
H. F. DYLLA

FIG. 6. AVS Presidents, 1984–1993. *Upper row:* Edward N. Sickafus, Donald M. Mattox, Jack H. Singleton, Paul H. Holloway, John W. Coburn. *Lower row:* Joseph E. Greene, David W. Hoffman, Lawrence L. Kazmerski, James S. Murday, H. Frederick Dylla.

time of the national symposium in Pittsburgh in October, she was already carrying a significant part of the workload for that meeting. So started a productive association with the Society that lasted for 21 years, providing a central clearinghouse for problems of the membership and governing groups alike, and maintaining an invaluable living calendar of events and deadlines to be met. In 1973 Betty Kelly started as a part-time employee, and continued until 1993. In 1979, a second full-time employee, Marcia Schlissel, was added; mechanization of the office was also started, with the installation of word-processing equipment, a Dictaphone product, affectionately named "Chumley."

The annual budget of the AVS has risen markedly over the years and it is fortunate that the Society has exercised fiscal prudence during this period. Much of the credit for the initiation of this policy of conservatism must be given to Dan Bills, who served as treasurer in the years 1968–70. Under his guidance, the board approved the

division of the budget into a series of cost centers with specific approval procedures and responsibilities for all expenditures, which have served the Society well. A major advantage that accrued, was a clearer definition of the cost of all prime functions within the Society. In later years, as the AVS office in New York has expanded, an attempt has been made to allocate the total cost of that office among the appropriate cost centers. The policy of maintaining the endowment of the Welch Award separate from general operating funds was started in 1973; currently, all the major award endowments are held in separate, "restricted" accounts. Reserve funds for publications, the short courses, and the annual symposium are also maintained, and by 1990 these approximated the annual cash flow in each category. Starting in 1987, a portion of the Society's reserves have been invested in the securities market.

Until the early 1980s, the work associated with the symposium was carried, to a very considerable extent, by

FIG. 7. The AVS board in Disneyland, taken during the 1974 Symposium in Anaheim. *Back row (L to R)*: Mars Hablanian, Charlie Duke, Peter Mark, Len Beavis, Eric Kay. *Front row (L to R)*: Roger Young, Lew Hull, Dorothy Hoffman (AVS President), Rey Whetten, Jack Singleton, Maurice Francombe.

the Local Arrangements and Program Committee chairs, and required an enormous commitment of time. The short course program was also expanding rapidly, and there was considerable pressure developing to transfer more of this load to the New York office. In June 1981 the board approved the hiring of a meetings manager to take care of the above concerns; one major additional consideration cited in the final decision, was to make the Society less dependent upon the expertise of a single person, a point brought home by the absence of the executive secretary from the meeting, due to surgery. Marion Churchill was engaged in October, and attended the Anaheim symposium just 2 weeks later.

The use of a computer for handling the abstracts and associated correspondence for the symposium was started in 1981. At its December 1984 meeting the board authorized the transfer of this task to the New York office. In February 1985 a fourth full-time employee, Diana Beasley, was added to the office, two IBM personal computers were purchased, and a consultant started development of an appropriate software package. Many AVS members, including Jerry Woodall, Jim Harper, Roger Stockbauer, and Mike Slade, were deeply involved in the implementation of this system, which was in full operation by May 14, logging abstracts for the fall symposium; the entire procedure was introduced without serious problems.

One major consideration mandating the installation of the computers at this time, was to prepare for the very large increase in the number of abstracts expected for the IUVSTA Congress and associated meetings, which the AVS was scheduled to host in the following year. The size of that meeting turned out to be fully as large as had been expected, and the new system was most effective. Quite unexpected, however, was the resignation and immediate departure of the newest employee, who had been hired and trained for this enterprise. This occurred at the peak of the abstract processing period; she was replaced by Margaret Banks in June, but this was not in time to help with the flood of abstracts. The problem was only solved by the dedicated effort of Marcia Schlissel, who worked very long hours in her customary meticulous fashion to complete the process, thus avoiding what could have been a serious slippage in the timetable. Starting with the 1990 symposium in Toronto, the abstract processing has been handled by Lynn Pizzo, an independent consultant in Rochester, N.Y.

In June 1989 Nancy Hammond retired from the position of executive secretary, and Marcia Schlissel resigned in January 1990. Angela Mulligan became membership and scholarship secretary in April 1989. Margaret Banks now handles the national short course registration and the national and divisional elections. In May 1992 N. Rey Whetten became technical director for the Society.

During 1986, with a great deal of help from Mike Slade, the use of the computer system was extended to handle the registration for the short course program.

The editorial functions for *JVST* were originally conducted by the editor of the *Journal*, often making use of the secretarial help at his place of work; but in 1980, when G. Lucovsky succeeded Peter Mark, a dedicated editorial office, initially staffed by one part-time employee of the AVS, was established in North Carolina to handle the increasing load.

In April 1976 the AVS was elected a full member of the American Institute of Physics, the culmination of an original application made in 1972, and a result largely due to the efforts of C. B. Duke. With that membership came the advantage of recognition by the largest physics societies in the country, and a role in the governance of the Institute. Full membership provided reduced cost for the AIP services the Society was already using, and permitted members access to several AIP services, such as the employment placement service and the receipt of *Physics Today*.

The overall result of the AVS involvement with AIP has been very advantageous from the start, major examples being in the sustained quality of the publication of the *Journal*, and of the management of the equipment exhibit at the annual symposium. But it is not unexpected that there have been some problems accompanying this association. In the early days these were often associated with membership and subscription fulfillment, mostly due to problems with the basic system, and in phasing in new computers and operating software, and partly because the AVS was unique among the AIP Societies in having many more subdivisions—particularly the large number of chapters. Of even greater consequence to the Society, in 1987 the AIP began seriously to consider the options available to expand from the overcrowded building in New York, and it quickly developed that the major thrust was to leave the city and move closer to Washington, D.C. This move was, from the first, not considered to be in the best interest of the AVS, since it was anticipated that the primary result would have been the loss of most of our staff. In 1990, the AVS contracted with SLACK Inc. to handle the membership records, billing, and mailing services, which had previously been done by the AIP, starting with the 1991 membership year. Currently the AVS is planning to move into its own office facilities in New York City.

XI. INTERNATIONAL INTERACTIONS

The AVS and its members have played a strong part in the operations of the International Union of Vacuum Science, Techniques and Applications (IUVSTA) since its founding in 1958. In turn, three distinguished AVS members have served as president of IUVSTA; M. W. Welch (1962–65), Luther E. Preuss (1971–74), and James M. Lafferty (1980–83); Theodore E. Madey assumed that position in 1992. Many other AVS members have also served the organization. For example, three members of the AVS are currently involved in the administration of the international Welch scholarship. IUVSTA has to some extent followed the pattern of AVS in establishing divisions.

The Society has also maintained strong interactions with a number of comparable organizations abroad, including those in Brazil, the People's Republic of China, and Mexico. The presentation of short courses, discussed earlier, and the invited participation of AVS members at meetings in these countries, are manifestations of this activity. In September 1987 the AVS was joint sponsor, with the Chinese Vacuum Society, of a symposium on Vacuum and Surface Science. In the 2 1/2-day meeting in Beijing, 11 AVS and 10 Chinese scientists presented papers. The meeting was attended by about 90 scientists. A reciprocal meeting, attended by both Chinese and Japanese scientists, was held in March 1989 in San Jose, California, with the cooperation of the Northern California Chapter.

XII. COLD WAR PROBLEMS

The Society has had a number of interactions with agencies of the United States government in the past 18 years, as a consequence of strained East-West relationships. In 1980 a major problem involved a conference, the First International Conference on Bubble Memory Materials and Process Technology, which was organized by the Society. Shortly before the meeting started, the AVS was *ordered* to impose severe limitations on attendance, excluding some participants (including those from the Soviet bloc) and requiring all non–U.S. participants to sign a commitment to limit transfer of any information acquired at the meeting. This affair has been discussed in detail by Vossen (2), who gives a thorough analysis of the problems involved in the control of information transfer. This incident was the beginning of a confrontation between government agencies and the scientific community, in which the government sought to hold the sponsors of conferences responsible for controlling both the attendance and the content of the papers presented.

In November 1983 at the annual symposium in Boston, the Society suffered yet another unwanted interaction with a government agency. Late on the afternoon of Thursday, November 3, an East German physicist, Alfred Zehe, was arrested by the FBI on espionage charges. Zehe, an exchange scholar at the University of Puebla, in Mexico, was said to have been under surveillance for some time, and was alleged to be engaging in espionage. Subsequent information suggested that he was the victim of a "sting" operation, in which a civilian employee of the U.S. Navy had offered to sell classified information to G.D.R. embassies, first in Washington,

D.C., and later in Mexico City. Zehe was reportedly asked by the G.D.R. embassy in Mexico City to evaluate the information as a technical expert; his presence at the symposium appeared to be totally unconnected with the espionage charges. Although Zehe could have been picked up at any time, since he had been under surveillance, the FBI perhaps chose the symposium as a high-profile site to highlight the dangers of espionage. The fact that local television cameras were on hand to record the event adds some credence to this interpretation. Details of a number of somewhat bizarre incidents surrounding the arrest have been published (6,7). After the arrest, Zehe was first held without bail for 7 months and then was allowed to remain in an apartment in Boston while awaiting trial. He was finally allowed to leave the country without trial.

XIII. LIFE BEGINS AT 40?

As the Society approaches its 40th anniversary, there are thoughts of changing its name to better reflect the broader range of topics in which it is now involved. But what's in a name? Rather, it is the general objectives and purpose of the Society that must be maintained. Over the past 40 years, the vitality of the Society has been continuously sustained by the active involvement of many, many people, in all areas of the operations, and by a willingness to bring in new people and the ideas that they so fervently espouse. The creative interaction between the various disciplines provides a potent synergism in the solution of problems, requiring continuous and cooperative interaction. Those of us who have had the great good fortune to have been allowed, over the years, to contribute in some small measure to various areas within the Society, and who, in doing so, have gained far more than we have given, can have little doubt as to the future of the Society, provided that it continues to be controlled, as far as is possible, by the members themselves, rather than by surrogates, for the benefit of our combined disciplines. We cannot expect the seas to be calm, which would be too boring, but at least we can hope that the voyage will be intellectually stimulating!

ACKNOWLEDGMENTS

The assistance of Marion Churchill and Ed. Greeley in collecting data for this history is gratefully acknowledged. Bill Buckman provided information on the Vacuum Metallurgy Division. Sources used include the previous histories (1,2,3,4), the AVS newsletter, the AVS board minutes, and the original reports from many AVS committees, especially those of the symposium Local Arrangements and Program, Committees. Where discrepancies exist, particularly in attendance figures, the original (contemporary) sources have been used. I particularly thank Collin Alexander, Marion Churchill, Fred Dylla, Dorothy Hoffman, John Noonan, Paul Redhead, Don Santeler, Bill Westwood, and Rey Whetten for reading an entire manuscript at some stage in the writing, providing many valuable suggestions and insights, and removing at least some of the prejudices of the author. Although the names of many key contributors to the development of the Society have been included, I must apologize to the many others who have been omitted—in particular those involved in the early years, which are more fully covered by Schleuning (1).

REFERENCES

1. Schleuning HW. The first twenty years of the American Vacuum Society. *J Vac Sci Technol* 1973; 10: 833.
2. Vossen JL, Hammond NL. The American Vacuum Society—1973–1983. *J Vac Sci Technol* 1983; A1:1351.
3. Lafferty JM. History of the American Vacuum Society & the International Union of Vacuum Science, Technique, and Application. *J. Vac Sci Technol* 1984; A2:104.
4. Harwood, VJ, Czanderna AW. AVS short course for scientific engineers and technologists: Past and future. *J Vac Sci Technol* 1982; 20:1412.
5. Minutes of the Board of Directors of AVS, June 4, 1982, p. 2.
6. Government bars Soviets from AVS and OSA meetings. *Physics Today* 1980; April: 81.
7. To catch a spy. *Science* 1983; 222:904.

SECTION I
Some Pioneers

Many outstanding scientists and engineers made major contributions to the development of vacuum technology in the first part of the 20th century. Biographical notes on a select few of these pioneers are contained in this section. They come from many countries, including Denmark (Knudsen), France (Holweck), Germany (Gaede, Jaeckel, Pirani), Great Britain (Burch, Yarwood), Holland (Clausing), and the United States (Dushman, Langmuir).

Cecil Reginald Burch *(1901–1983)*

W. Steckelmacher

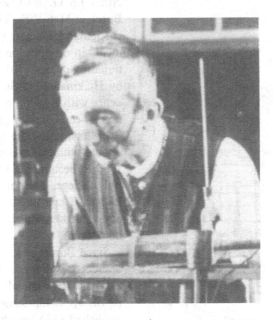

Burch at the age of 48 shown at work at his original pot-still of 1926, wearing the clothes of the earlier period.

Burch was born in 1901, educated at Oundle School and Gonville and Caius College, Cambridge. He took the Natural Sciences Tripos and joined the Metropolitan–Vickers Electrical Co. Ltd. as a graduate apprentice in 1923. The important effect of his work at Met–Vick from 1923 to 1933 was reported by J. Blears in 1981 (1) following his presentation of the Burch prize and also indicated in a "Met–Vick History 1899–1949" by J. Dummelow (2).

An indication of the invention of oil diffusion pumps was first given in 1928 in a short paper by Burch (3) in which he states: "In the course of some work. . .on the distillation of petroleum derivations, I became aware of the possibility and advantages of using oil in place of mercury as working fluids in condensation pumps." This concept and further experimental work then led to the invention by C. R. Burch and F. E. Bancroft (4) of the oil-operated diffusion (or condensation) pump. Related to this, Burch (5) carried on with vacuum distillation.

One important application described in detail by Burch and Sykes (6), involved the development of continuously evacuated transmitting tubes using oil diffusion pumps. The low vapor pressure materials produced in the molecular still for this work were given the name "Apiezon products."* The design and properties of the pumps shown in Fig. 1 (from Burch and Sykes, 1935 ref. 6) are clearly very similar to present-day pumps. In this paper the wide-ranging applications of these pumps are also emphasized, both for high-power long- and short-wave radio transmissions and high-frequency furnaces (for the production of special alloys and hard metals). Also mentioned is the development of continuously pumped multi-electrode tubes, such as the design (by his brother F. P. Burch[†]) of a large-screen grid tube with exceptional

---

*From the Greek α (privative) and *piezon* (pressure).
[†]F. P. Burch unfortunately died in 1933. As mentioned by Gibbs (7), "His brother's sad and painful death in 1933 affected him very deeply and precipitated the change in his career," as noted herein.

stability to be used in a short-wave telegraph transmitter at Leafield Radio Station. A 60 kW tetrode tube, another outcome of C. R. Burch's work, was in operation with the Post Office when Watson Watt was starting his experimental station at Oxford for radio direction finding (r.d.f.) later known as "radar," reported in the review by Dummelow (2). This review also mentions the construction in 1931 of a 500 kW continuously evacuated tube for the Post Office radio station at Rugby, to replace the then-existing bank of 54 sealed-off tubes, which was the largest thermionic tube in the world at the time.

The Apiezon oils and greases for vacuum applications were manufactured under Met–Vick patents and marketed by Shell–Mex Ltd. The development of Apiezon products was mentioned in the "Biographical Notes on G. Burrows" (8) as joint work with C. R. Burch at Met–Vick. This stimulated several others, but in particular Hickman (9), to carry out further improvements in oil-diffusion pumps, leading eventually to fractionating pumps (10). A good summary of this was given by Hickman (11), including some details of correspondence with Burch. For some interesting brief biographical notes, see Hickman (12).

Following the death of his brother,* a new interest in optical studies developed when Burch left Met–Vick in 1933 to become a Leverhume Fellow in Optics at Imperial College, London. Then in 1935 he moved to the Physics Department at the University of Bristol, with many further contributions (and publications) on these new topics.

Since Burch was so closely associated with the development of modern high-vacuum equipment, he was awarded the Duddell Medal by the Physical Society on April 9, 1943; some brief biographical details were published (13) together with one of his contributions to optics "on aspheric systems." An abridged version of his address "A Technologist Looks at the Future," delivered on the occasion of the presentation of the medal, was published (14). This was again full of interesting anecdotes, and just one is quoted here, since it indicates so clearly his practical approach: "Let us do all in our power to make it possible for the embryo physicist of the future to satisfy fully any desire he may have to acquire knowledge of crafts and skills in the use of tools. I cannot assess too highly in this respect the education I received at Oundle School. I use deliberately the phrase 'education in the use of tools' rather than 'training . . . .'" When presenting this medal, it was pointed out that Burch was not only a first-class physicist and instrument designer, but he also possessed considerable mathematical ability and mechanical skill, which he used to construct for the most part his own instruments and apparatus. He was also elected a Fellow of the Royal Society in 1944. At the High Vacua Convention arranged by the Society for Chemical Industry in October 1948, a paper by Burch (15) also dealt with "The First Oil Condensation Pump." His presentation of the first Burch award in 1981 (16) and the publication of his speech‡ was mentioned above.

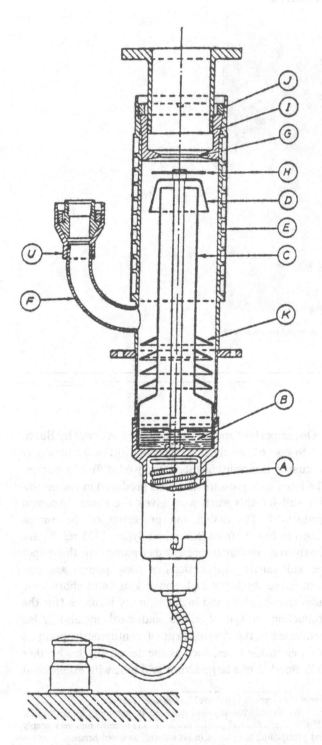

FIG. 1. Oil diffusion pump designed by Burch in 1935(6). **A:** Electric heater. **B:** Boiler. **C:** Uptake pipe. **D:** Cowl. **E:** Water-cooled surface. **F:** Forepump connection. **G,H:** Baffles.

‡At the Interdisciplinary Surface Science Conference, held in Liverpool, April 1980.
*F. P. Burch unfortunately died in 1933. As mentioned by Gibbs (8), "his brother's sad and painful death in 1933 affected him very deeply and precipitated the change in his career," as outlined below.

After he died in July 1983 at the age of 82, a brief appreciation by Gibbs (8) gave a really good view of his character and personality.

REFERENCES

1. Blears J. A tribute to the work of Dr. C. R. Burch, Metropolitan-Vickers Research Laboratory on the occasion of the presentation of the first C. R. Burch prize. *Vacuum* 1981; 31:725.
2. Dummelow J. *1899–1949, Metropolitan-Vickers Electrical Company Ltd., Manchester.* Manchester, England: Rowlinson-Broughton, 1949.
3. Burch CR. Oils, greases and high Vacua. *Nature* 1928; 122:729.
4. Burch CR, Bancroft FE. British patent 346, 293, applied for Jan. 6, 1930, granted Apr 7, 1931.
5. Burch CR. Some experiments on vacuum distillation. *Proc Roy Soc* 1929; 123A:271.
6. Burch CR, Sykes C. Continuously evacuated valves and their associated equipment. *J. I.E.E.* 1935; 77:129.
7. Burrows G. Biographical note. *Vacuum,* 1957; 8:2.
8. Gibbs DF. Appreciation of C. R. Burch. *Phys Bull* 1983; 34:422.
9. Hickman KCD, Sanford CR. A study of condensation pumps. *Rev Sci Instr* 1930; 1:140.
10. Hickman KCD. Trends in the design of fractionating pumps. *J App Phys* 1940; 11:303.
11. Hickman KC. Vacuum pumps and pump oils. Part I: some fractionation pumps. Part II: Comparison of oils. *J Frank Inst* 1936; 221:215, 383.
12. Hickman KCD. Biographical notes. *Vacuum* 1952; 2:2.
13. Burch CR. Nineteenth Duddell Medalist. *Proc Phys Soc* 1943; 55:444.
14. Burch CR. A technologist looks at the future (address for Duddell Medal presentation). *Nature* 1943; 152:523.
15. Burch CR. The first oil condensation pump. *Chemistry & Industry* 1949; 6:87.
16. Burch CR. The British Vacuum Council 'C. R. Burch' Prize. *Vacuum* 1981; 31:723.

Pieter Clausing *(1898–Present)*

H. Adam and W. Steckelmacher

Pieter Clausing age 57.

Pieter Clausing was born in 1898 in the town of Haarlem in Holland. At the appropriate age he joined the first Haarlem High School (Eerste Haarlemsche Hoogere Burgerschool) and continued with further studies in mathematics and physics at the faculty of Natural and Physical Science at the Universities of Amsterdam and Leiden. He publically acknowledged the practical help and guidance he received there, particularly from such famous and highly regarded professors as Sissingh, Kuenen, Kamerlingh Onnes, and Lorentz. Furthermore, Clausing indicated that he received much help from both Prof. Ehrenfest (with respect to all the basic physical problems) and Prof. de Haas. He graduated with an equivalent to the M.Sc. In 1923 Clausing joined N. V. Philips' Gloeilampen fabrieken (incandescent lamp factory) at Eindhoven (Holland) and started his professional career with the Philips Research Laboratories. The director was Prof. Holst, with whom he worked on a number of joint projects.* He soon got involved with vacuum technology and related problems, and at an early stage of these research activities published several papers (2–8). These investigations also included an analysis of the stationary flow of rarefied gases through cylindrical tubes, thus extending the earlier investigations by Knudsen. Prof. Holst soon suggested that Clausing should also work for a Ph.D., which led to his thesis entitled. "On the adhesion time of molecules (on a surface) and the flow of very rarefied gases."

The subject was very much in vogue at the time, kept alive by the still-fierce controversy whether specular or diffuse reflection was the basis of many phenomena

---

*The young Holst was particularly keen on fundamental scientific work, so that his activities did not lead to the growth of an ordinary industrial and development laboratory, but to the founding of a real research laboratory. In 1923 Holst's research group, which consisted of 15 graduates and about 20 assistants, moved into a new laboratory building of their own, just at the time when Clausing joined them. Holst retired as director of the research laboratories in 1946 (1).

observed in the molecular flow regime. In addition, Prof. Holst was keen to put research work at Eindhoven in a favorable relationship to similar work carried out over many years at the General Electric Company in the US. The thesis was submitted to the University of Leiden, defended and accepted in May 1928 (9). Clausing continued his work at Philips until 1960, when he retired. He was in charge of the section "Testing of Materials" within the Philips Research Laboratories and was soon appreciated as a scientist of extraordinary scientific capability. In particular, Clausing became known as a very critical scientist and thus, for some of his colleagues, not easy to get along with. Early in his life he became very interested in religion, and the Philips authorities allowed him to read theology at the University of Utrecht. He finished his theological studies and, in his later years, this resulted in the publication of several books on religious subjects.

Clausing's scientific work, and hence his publications (mainly in Dutch and German, but some of his papers were translated into English) were on the one hand concerned with the more specific topics on special materials used in the electron tube and lamp industry and, on the other, the more general aspects of high-vacuum technology in the development and production of electron tubes and gas-filled discharge devices. These developments were at the time the highlights of modern technology, headed by Langmuir and his school within the Research Laboratories of the General Electric Company in Schenectady. Clausing's highly critical mind, mathematical genius, and experimental skill, a very rare combination indeed, led to a thorough scrutiny of previously published and more or less accepted (and applied) results on the topic of the fundamental aspects of molecular flow of rarefied gases in vacuum systems (10–15). This relates to the earlier studies, as elaborated by such famous scientists as Knudsen and Smoluchowsky, around 1910, followed by Gaede, Langmuir, and Dushman about 1913.

Clausing's research efforts culminated in his formulae for the molecular flow through tubes of any length (10,14,16), including his famous table of Clausing factors, still widely used today.

Another highlight of his work was his derivation of the jet pattern, i.e., beam shape and angular distribution of gas flowing in the molecular mode from the exit of tubes and thick-walled apertures (17), which showed a considerable deviation from the cosine law, which would apply only to very thin-walled apertures and the emission from surfaces (Fig. 1). The latter was the subject of further studies (12,13,18,19). Both the evaluation of flow conductances and molecular beaming effects became important bases for numerous further studies and publications, as indicated in the reviews (20,21). The gas flow pattern at the entrance and exit of cylindrical tubes, following that by Clausing in 1930 to 1933, was analyzed again some 25 years later by Dayton (22), thus updating

this earlier contribution. In a review (23) dealing with the flow of rarefied gases in vacuum systems and the associated problems of standardization of measuring techniques, the importance of the basic calculations by Clausing are also emphasized. These calculations and concepts also became very important in the evaluation of pumping speeds and associated measurements as reviewed by Dayton (24) and Venema (25).

One problem for many who were interested in these early publications (in Dutch and German) was possibly the language and even the availability of some of the journals. It is therefore interesting to note that, with the wide application of high-vacuum technology and the realization of the importance of Clausing's studies, many of his basic papers were translated in the 1950s by the Atomic Energy Commission (U.S.) and became available (through the clearing house for technical information) to technical and university libraries. Furthermore, in order to make his 1932 paper, discussing the molecular flow through a short tube, more readily accessible, an English translation was published in 1971 (14). This stimulated an important contribution in 1973 by Venema (from the Philips Research Laboratories), which gave an interesting historical background to Clausing's analysis of these molecular flow problems and relating them to developments, such as the analysis by De Marcus (26,27) of the integral equations involved, the molecular flow or transmission probabilities for tubes in series as investigated by Oatley (28) (see also subsequent correspon-

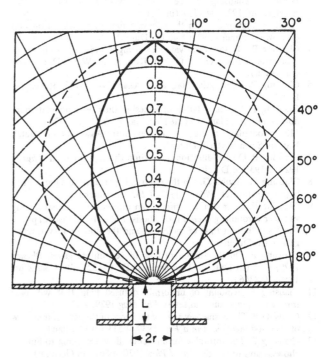

FIG. 1. Polar diagram of gas flow issuing from a short cylindrical tube into a vacuum (for the case $L=2r$) compared to flow through an orifice (dashed line) calculated by Clausing (17).

dence of Steckelmacher [29]) and Monte Carlo methods as developed for the analysis of more complex structures.

There was one small problem that took some time to resolve. In paragraph 19 of his Ph.D. thesis, Clausing (30) treated the stationary molecular flow of a gas with adsorption of the molecules on the wall of the tube described by the Langmuir isotherm. Soon after the publication of the thesis, Langmuir visited the Philips laboratories [on one of his many trips to Europe] and mentioned his suspicion of Clausing's results on this topic; although, as Clausing mentioned, they were derived step by step on the blackboard and Langmuir could not find a mistake. Nevertheless, Langmuir departed quite unsatisfied. It was only a long time afterward that it turned out that a wrong diffusion coefficient had been used, and much later, this interesting mistake was analyzed and the results corrected by Clausing (31) in a brief publication. This was, after a long pause (since 1933), the last of Clausing's vacuum-related publications.

ACKNOWLEDGMENT

The authors are indebted to A. Venema for his contribution to relevant details of Clausing's life and achievements.

REFERENCES

1. van Santen, JH. The Philips Research Laboratories at Eindhoven, The Netherlands. *Chem and Ind* 1964; 37:1564.
2. Holst G, Clausing P. Versl. Afd. Nat. Kon. Akad. Wet., Amsterdam. 1925; 34:1137 (in Dutch).
3. Holst G, Clausing P. On the adhesion time of metallic atoms on glass walls. *Physica* 1926; 6:48 (in Dutch).
4. Clausing P. On the mean values of straight and curved chords of geometric bodies, Versl. Afd. Nat. Kon. Akad. Wet. 1926; 36 860 (in Dutch).
5. Clausing P. Versl. Afd. Nat. Kon. Akad. Wet. 1926; 36:863 (in Dutch).
6. Clausing P. On the relation of adhesion time with nonstationary flow conditions, Handel 21 ste Ned. Natuur & Gen. K. Congress p. 113. (Proceedings of the 21st Congress of the Association of Dutch Natural Scientists) (in Dutch).
7. Clausing P. On the diffusion of thorium through tungsten. *Physica* 1927; 7:193 (in Dutch).
8. Clausing P, Moubis G. On the electrical resistance of Ti and Zr at high temperatures. *Physica* 1927; 7:245 (in Dutch).
9. Clausing P. Over den verblijfdijd van moleculen en de strooming van zeer verdunde gassen. Ph.D. Thesis, 1 May 1928, Leiden, p. 175 (on the adhesion time of molecules and the flow of very rarefied gases) (in Dutch).
10. Clausing P. On the stationary flow of very dilute gases. *Physica* 1929; 9:65 (in Dutch). Translation: AEC-tr-2525, *Nucl Sci Abstr* 1956; 10:8346.
11. Clausing P. Amount of uniformly diffused light that will pass through two apertures in series. *Phil Mag* 1929; 8:126.
12. Clausing P. The cosine law as a consequence of the second main law of thermodynamics. *Ann d Phys* 1930; 4:533 (in German).
13. Clausing P. The formulae for molecular flow according to Smoluchowski and to Gaede. *Ann d Phys* 1930; 4:567 (in German).
14. Clausing P. The flow of highly rarefied gases through tubes of arbitrary length. *Ann d Phys* 1932; 5, 12:961 (in German). Translation: AEC-tr-2447, *Nucl Sci Abstr* 1956; 10:3919. Also translated in *J Vac Sci Tech* 1971; 8:636.
15. Clausing P. A comment on Gaede's experiments on molecular flow. *Ann d Phys* 1932; 14:134 (in German).
16. Clausing P. On the stationary flow of very dilute gases through cylindrical tubes of any length. (in Dutch), Versl. Afd. Nat. Kon. Akad. Wet., Amsterdam 1926; 35: 1023. Translation: AEC-tr-1744, *Nucl Sci Abstr* 1954; 8:792.
17. Clausing P. On the formation of beams in molecular flow. *Zeit Phys* 1930; 66:471 (in German). Translation: AEC-tr-2446, *Nucl Sci Abstr* 1956; 10:3918.
18. Clausing P. Life of an adsorbed molecule and its measurement from studies of molecular streaming. Part 1. *Ann d Phys* 1930; 7:489. Part 2 *ibid*. 7:521 (in German). Note; these papers describe some of the experiments already reported in the Ph.D. Thesis, 1928, but give a considerable extension of these.
19. Clausing P. A measurement of the molecular velocities and a test of the cosine law. *Ann d Phys* 1930; 5, 7:569 (in German).
20. Steckelmacher W. Review of the molecular flow conductance for systems of tubes and components and the measurement of pumping speeds. *Vacuum* 1966; 16:561.
21. Steckelmacher W. Knudsen flow 75 years on: the current state of the art for flow of rarefied gases in tubes and systems. *Rep Progr Phys* 1986; 49:1083.
22. Dayton BB. Gas flow patterns at entrance and exit of cylindrical tubes. *Trans 3rd AVS Nat Vac Symp*, Chicago, 1956. Oxford Pergamon 1957:5.
23. Steckelmacher W. The flow of rarefied gases in vacuum systems and problems of standardization of measuring techniques. Proc. 6th. Int. Vac. Congr. 1974. *Jap J Appl Phys* Suppl. 2: pt. 1, 117.
24. Dayton BB. Measurement and comparison of pumping speeds. *Ind Eng Chem* 1948; 40 (5):795.
25. Venema A. The measurement of pressure in the determination of pump speed. *Vacuum* 1957; 4:272.
26. De Marcus WC, Hopper EH. Knudsen flow through a circular capillary. *J Chem Phys* 1955; 23:1344.
27. De Marcus WC. The problem of Knudsen flow, parts I–VI. U.S. Atomic Energy Comm., Oak Ridge, Report K1303, AD 124579.
28. Oatley CW. The flow of gases through composite systems at very low pressures. *Br J Appl Phys* 1957; 8:15, 495.
29. Steckelmacher W. Comment. *Br J Appl Phys* 1957; 8:494.
30. Clausing P. On the relation of adhesion time with adsorption. *Physica* 1928; 8 289.
31. Clausing P. On the molecular flow with Langmurian adsorption of the molecules on the wall of a tube; a correction. *Physica* 1962; 28:298.

SUGGESTED READINGS

1. de Boer JH, Clausing P, Zerker G. A simple method for the production of small amounts of K-, Rb- or Cs- metals. *Z. angew, anorg. allg. Chemie* 1927; 160:128 (in German).
2. de Boer JH, Clausing P. On the electrical resistance of titanium, zirconium and their mixed crystals. *Physica* 1930; 10:267 (in Dutch).
3. Clausing P. On the entropy reduction in a thermodynamic system on attack by intelligent creatures. *Z Phys* 1929; 56:671 (in German).
4. Clausing P. Radiation of a filament surrounded by an incandescent tube. *Rev Opt* 1931; 10:353 (in French).
5. Clausing P. A comment on the molecular flow (at low pressures). *Ann d Phys* 1932; 14:129 (in German).
6. Clausing P. On the melting point of zirconium oxides and hafnium oxides. *Z angew anorg allg Chem* 1932; 204:33 (in German).
7. Clausing P. On the electrical resistance of titanium and zirconium nitrides and a new resistance effect. *Arch Ned Sci* 1932; A14:53 (in German).
8. Clausing P. Concerning the special application of initial values in diffusion problems. *Physica* 1933; 13:225 (in Dutch).
9. Clausing P. On the evaluation of the composition of small amounts of a neon-argon-mixture. *Physica* 1933; 13:320 (in Dutch).

10. Clausing P, Ludwig JB. On the total radiation of oxide cathodes. *Physica* 1933; 13:193 (in German).

11. Rosenfeld A. *The Quintessence of Irving Langmuir.* Oxford: Pergamon, 1966.

12. Steckelmacher, W. The effect of cross-sectional shape on the molecular flow in long tubes. *Vacuum* 1978; 28:269.

13. Venema A. The flow of highly rarefied gases. *Philips Tech Rev* 1973; 33 (2):43.

Saul Dushman *(1883–1954)*

J. M. Lafferty

Portrait of Dr. Saul Dushman taken in 1918 at the age of 35. Courtesy Hall of History Foundation.

INTRODUCTION

It is indeed a pleasure to write about Saul Dushman, a former associate and friend at the General Electric Research Laboratory (now GE R&D Center). He was a man I highly respected for his scientific knowledge and human kindness. I met Dushman for the first time in 1940 when I came to the Laboratory as a "summer man." I well remember walking into his office that first day when, after a brief discussion, he asked me "Well, Lafferty, what would *you* like to do this summer?" Still a graduate student at the University of Michigan, I was completely unprepared to answer such a question. I had expected to be *told* what to do. What a refreshing atmosphere in which to start one's work in the laboratory—an environment seldom encountered today even for the postgraduates! But this was characteristic of the GE Research Lab at that time, thanks to the influence of its founder, Willis R. Whitney, and its associate directors, Saul

Dushman and Albert W. Hull. More about that later. As it turned out, I told Dushman of my interest in electronics and he assigned me to Kenneth H. Kingdon to work on an instability problem associated with the FP-54 electrometer tube (1) that he was using in his mass spectrometer to separate uranium isotopes for Columbia University.

Later, when I came to the Laboratory permanently in 1942, I became well acquainted with Saul Dushman, and what I have written here is based on my personal contacts with him and liberal use of a wealth of information on his early life made available to me from the files at the GE R&D Center.

Dushman's scientific interests and contributions covered a wide range of subjects, including the production and measurement of high vacuum, vacuum tubes, unimolecular reactions, electron emission phenomena, phototubes, electrical discharges in gases at low pressures, light sources, and quantum mechanics. The list of

32

his publications in the Appendix shows the breadth of his interests. However, in this paper I will only describe his contributions related to vacuum science and technology, which earned him a worldwide reputation, and about his life in the laboratory, where he was even more highly regarded by his associates as a beloved friend than as a scientist.

EARLY LIFE

Saul Dushman was born in Rostov, Russia, on 12 July 1883, the son of Jewish parents, Samuel and Olga (Hurwitz) Dushman. An uncle who was a chemist for an oil refinery on the Caspian Sea aroused Saul's interest in science at an early age. When Saul was nine, his family emigrated to Toronto, Canada, chiefly for the purpose of providing educational opportunities for Saul. There he entered public school unable to speak a word of English. Nonetheless he graduated from high school in 1900 with the highest scholastic record ever achieved in the Province of Ontario and won the Prince of Wales Scholarship consisting of four years of college at the University of Toronto and a cash award of $110. Dushman later said that no money he ever earned meant nearly as much to him as that cash and scholarship award. Torn between the classics and science, in both of which he excelled, he chose the latter and joined the university's department of physics and chemistry. In 1904 he graduated with an A.B. degree. Dushman remained at his alma mater for eight more years, first as a demonstrator for five years and then as a lecturer in the newly established department of electro-chemistry.

During the last three years he also did graduate studies in physical chemistry under Prof. W. Lash Miller, who, as Langmuir (2) has said, was an ardent disciple of Wilhelm Oswald at Leipzig. Oswald taught that atomic theory, even if it is sometimes a convenient hypothesis, "should be employed as sparingly as common usage will permit." Dushman completed his graduate studies in 1912, earning a Ph.D. degree in physical chemistry.

This same year Dushman was invited to give a lecture at the General Electric Research Laboratory in Schenectady by its director, Willis R. Whitney. This meeting is best described by Dushman himself in *An Album Of Memories* (unpublished). He writes:

It was on a relatively balmy day in February that I walked over from the Edison Hotel down Dock Street, a cobble-paved road along the Erie Canal. The *Kitty West* (a boat), used during the summer for hilarious Saturday night carousals, was moved to State Street as the canal was still covered with ice. At the GE main gate I was taken to Building 6 where the lab was located.

It was an old brick building with a wooden front and consisted of two stories. The floors were of rough boards and, according to modern standards the structure would have been condemned as a veritable fire-trap. Ralph

Robinson, subsequently manager of the Vacuum Tube Division, met me at the door. He was in charge of colloquies which were then held on every Saturday morning at 11:00. It is interesting to note that these gatherings were held in a room designated the library, which was also the office of Dr. Coolidge. Most of those who attended brought in their own chairs or stools or sat on the tables.

My recollections of the colloquy which I gave are rather amusing. On the advice of my professor at the University of Toronto, where I was at that time a lecturer, I presented a learned, mathematical discussion for diffusion at the electrodes in an electrolyte—the subject of my thesis for a degree. I doubt that more than half a dozen understood what I said. I spoke for about 30 minutes and then Langmuir discussed the talk for pretty nearly the same length of time, pointing out many defects in my arguments and also indicating what further experiments I ought to carry out. It was not to be the only occasion on which he taught me an invaluable lesson in the application of the scientific method. I still consider it a miracle that after this colloquy I was engaged by Dr. Whitney to come to the laboratory the following summer.

It was clear that Langmuir's nearly 30-minute, incisive critique of Dushman's lecture had persuaded him that an industrial laboratory could apply the same high critical standards to scientific work that he was used to at the University of Toronto. Langmuir (2) remembered during that visit showing Dushman shadow casting resulting from rectilinear propagation of atoms evaporated on glass from a tungsten filament in high vacuum. Langmuir explained that the deposit on the glass, which absorbed 20% of the light, consisted of a layer of tungsten about 14 atoms deep. Dushman smiled and asked Langmuir whether he regarded that statement as more than "a figure of speech." Langmuir further recalled that, after a few months at the lab experimenting with electron discharges in high vacuum, it came to Dushman as a marvelous revelation that atomic physics was more than a "convenient hypothesis."

In accepting a job at the GE Research Laboratory, Dushman was to become one of the world's leading experts in high vacuum research. In his *Album of Memories*, Dushman writes more about his early days at the laboratory:

As far as I can recollect rather dimly, the total laboratory staff consisted of about 75 people of whom about 25 or 30 were technically trained and the rest were machinists, assistants and glass blowers. We had only about two of the latter and one of them was accustomed to celebrating Saturday night so strenuously that many times he was not able to come in on Monday morning.

I can still remember that during the following winter, whenever I worked in the lab at night I would see Dr. Collidge working a Toepler pump to exhaust an x-ray tube and operating a rectifier which made a terrific din.

Gilson, my first "boss," was a clever mechanical engineer and extremely practical. His pet aversion was a differential equation. It became obvious very soon that we could not be in agreement upon the method of carrying on research. My first year in the lab was therefore, far from pleasant and I welcomed the opportunity when Langmuir asked me to join him in the development of his extremely stimulating ideas on the application of incandescent cathodes in highly evacuated devices.

Thus began a lifelong friendship between the two researchers (Fig. 1). It was this work in the field of electron emission from hot filaments that provided basic information in the development of electron tubes and stimulated Dushman's interest and investigations in high vacuum.

In 1915 Dushman published an important paper that described the development of a high-voltage vacuum diode (Fig. 2) that could rectify as much as 10 kW of alternating current power at 100,000 volts. A new name was needed for this device to distinguish it from former hot-cathode rectifiers in which positive ions played an essential role. Dushman and Langmuir spent a Sunday morning in 1914 at the home of Dr. Bennett, professor of Greek at Union College, discussing Greek roots that might be of help. This marked the birth of the -tron family of vacuum devices. Dushman explained in his paper that the designation *kenotron* was especially coined from the Greek *kenos* meaning "empty," and the suffix *-tron* signifying an instrument or appliance. The -tron family then grew to include pliotron, radotron, dynotron, magnetron, thyratron—and, much later, cyclotron, betatron, klystron, and many others.

CONTRIBUTIONS TO VACUUM SCIENCE

Langmuir (2) recalled that in 1915 Dushman called his attention to a paper that had just been published in Germany describing Gaede's diffusion pump, which had a maximum speed of 80 cm^3/sec. Dushman then constructed one of Gaede's pumps and studied it with great care. It was during their discussion of these experiments that the ideas were developed for eliminating the need for the narrow slit in the Gaede pump and thus removing the speed limitation. Dushman then built and tested this condensation pump, obtaining in the first model a speed of 3,000 cm^3/sec.

Dushman's work with Langmuir on vacuum tube devices, thermionic emission, incandescent filaments for lamps, and low-pressure gas discharges all exposed him to the important need for high vacuum, as well as the problems associated with producing and measuring it (Fig. 3). Over the years he published a series of papers and books on the theory and use of vacuum gauges and vacuum technique (see Appendix, papers 18, 26 35, 55, 56, 57, 59, 60, 61 and books 63 and 66). Langmuir's work on surfaces led him to some new ideas about the structure of atoms and the forces that held them together. This fascinated Dushman and he joined in the quest. Long stretches of his notebooks was filled with attempts to account for the arrangements of electrons in atoms.

Dushman fancied himself as becoming a theorist. He looked the type, too (Fig. 4), with his bald head, round-rimmed glasses, and ever-present pipe in defiance of Whitney's smoking ban at the lab. In the early 1920s he was even a step beyond Langmuir and worked with quantum theory. One area where Dushman's knowledge of quantum theory would have immediate application was in gas discharge light sources. The Laboratory had built an impregnable patent protection for the incandescent lamp. There was concern that the efficient process used by the mercury arc and other discharge lamps to turn electrical energy into light would erode the patent protection that GE had on light production. Whitney chose Dushman, with his knowledge of vacuum science and quantum theory, to be the director of research for a new lighting laboratory established in August 1922 at the GE lamp plant in Harrison, New Jersey. This lasted until 1935, during which time Dushman would commute between Schenectady and Harrison.

In 1922 Dushman published a small volume entitled *The Production and Measurement of High Vacua* (Appendix 16 and 63). This work consisted of a compendium of articles he had written for the *General Electric Review* during 1920–21. This book was written primarily for his fellow workers and others who were engaged in the field of physical electronics and lamp production. It began with a review of the kinetic theory of gases as a basis, and contained all of the information known about vacuum in the GE Research Laboratory at that time. It became a standard reference on the subject for many years.

Over the years Dushman became one of the most highly regarded scientists in the Laboratory, yet he was not a prolific discoverer or inventor. He obtained only eight patents during his 40 years of service. But he had the unique ability to read and digest scientific literature with great rapidity and then interpret and express it clearly in simple language that one could understand without being an expert in the field. This ability led to a long series of publications of papers and books that earned him a worldwide reputation. He became a compendium of information. It is no surprise that his associates in the Laboratory consulted with him on a wide range of subjects.

Dushman has admitted that throughout his life he was vicariously a teacher (Fig. 5). Those of us who had the privilege of hearing him lecture were struck by the fact that the teaching profession lost a gifted teacher when he decided upon an industrial career at GE.

Perhaps Dushman's most important work was his book *Scientific Foundations of Vacuum Technique* (Wiley, 1949). This book was completed just before his retirement. It was the culmination of all his knowledge and experience over a lifetime in the field of vacuum science and technology. It was a comprehensive survey of all phases of achieving, maintaining, and measuring low gas

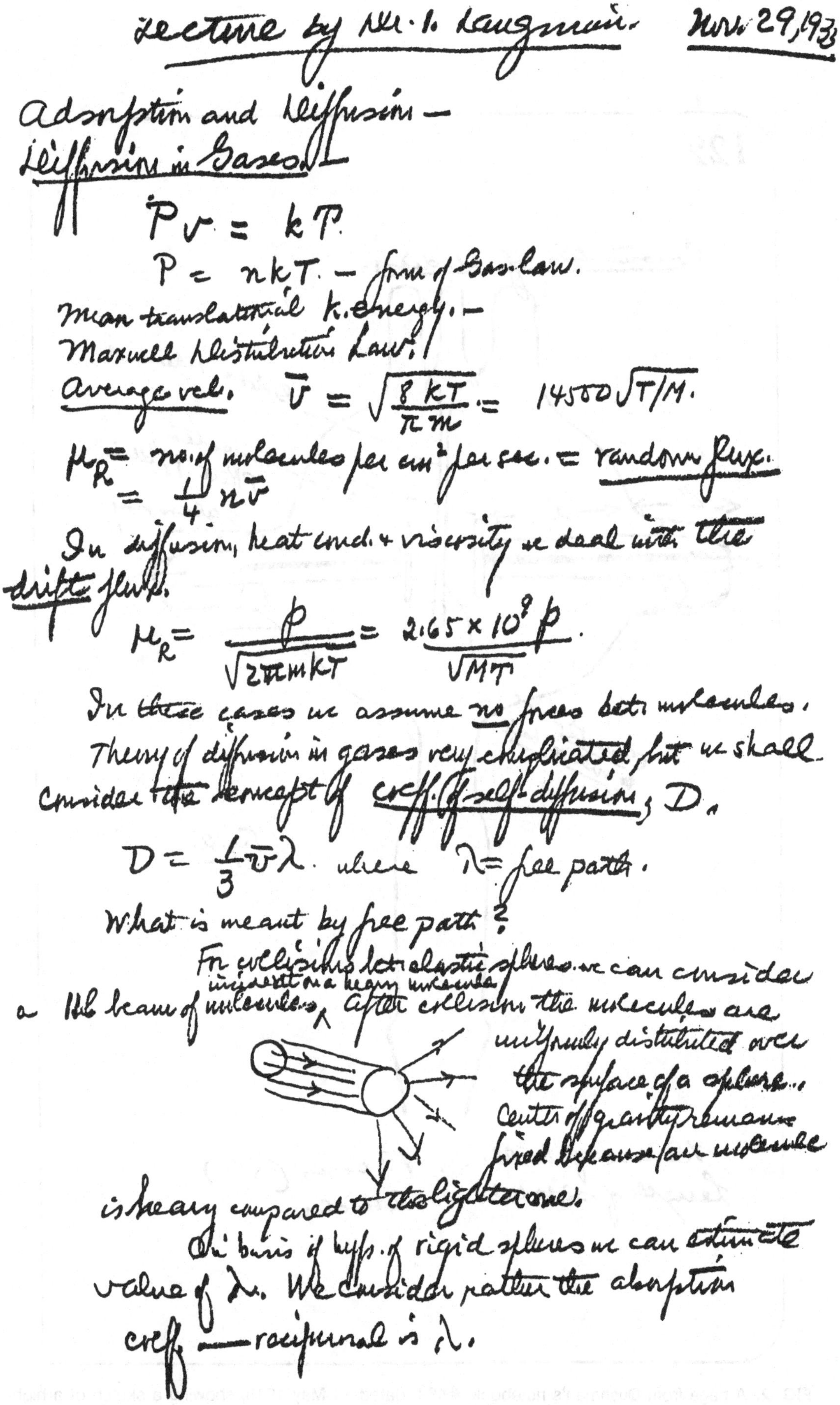

Lecture by Dr. I. Langmuir. Nov 29, 1933

Adsorption and Diffusion —
Diffusion in Gases.

$$Pv = kT.$$

$$P = nkT. — \text{form of Gas law.}$$

Mean translational k. energy. —
Maxwell distribution Law.

Average vel. $\bar{v} = \sqrt{\dfrac{8kT}{\pi m}} = 14500 \sqrt{T/M}.$

$\mu_R =$ no. of molecules per cm² per sec. = random flux.

$\quad = \dfrac{1}{4} n\bar{v}$

In diffusion, heat cond. + viscosity we deal with the drift flux.

$$\mu_R = \dfrac{p}{\sqrt{2\pi mkT}} = \dfrac{2.65 \times 10^9 \, p}{\sqrt{MT}}.$$

In these cases we assume no forces betn molecules.

Theory of diffusion in gases very complicated, but we shall consider the concept of coeff. of self-diffusion, D.

$$D = \dfrac{1}{3} \bar{v} \lambda. \quad \text{where} \quad \lambda = \text{free path.}$$

What is meant by free path?

For collisions between elastic spheres we can consider a like beam of molecules, impinging on a heavy molecule. After collision the molecules are uniformly distributed over the surface of a sphere. Center of gravity remains fixed because one molecule is heavy compared to the lighter one.

On basis of hyp. of rigid spheres we can estimate value of λ. We consider rather the absorption coeff. — reciprocal is λ.

FIG. 1. A page from notes taken by S. Dushman at a lecture given by I. Langmuir in 1933.

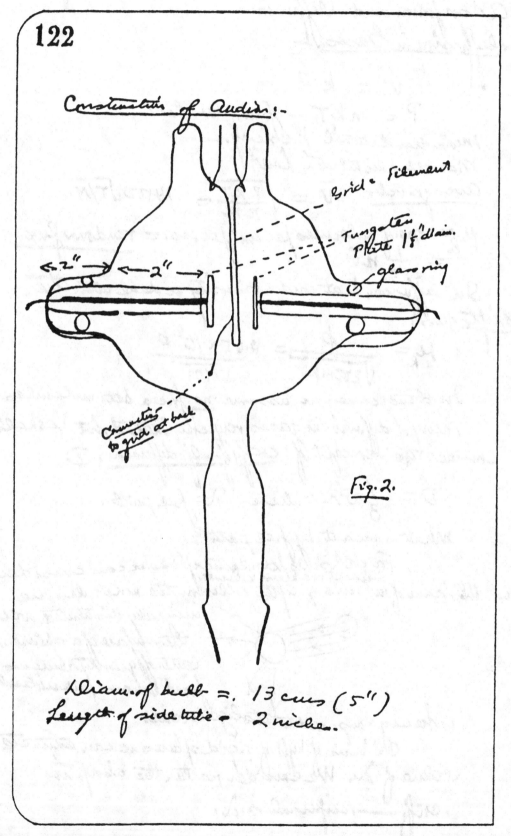

FIG. 2. A page from Dushman's notebook #464, dated 14 May 1913, showing a sketch of a high-voltage full-wave vacuum tube rectifier.

pressures. The book was unique because it presented the scientific basis for the subjects covered in the clear, precise way Dushman was noted for. The book immediately enjoyed unprecedented success and soon became the "bible" in its field, earning Dushman world-wide acclaim. He was in the process of writing a second volume of this work when he was stricken with his final illness in 1954.

By the time the book went out of print ten years later, over 9000 copies had been sold and the annual sales were nearly constant. The book had become a classic in its field, and there was obviously a need for a revised (second) edition. There were many new subjects to be covered. I undertook the responsibility for doing this, but the revision of such a comprehensive treatise was a job larger than I wanted to undertake alone. I was fortunate to find within the GE Research Laboratory at that time a number of scientists, many of whom had been associates of Dushman, with the highly specialized knowledge needed to revise such a book. We undertook the task and published a second edition in 1962, retaining wherever

possible the original plan of the book. This edition was also very successful, selling over 13,000 copies before it went out of print in 1991. (I have now been asked to produce a third edition of the book. I am giving it serious consideration, but no promise at this time.)

A HUMAN CATALYST

I borrow the words for the title of this section from the title of a memorial to Dushman written by Irving Langmuir (2) because they best describe the character of a man dedicated to interacting with his fellow humans and promoting interaction among them. To fully understand Dushman's deep interest in the human side of the GE Research Laboratory, it is important to understand the unique "spirit of the laboratory" inculcated by its founder, Willis R. Whitney, in 1900. This was a spirit of strict honesty, friendship, informality, and cooperation—a spirit that persisted well into the forties

FIG. 3. Photograph taken in 1920 of Saul Dushman (*front, center*) and his staff of high vacuum investigators. Courtesy Hall of History Foundation.

and early fifties. It made a favorable and lasting impression on me when I first came to the lab. This "spirit of the laboratory" can best be described by Dushman himself in a talk he gave on the subject on 12 December 1951. I quote:

> Let us now consider briefly what we mean by the Lab Spirit. How may we describe in mere words that which is intangible, a product of the emotions of men working together harmoniously through both "thick and thin" for a common cause?

> When I consider the characteristics of that relatively small group of about 185 that constituted the technical staff during the late thirties there seem to be *three* principal factors that brought about a striking geniality and sense of family unity that permeated practically all of us, and which has remained with us right up to the present.

> The first and most important characteristic has been that of mental integrity. To be honest with oneself and to be honest with others are the prime necessities in the search for truth. Mental integrity is also the basic principle by means of which men are able to work together.

> The second characteristic, which follows logically from mental integrity, is complete and unselfish cooperation in furthering those ends which are of greatest service to the lab and, consequently, to the company as a whole. Only lack of time prevents me from mentioning the numerous occasions on which a major portion of the Lab staff cooperated in this manner in order to achieve the success of a certain product, process, or idea.

> Lastly, there is a third characteristic of the Lab Spirit, that which I would designate as *faith*. By this I do not mean a blind emotional belief that accompanies men's thinking and action in other than scientific fields. Rather, I am thinking of that faith which men must have both in themselves and in each other in order to ensure successful cooperation and achievement. To be able to trust each other to the utmost, to know that a statement made by any one member is the truth to the best of his knowledge, to have confidence that the management is always trying to be fair with those for whom it is responsible—these constitute the kind of faith that will always unite all of the members of our staff in whatever difficulties that may arise and most ensure ultimate success. The spirit of the Lab is, in my view, a characteristic of an institution or organization such as could develop only in our country. It is in accord with all our democratic traditions of cooperation and fair play. It embodies, in a sense, a "bill of rights" for all the members of our Lab.

During his entire career at GE, Dushman considered it his personal responsibility to perpetuate the "spirit of the

FIG. 4. Rendition of a portrait of Saul Dushman by Eugene A. Montgomery.

laboratory." Known unofficially as the Laboratory's "dean of men," he made a special effort to welcome new members of the staff and make them feel at home. His associates referred to him as the Laboratory's "greatest morale builder."

Saul had a sense of humor that endeared him to his associates. His presence always had a dominating influence in group meetings, both social and scientific. He could also take a joke as well as the next one, but for 30 years he searched for the fellow who evacuated his bicycle tires and filled them with water. After lugging the suspiciously heavy bike up a long flight of stairs every day for six months and pondering on the inexorable law of gravity, he one day noted a tiny puddle of water where no puddle should have been.

I well remember having dinner with Saul and two other new young associates of mine one evening in Boston in the early 1940s, during one of the famous Nottingham conferences at M.I.T. on physical electronics. Early in the meal the subject of expense accounts came up. Saul explained that in his view when one traveled at company expense one should not buy the most expensive meal on the menu, but should purchase a meal as though he were paying for it himself—or, as he put it, "Live the way you

would at home... ." At this point he then bought a good bottle of wine for the four of us!

My "boss" for many years and also a great morale booster at the Lab, Albert W. Hull, has best summarized Dushman's genial personality and warm interest in others as follows (3):

> Dr. Dushman's door, like Dr. Whitney's, was always open, "rain or shine," to anyone who was in trouble. But his interest in the personal problems of his colleagues was not the conventional sympathetic ear; rather it was a hard-headed, though kindly, practical advice and help. Often he took the problems of his fellow workers home with him, worked and worried over them as though they were his own.

> Dr. Dushman's most outstanding quality, both in science and in life, was strict honesty. It was so much a part of him that his outspoken candidness, which was characteristic of him, never gave offense. His mild profanity, which never exceeded a "damn," was a lovable characteristic for which he is well remembered. It was not profanity with him, but an expression of his forthrightness of speech, as of thought.

My class mentor, A. D. Moore, professor of electrical engineering at the University of Michigan, wrote the

FIG. 5. Photograph of Dushman (third from left) taken in 1944 with GE Leak Detector training course students. Courtesy Hall of History Foundation.

following tribute to Dushman after he had completed two weeks of lectures in Ann Arbor at the Electronics Institute in the summer of 1937:

The Physicist Again

The other day, in speaking of the physicist in jest,
I use the implication that he never takes a rest
From being a pure scientist: that with his burning zeal,
He's married to research so much, the arts have no appeal.

Now science is a quest for truth, and if the truth be told,
There's plenty who have wandered from the scientific fold:
And, while remaining true and faithful to the research wife
They like to flirt around a bit with other things in life.

They tell me this man Dushman simply loves to sit around
And keep the conversation going anywhere he's found.
No matter what the subject of the people he's among,
He has some sound ideas and an ever-ready tongue.

He opened up the Institute in Grade-A, top-notch style,
And technically has made it something very much worthwhile.
But, otherwise and on the side, we dedicate this rhyme
To the hope that in Ann Arbor he had a damned good time!

In times of trouble, Dushman was a pillar of strength to the Laboratory, GE, and the community. During World War II, when government security regulations threatened to raise barriers between various sections of the Laboratory, he started a regular Wednesday luncheon at the local YWCA cafeteria that was invariably known as the "Dushman luncheon." Entirely informal, this was a forum for the free discussion of any topics that seemed timely.

Again, during that troubled day in 1946 when staff members were excluded from their labs during a GE strike, he became "dean" of the "University in Exile" at Union College. There he organized and taught with others a number of courses appropriate for the staff.

LIFE AT HOME

While still a graduate student, Saul Dushman was married to Amelia Gurofsky in Toronto on 1 May 1907. Mrs. Dushman died in May 1912, leaving one daughter, Beulah, who eventually joined the National Bureau of Standards in Washington, DC. Two years later Saul married Anna Leff in New York on 28 June 1914. They met through mutual friends during her visit to Schenectady. An insight into their home life was revealed through an interview with Anna in the late 1930s.

She knew that work was always on Saul's mind, but cited no examples of his getting hunches or inspirations at odd moments, either at home or during vacation. He didn't discuss his work very much at home, but he did work a lot at home and always became deeply absorbed in it. He liked to have Anna remain in the room with him while he was working, but gave no sign he knew she was there. She knew, though, that he would become aware of her leaving. He was not particularly interested in social activities, he enjoyed his home. He liked plays and good movies, but not just for something to do. He didn't care for music. He was handy with tools when he had the time and did most of the construction work on their camp on the Hudson River. Saul and Anna enjoyed gardening, both vegetables and flowers. Anna did little canning because they both preferred commercial canned food. However, Anna did like to make tomato relish and marmalade.

Saul's favorite hobby was reading technical magazines. At one point he translated a Russian technical book into English, working late hours on it and often taking it to bed with him.

Anna said that Saul might be called absent-minded, but *she* preferred to consider it deep concentration. He did remember birthdays and anniversaries. While wrapped up in work he would let a cigar burn down and mar the table. He sometimes fell asleep on the couch with a cigar in his mouth—burning his vest on one occasion.

One day he parked his car uptown in Schenectady and forgot to put a nickel in the meter. On returning he found a ticket. He took it to the police station and paid the fine. He decided to keep it a secret because he didn't want to admit his absent-mindedness. He arrived home late because of his visit to police headquarters. Anna asked if he had been detained at the Lab. He said yes. Then during dinner he couldn't resist telling her what had happened. The next day he told everyone in the lab.

Anna Dushman felt that wives of scientists must be sympathetic and understanding. They must realize how much work means to their husbands and that they have less time for wives than average men. Their minds are too active and concerned with the laboratory to consider household affairs.

Dushman retired from the Laboratory on 31 December 1948 after more than 36 years of service (Fig. 6). He and Mrs. Dushman then spent the winter in Florida for a well-earned vacation. However, in the spring he was back at the Lab as a consultant, where he continued to work until he suffered a stroke from which he never fully recovered. During his illness at home he wrote the following reflections under the stimulation of a book he was reading:

Reading for Relaxation

Why do you enjoy one book and dislike the other? It is not the finely written prose, the harmony of words and thought—although these invariably give you intellectual satisfaction—but it is the revelation you receive, as if by inspiration, of the philosophy of living, of what it means to be human. You read and inwardly glow with pleasure as the author expresses so well those thoughts which you, the reader, have been cherishing dimly but could never write down so clearly. A good book, one that stimulates you, that makes life worthwhile, is in a sense, a mirror of your soul; but it differs from a physical mirror in this very special characteristic that it is what is known scientifically as a selective reflector. It reflects only those radiations emanating from within your mind that enrich your thinking and inspire you with the thought that being alive is the greatest experience in creation.

He died at his home in Scotia, New York, on 7 July 1954 at the age of 70 after a prolonged heart ailment.

ACHIEVEMENTS AND HONORS

Dushman's awards started at an early age. As previously mentioned, he was a Prince of Wales scholar at the University of Toronto from 1900 to 1904. He became a naturalized United States citizen in 1917.

From 1923 to 1925 Dushman served as director of research for the Edison Lamp Works at Harrison, New Jersey, the first GE lamp factory. In 1928 he was made assistant director of the GE Research Laboratory, a post

he held until his retirement in 1948. Dushman took a major responsibility for the administration of the GE science fellowships, which were awarded every year to a selected group of high-school science teachers. He also served for many years on the committee of awards for the GE Coffin fellowships.

In June 1940 Dushman received the honorary degree of Doctor of Science from Union College. In November 1941 he was elected by the board of trustees of Princeton University to membership on the advisory council of the department of physics.

Dushman was especially proud of the invitation he received to give the opening address at the High Vacuum Conference held at Perthshire, Scotland, in September 1947.

Dushman was elected to Sigma Xi and was a member of the APS, the ACS, the AAAS and the AIEE (now IEEE).

Dushman continues to receive recognition posthumously. In 1977 the GE R&D Center established the Dushman Award. The annual award recognizes outstanding contributions to scientific projects or technologies by individuals or teams at the GE R&D Center. The

FIG. 6. Photograph of Dushman at his desk in the Laboratory taken at about the time of his retirement in 1948.

nominees are judged on the basis of outstanding scientific and technical contributions, and impact on the R&D Center, on GE, and on world science and technology.

To these achievements should be added the most important of all honors—the respect and love of those who knew him.

REFERENCES

1. Lafferty JM, Kingdon KH. *J Appl Phys* 1946; 17:894.
2. Langmuir I. *Vacuum* 1954; 3:113.
3. Hull AW. *Science* 1954; 120:686.

APPENDIX: PUBLICATIONS BY DUSHMAN

Papers

1. Heating of cable carrying current. *Trans AIEE* 1913; 32(1): 333–357.
2. Modern theories of light. *GE Review* 1914; 17:185–195.
3. Determination of *e/m* from measurements of thermionic currents. *Phys Z* 1914; 15:681–685. *Phys Rev* 1914; 4:121–134.
4. Recent views on matter and energy. *GE Review* 1914; 17:694–700, 901–908, 952–960, 1197–1203.
5. Theory and use of the molecular gauge. *Phys Rev* 1915; 5:212–229.
6. New device for rectifying high tension alternating current. *GE Review* 1915; 18:156–167.
7. Periodic law. *GE Review* 1915; 18: 614–621.
8. Kinetic theory of gases, Part I. *GE Review* 1915; 18:952–958; Part II, 1042–1049.
9. Absolute zero. *GE Review* 1915; 18:93–100, 238–248.
10. Factors affecting the relation between photo-electric current and illumination. *Astrophysical Journal* 1916; 43:9–35.
11. Theories of magnetism, Part I. *GE Review* 1916; 19:351; Part II, 666; Part III, 736; Part IV, 818; Part V, 1083.
12. Structure of the atom, Part I. *GE Review* 1917; 20:186–196; Part II, 397–411.
13. Theory of unimolecular reaction velocities. *J Frank Inst* 1920; 189:5.
14. Introductory remarks on present theories of atomic structure. *JAIEE* 1920; 39:119.
15. A theory of chemical reactivity, calculation of rates of reactions and equilibrium constants. *J Am Chem Soc* 1921; 43:397.
16. The production and measurement of high vacua (high vacuum). *GE Review* 1920; 23: June, July, August, September, October; 1921; 24: January, March, May, July, September, October.
17. The reaction between tungsten and Napthalene at low pressures. *J Frank Inst* 1921; 192:545.
18. Studies with the ionization gauge, Part I. *Phys Rev* 1921; 17:1.
19. Some recent applications of the quantum theory to spectral series. *J Optical Soc Am Rev Scien Inst* May 1922; VI (3):235.
20. A general relation for electron emission from metals. *Phys Rev* 1922; July: 111.
21. The diffusion coefficient in solids and its temperature coefficient. *Phys Rev* 1922; July: 113.
22. Atomic structure. *Albany Medical Annals*, April, 1922. Included in proceedings and papers presented at the meeting of the Interurban Surgical Society held in Albany and Schenectady, November 25–26, 1921.
23. Theory of electron emission. *Trans Am Electrochem Soc* 1923; 44: 101.
24. Graphs for calculation of electron emission from tungsten, thoriated tungsten, molybdenum and tantalum. *GE Review* 1923; 26: 154–160.
25. Electron emission from metals as a function of temperature. *Phys Rev* 1923; 21:623.
26. Studies with ionization gauge, Part II. *Phys Rev* 1924; 23:734.
27. Diffusion of carbon through tungsten and tungsten carbide. *J Phys Chem* 1925; 29:462.
28. Electron emission from tungsten, molybdenum and tantalum. *Phys Rev* 1925; 25:338.
29. Electron emission from thoriated tungsten. *Phys Rev* 1927; 29:857.
30. Line spectra and the periodic arrangement of the elements. *Chem Rev* 1928; 5:109–171.
31. Cohesion and atomic structure, Part 2. *Proc ASTM* 1929; 29:7–24.
32. Thermal emission of electrons. *Int Critical Tables* 1929; 6:53.
33. Thermionic emission. *Rev Mod Phys* 1930; 2(4):381–476.
34. Modern physics—A survey. *GE Review* 1930; 33:328, 394.
35. Recent advances in the production and measurement of high vacua. *J Frank Inst* 1931; 211:689–750.
36. Quantum theory. *J Chem Education* 1931; 8:1074–1113.
37. The lamp of the future? *Electrical News Engineering* 1931; 40:51, 52, 68.
38. A symposium of some activities of the research laboratory of the general electric company. *Trans Am Inst Chem Engrs* 1933; 28: 31–55.
39. Production of light from discharges in gases. *GE Review* 1934; 37: 260–268.
40. Electron emission. *Elec Eng* 1934; 53(7):1054–1062.
41. Low pressure gaseous discharge lamps. *Elec Eng* 1934; 53(8): 1204–1212; 1934; 53(9):1283–1296.
42. The story of the Charles A. Coffin Fellowships. *GE Review* 1934; 37:541–547.
43. The evolution of physical concepts. *The Scientific Monthly* 1936; 42(5):387–395.
44. Exploring the atomic nucleus. *Elec Eng* 1936; 55(7):760–767.
45. The search for high efficiency light sources. *J Optical Soc Am* 1937; 27:1–24.
46. Educating physicists for industry. *J Appl Phys* 1937; 8(1):59–67.
47. The quantum theory of valence. *J Phys Chem* 1937; 41:233–248.
48. Evolution of concepts in nuclear physics. *GE Review* 1937; 40(11): 503.
49. Recent developments in gaseous discharge lamps. *J Soc Motion Picture Engineers* 1938; 30(1):58–80.
50. Contributions of mathematics to research in physics and chemistry. *J Eng Education* 1939; 29(9):716–724.
51. Creep of metals. *J Appl Phys* 1944; 15(2):108–124.
52. Postwar training of physicists for industry. *Am J Phys* 1944; 12(4): 219–224.
53. Mass-energy relation. *GE Review* 1944; 47:6–13.
54. Application of theory of absolute reaction velocities to creep of metals. *Rev Modern Phys* 1945; 17:1276.
55. Calibration of ionization gauge for different gases. *Phys Rev* 1945; 68:278.
56. Proposed unit for high vacuum. *Science* 1945; 102:383.
57. Manometers for low pressures. *Instruments* 1947; 20:3.
58. Thermionics. 1947 *Encyclopedia Britannica*.
59. Scientific aspects of vacuum technique. *Chem Indus* 1948; 67, S3.
60. Development of high vacuum technique. *Indus Eng Chem* 1948; 40: 778.
61. Use of characteristic curves for pump-speed vs. pressure. *Rev Scien Inst* 1949; 20(2):139.
62. The new chemical elements. *GE Review* 1951; 54(4).

Books

63. *Production and measurement of high vacua.* General Electric Review; Schenectady, 1922.
64. Quantum theory of atomic spectra and atomic structure. In: Taylor HS, Glasstone S, eds. Volume 1: *Atomistics and thermodynamics: Teatise on physical chemistry.* 3rd ed. New York: D. Van Nostrand; 1942.
65. *Elements of quantum mechanics.* New York: John Wiley & Sons; 1938.
66. *Scientific foundations of vacuum technique.* New York: John Wiley & Sons; 1949.
67. *Fundamentals of atomic physics.* New York: McGraw-Hill; 1951.

Wolfgang Gaede *(1878–1945)*

Günter Reich

(Translated from the German by G. Lewin)

Wolfgang Gaede at an age of 39 years (1917).

INTRODUCTION

During the first half of this century, Gaede contributed substantially to the understanding of the processes occurring in very rarefied gases; he conceived and applied important methods for the production of low pressures; he developed new pumps and improved existing ones. This was done in cooperation with the company E. Leybold's Nachfolger during the years from 1906 to 1944.

Alfred Schmidt became a partner of Leybold in 1894 and owner in 1910. He signed a contract with Gaede in 1906 which was the beginning of a very productive relationship, characterized by the economic constraints of a commercial enterprise on the one hand and, on the other hand, by the ambition of a scientist and inventor who wanted to find optimal solutions to the needs for new products. Market research and product management

were not known at this time; nevertheless, this cooperation did not result in any major errors. It was ended in 1931 by the death of A. Schmidt. His son-in-law Manfred Dunkel became the new manager. Due to the difficult economic situation, he had to modify Gaede's contract to the latter's disadvantage; it was also impossible to turn each of Gaede's ideas into a new product. Nevertheless, this did not detract from his productivity; on the contrary, the chemist Dunkel stimulated Gaede's interest in applications. During this time Gaede wrote his last important report, "The vacuum technique of vapors and the gas ballast pump."

Gaede's life prior to his meeting Schmidt will be described first, and then the growth of Leybold's up to that time. The early cooperation and the financial agreements follow. His molecular pumps, diffusion pumps, oil-sealed forepumps and vacuum-measuring devices are dealt with in subsequent chapters. Finally the author

43

attempts a subjective appraisal of Gaede's accomplishments.

Beside the literature listed at the end, the author had at his disposal all the German patents of Gaede, and catalogs, some other notes, letters, and official Leybold documents.

THE EARLY DAYS

Max Paul Wolfgang Gaede was born on May 25, 1878, in Lehe, now part of Bremerhaven. His father was a Prussian artillery officer, which necessitated a frequent change of residence; he was last stationed in Strassburg and, after his retirement in 1891, moved with his family to Freiburg (Breisgau), his wife's former home. Gaede graduated from high school in 1897 and studied first medicine and later physics at the university in Freiburg. He graduated in 1901 (under Himstedt) (1) and took an assistantship. He measured the contact potentials of metals, but the results were poorly reproducible, and he discussed the effect of water films on the surfaces. His publication (2), submitted on 26 February 1904 ended with the statement: "I am now investigating the influence of electrical forces on the Volta effect in vacuum and I hope to report about it soon." He had available only a Sprengel-Kahlbaum pump with a pumping speed of about 0.01 l/s, much too small to remove the surface gases. His sister states in her biography of Gaede (ref. 22a, p. 23): "The only solution left to him was to try to construct a pump of much greater speed than those available at the time (3)."

Shortly afterward Gaede designed and built a rotating mercury pump. Unlike the other pumps made of glass, it contained a rotating drum in a metal housing. The drum was made of porcelain, and he had to convince the *Königliche Porzellanmanufactur* (Regal Porcelain Manufacturing Co.) in Berlin that a drum of this shape could be made of porcelain (Fig. 1). This pump was of relatively small size, since the gas was ejected below atmospheric pressure. It contained 26 kg of mercury and required a forepump. A pumping speed of 0.1 l/s was attained at a rotational speed of 20 min^{-1}. Very low "partial air pressures measured with a McLeod gauge" were reached. His sister reports further (ref. 22a, p. 23) that the pump was assembled in early September 1905 (3). The patent was applied for on 7 September 1905. The pump performed properly right away and no major modifications were required for industrial production. Gaede travelled at once, with the pump and the mercury in his baggage, to the 77th meeting of the Natural Scientists (*Naturforscher*) and Physicians in Meran. He was permitted to demonstrate his pump on September 26 between two lectures. It must have made a major impression. When asked during the discussion "How much does the pump cost?" he replied "I have not yet contacted any potential manufacturer. If somebody wanted one, I would supply it to institutes as cheaply as possible, I guess 200 marks without mercury (3)." He received numerous orders, which he could not help accepting. He assembled and tested about 60 pumps at his workplace until the manufacture started at Leybold in 1907 (ref. 22a, p. 30). He described a modified design with a cast-iron drum a few years later (4).

ALFRED SCHMIDT AND THE LEYBOLD COMPANY IN COLOGNE

E. Leybold's Nachfolger has been founded in Cologne in 1850. It was a typical local commerce and service estab-

FIG. 1. Rotary mercury pump by Gaede, 1905, the inlet port *R* is a ground joint made of glass and covered with mercury from Leybold Catalog 1907.

lishment and was taken over by Ernst Leybold in 1851. In 1854 it became a "manufacturing and supply company for glass medicine bottles and other containers, pharmaceutical, chemical, and physical utensils and apparatus." This company was taken over by Emil Schmidt in 1870. A partial catalog, published in 1878, lists 1333 "physical apparatuses for high schools manufactured in the machine shops of E. Leybold's Nachfolger, Köln," indicating already the direction of later specialization.

While studying chemistry in Berlin, Emil Schmidt's son Alfred (born in 1867) (22) had the opportunity in 1889 to watch the manufacture of incandescent lamps at the Siemens-Schuckert Co. The lamps were still evacuated with manually operated Töpler pumps. Subsequently he observed how the production manager A. Raps, a friend of his, automated the pumping process by inventing a mechanism that was connected to the water supply and actuated the rise and fall of the mercury in the pumps (Fig. 2). This permitted the worker to operate six pumps instead of only one. After joining his father's company, he obtained a Rasp pump and tried unsuccessfully to find customers for it. "The pump became a success a few years later, after Röntgen had discovered x rays and we were manufacturing a substantial number of Röntgen tubes. They were evacuated with a number of these pumps (22)."

Alfred Schmidt became a partner of Leybold in 1891 and continued his efforts to sell vacuum pumps, but without much success. Hence it was a great disappointment for him to learn in 1904 that his competitor (A. Pfeiffer Wetzlar) had received a license to manufacture and market the "Geryk" vacuum pump. This pump was manufactured by the British Pulsometer Engineering Co. from a design based on the Fleuss patents and was an oil-sealed piston pump, capable of reaching pressures below 1 Torr. While looking for a substitute, Schmidt found by accident Gaede's report about his new mercury pump (3). He took the next train to Freiburg in February 1906 to acquire the rights for this pump. Negotiations ensued with the participation of a patent attorney from Berlin and a licence agreement was signed on 23 April 1906. It was the beginning of a very unusual cooperation lasting many years.

WOLFGANG GAEDE AND LEYBOLD, 1906–1945

The German patent application G21834 covering Gaede's rotating mercury pump was rejected in 1910 because of objections by competing companies. The license agreement between Gaede and Leybold was similar to those of today, except that the amount of the royalty was very high: 25% of the sales price (without mercury). At the time, Leybold badly needed a competitive product, but it is doubtful whether Alfred Schmidt realized that "the agreement did not give him only a license to manufacture a certain pump, but also the cooperation of an ingenious physicist who would fundamentally change Leybold's position in the world" (ref. 23, p. 33). There is no indication that Gaede was obligated to continue this cooperation. "After the first prospectus of the rotating mercury

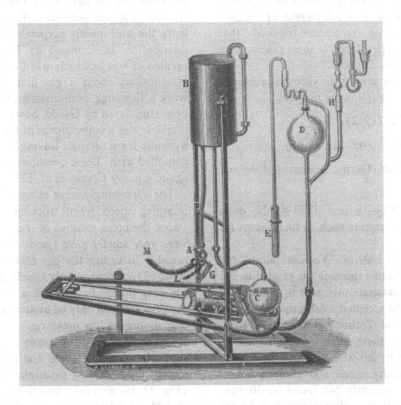

FIG. 2. Automated Töpler pump by A. Raps from Leybold Catalog 1904.

pump of Gaede had been published in March 1907, the sales increased suddenly and likewise the profits of Leybold, without, however, matching the royalties Gaede received (23)." The pump was produced in large quantities until 1926. It is remarkable that there was no need at any time to modify the design.

Another contract with similar conditions was signed on 3 December 1906: "Based on his experience with the processes occurring in high vacuum, Gaede makes Leybold the following proposals for improvement of evacuation on an industrial scale:

1. the manufacture of a larger rotating mercury pump...

2. the manufacture of a twin piston pump..."

The wording of this contract suggests that Gaede proposed these developments to Leybold, not vice versa. In addition, his sister states (ref. 22a. p. 85) that Gaede mentioned frequently that "he never invented anything in compliance with somebody's request."

Gaede quit his assistantship at the university and started his own laboratory with a machine shop in 1907. Leybold provided the lathes and milling machines. He employed two assistants and two mechanics. The laboratory was made available to the university from 1912 to 1921 as the "Technical-Physical Institute." There is little known now how the operating expenses of this institute and, later on, of Gaede's laboratory in Karlsruhe, were split between Gaede and Leybold. Apparently, the arrangements were changed frequently, Gaede received large royalty payments and was able to partially support the laboratories, but Leybold always shared the expenses, especially when investigations of the physics of vacuum were conducted. The license agreements required "that all inventions which were developed in these places, were produced and marketed by Leybold."

Gaede did not limit his activity to vacuum problems. There are references to the following projects:

Refrigerators to make ice (1912),

Wireless telegraphy (1912) (ref. 22a, p. 45),

Safety devices for railroads (German patents 464566 and 466444; 1926).

During the wars of 1914–18 and 1939–45 he dealt frequently with military projects such as fuel pumps for airplanes.

Gaede's first lecture in 1909 on "Vacuum pumps and their importance for scientific research" is printed in ref. 22a, p. 93. He became associate professor in Freiburg in 1912 and was offered and accepted a full professorship of physics at the Institute of Technology in nearby Karlsruhe in 1919. He had there at his disposal a well-equipped laboratory where his assistants and mechanics would work. His university career was abruptly terminated when he was prematurely pensioned on 30 June 1934 as a result of an utterly unfair persecution by the Nazi authorities. He suffered tremendously under the emotional strain of the proceedings, which included a "political examination."

Gaede installed a laboratory in his house in Karlsruhe and Leybold supplied the shop facilities. This enabled him to continue working in the accustomed manner, which included the cooperation with Leybold. As indicated in a letter he had written to Dunkel, Gaede was very touched by the presentation of the Siemens Ring on December 13, 1934, bestowed every third year on the birthday of Werner von Siemens. (The award lecture is printed in ref. 13b.)

Laboratory and shop were transferred to Munich in 1940 in compliance with a request of Gaede. Leybold paid all expenses thereafter. Gaede received a generous compensation and Leybold a license for all patents without royalty payments.

The laboratory and shop were completely destroyed by bombing in 1944. Gaede, who had never married and had lived with his sister Hannah for many years, died in Munich on 6 June 1945, a few weeks after the end of the war. He was 67 years old. He had been suffering from diphtheria aggravated by the lack of proper nourishment.

THE MOLECULAR PUMP[48]

Gaede discussed M. Knudsen's paper (24) about the laws governing molecular gas flow in his paper (6) "The external friction of gases" (1913). He confirmed Knudsen's results for pressures below one millitorr; however, he states that the gas film on the wall affects the external friction at higher pressures; this causes the molecules to leave the wall mainly perpendicularly, not with a cosine distribution as surmised by Knudsen. He shared the opinion of von Smoluchowki (25) in connection with his assumptions about a gas film. Clausing gave Gaede's work a thorough examination (26) and appears to have been stimulated by Gaede, however he insisted that "the cosine law is a consequence of the second law of thermodynamic for molecules leaving a wall," hence it must be complied with. These questions have recently been raised again, e.g., by Comsa et al. (27).

The surprisingly large effect of the gas friction on the pumping speed when working with a mercury pump (since the cross sections of the connecting tubes [Fig. 1] were very small) gave Gaede the idea that "it must be possible to utilize the gas friction to attain a pumping effect (7)." Consequently Gaede invented the principle of the drag pump; he called it a "molecular pump," since the principle can only be understood by analyzing molecular motion at low pressures.

Gaede developed the theory of drag pumps in order to maximize pumping speed and compression, and demonstrated the operating principle in a simple arrangement (Fig. 3). He also described the measurement of pumping speed by the determination of the rate of pressure drop in

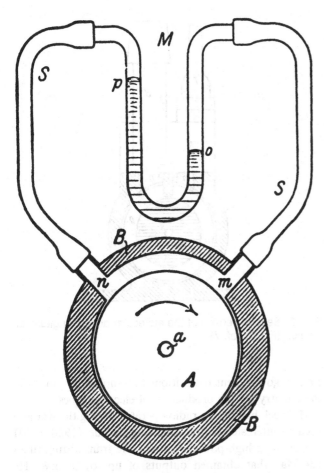

FIG. 3. Principle of molecular-drag pump. Rotation of cylinder *A* in the direction of the arrow produces a pressure difference between *n* and *m* thereby causing a pumping action in the same direction (Fig. 9, ref. 21).

a vessel of known volume and calculated the effect of the "conductance" of the connecting tubes on the "effective pumping speed." Essential parts of Gaede's theoretical considerations had already been included in his treatise (6a) submitted in 1909 when he joined the faculty of the University of Freiburg.

First he built a "screw thread pump" (*Gewind-epumpe*). A smooth, solid cylinder rotates in an internally threaded hollow cylinder (Fig. 4). "Left handed and right handed threads are cut beginning in the middle.

Hence the pressure is lowest in the center" (ref. 7, p. 348). He demonstrated "that an appreciable pressure reduction can be accomplished." When he tried to increase the effectiveness by the use of more threads, he reached a practical limit. Also, strong vibration occurred at higher rotational speeds. Thus Gaede gave up this idea, which was later revived by Holweck (28) in 1921 [see ref. 21 (1931), pp. 592–93].

Gaede then designed a second version, where the moving surface area was larger than the stationary one, his theoretical considerations indicated that "this increases the efficiency of a pump of the same external size." This pump contained several stages, arranged in series but of different dimensions. The inlet port was in the center again and two symmetrical pumping sections were located on either side (Figs. 5 and 6). It was already recognized and specifically mentioned that oil vapor emanating from the bearings could not penetrate to the high vacuum side and that the compression ratio for heavy gases was very high.

Leybold manufactured about 300 molecular pumps from 1912 to 1923. It is known (from ref. 22a, p. 30) that in November 1907 Gaede stayed at Leybold's for some time to prepare the manufacture of his new pump; but it was not until 1912 that the first 60 pumps could be made. The pump was first reported (5) at the 83rd convention of the Natural Scientists in Karlsruhe 24–29 September 1911. One year later, at the next meeting in Muenster, on September 16 the pump was demonstrated for the first time by his assistant K. Goes (29) (Fig. 7). The peripheral velocity of the rotor was 35 m/s at the specified 8000 rpm; this is very small compared to today's typical speeds of about 200 m/s. The compression ratio for air was 2×10^5 and the pumping speed was 1.5 l/s below 10^{-4} Torr, a great improvement over the rotating mercury pump's speed of 0.08 l/s. The speed was still as high as 0.6 l/s at 10 Torr. (The forepump was probably a rotary vane pump.) The ultimate pressure, measured with a McLeod gauge was below 10^{-6} Torr. Pressures of condensable vapors could not be measured at the time; the out-gassing of the grease in the ground joints must have been appreciable. Indicative of the high performance

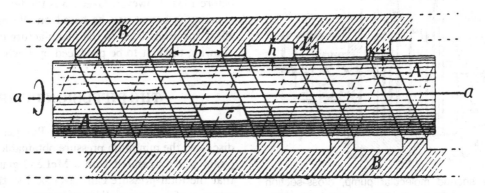

FIG. 4. Principle of the screw thread pump (Fig. 12, ref. 6a and Fig. 2, ref. 7).

of this pump was the relatively short down time and the "quality" of the vacuum in x-ray and canal ray tubes (e.g., a "pure electron discharge" could be obtained).

Gaede noted that the outstanding feature was "that the molecular pump exhausts not only gases but also vapors, contrary to the other pumps known so far." The mercury pumps in use at the time could pump permanent gases only, not water vapor, in analogy to the McLeod gauge, which does not measure the partial pressure of the water vapor. This advantage was important for many applications—in particular for the evacuation of electron tubes. W. D. Coolidge of General Electric reported in December 1913 (30), that the manufacture of a high-power x-ray tube with a pure electron discharge was rendered possible only by the use of Gaede's molecular pump. Langmuir suggests in his U.S. Patent 1558436 to evacuate an electron tube with Gaede's molecular pump. Dushman, after having investigated the pump, confirmed Gaede's results (30a). H. D. Arnold of Western Electric Co. reported (31) that he received the new pump on 22 April 1913. The first electron tubes for telephone amplifiers exhausted with this pump were installed between New York and Baltimore half-a-year later, and across the continent in June 1914. In recognition of this success Gaede was awarded the golden Elliot-Cresson Medal of the Franklin Institute in Philadelphia. Telefunken of Berlin began in 1915 to evacuate transmitting tubes with molecular pumps (32). During the war, in 1915 (33), the

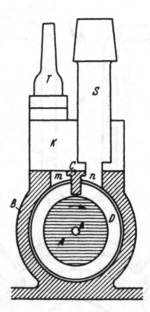

FIG. 6. Second version of the molecular pump, longitudinal-section (Fig. 7, ref. 7).

French government requisitioned all molecular pumps in the country for the production of electron tubes.

Holweck had been a radio operator during the war and became interested in the following decade (1920–1930) in designing high-power demountable transmitting tubes (34,34a) that obtained outputs of up to 150 kW. He started with Gaede's screw-thread drag pump and successfully overcame his difficulties. His pump was larger and heavier than Gaede's second version, but it has a pumping speed of 3.5 l/s and a compression ratio of several million for air (35). The peripheral speed was likewise 35 m/s at the specified 4500 rpm.

The theory of the molecular drag pump as developed by Gaede is still the basis for the calculation of all drag pumps (36). It is tragic that Gaede, in his endeavor to optimize the pump based on his theory, pursued a risky design. The vacuum connection of the stages was a difficult problem, and more than 4 stages are hardly possible. Hence this pump did not have a chance in the long run. It is not known whether Gaede continued its further development and participated in the engineering of the Siegbahn molecular pump, which Leybold manufactured before 1931. However, Gaede was the first to realize that a pumping action can be based on the molecular drag effect and lends itself to the manufacture of pumps, thus allowing the first to be produced by Leybold.

THE DIFFUSION PUMP

The physics text book by Müller-Pouillet in 1906 (37) discussed the minimum pressure obtainable by a mercury pump when measured with a McLeod gauge. It implied that the total pressure can never be less than the vapor pressure of mercury at the ambient temperature, even if a

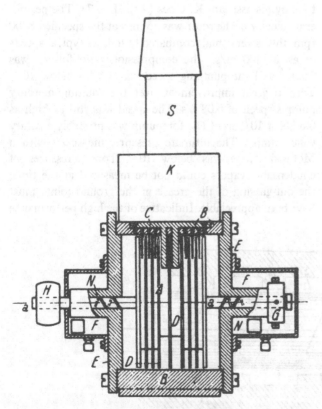

FIG. 5. Gaede's second molecular pump, cross-section (Fig. 6, ref. 7).

cold trap is inserted between the pump and the vacuum vessel, because the theory of diffusion requires that the sum of the partial pressures is the same everywhere in the system. Hence the lowest obtainable pressure must equal the mercury vapor pressure in the pump, even in the absence of mercury in the vessel. The validity of this statement at low pressures could not be determined at the time, since the pressure could only be measured with a mercury-filled McLeod gauge. The residual pressure in the absence of mercury could not be measured.

Gaede attacked this problem theoretically and experimentally. He wanted to show that the ultimate partial pressure obtained with his rotating mercury pump is the one measured with the McLeod, with or without a cold trap between the McLeod and the vacuum vessel or the pump. Summarizing his results in his paper "The Diffusion of Gases through Mercury Vapor at Low Pressures and the Diffusion Pump (9)," and in ref. 6a, Chapter 2, he writes that the above statement has been refuted by his experimental results. He showed that particle diffusion depends only on their partial pressure gradient if their mean free path in the mercury vapor is not small compared to the diameter of the tube connecting the mercury pump with the cold trap (molecular flow).

Consequently the partial pressure of air in the pump is always lower than in the vessel while pumping. Gaede's statements were revolutionary at the beginning of the century and it is not surprising that some of his arguments are difficult to comprehend. In addition, when speaking about a diffusion pump, he meant something different from what was realized later on. Figure 8 shows some details to elucidate the situation.

The vessel on the left (subscript 2) is initially filled with air (subscript g). The vessel on the right (subscript 1) contains mercury (subscript v). According to Müller-Pouillet (37),

$$p_{v2}+p_{g2}=p_{v1}+p_{g1}$$

irrespective of the condition of the cold trap (cooled or at ambient temperature); however, it should be obvious that this does not apply to molecular flow. We now assume that a mercury pump (e.g., a Töpler pump) is connected to vessel 1 and, therefore, $p_{g1} \ll p_{v1}$. If the cold trap is empty, after a sufficient pumping time $p_{g2} \ll p_{v2}$, and the total pressure in 2 is $p_{t2}=p_{v2}=p_{v1}$. But if the cold trap has been filled, $p_{v2}=0$ and it should follow, according to Müller-Pouillet, that $p_{t2}=p_{g2}=p_{v1}$.

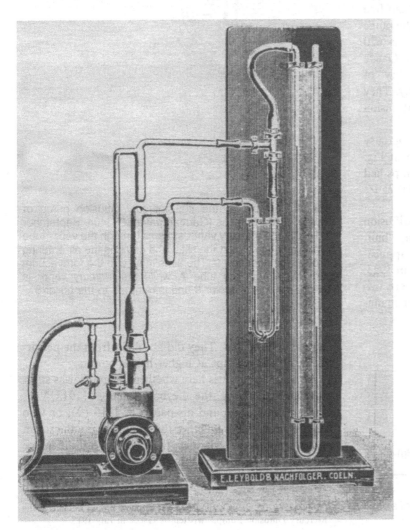

FIG. 7. Demonstration of the pressure difference between inlet and outlet of Gaede's molecular pump (Fig. 1, ref. 29).

Contrary to this reasoning, Gaede found for conditions as defined above that

1. ... p_{g2} can be smaller than p_{v1}, but not equal to p_{g1}, for the mercury vapor flowing to the cold trap causes p_{g2} to be always larger than p_{g1}. This means that the pressure measured by a McLeod gauge connected to the right (1) vessel is too small with respect to the true pressure in (2).

2. gas molecules can diffuse counter to the mercury vapor flow through the connecting tube and through vessel 1 into the mercury pump, the rate of diffusion increases as their mean free path in mercury vapor increases. "This arrangement acts like a pump. We can call the conductance of the connecting tube the pumping speed of the diffusion pump."

These considerations apply equally to the diffusion slit 3 in Fig. 9 or q in Fig. 8. Hence it must be quite narrow, and the pumping speed is quite small. Consequently Gaede's first diffusion pumps have narrow slits through which the gas enters into the mercury vapor jet.

The measurements plotted in Fig. 10 show that the pumping speed initially increases as the mercury vapor pressure is increased, reaches a maximum, and drops thereafter. This is to be expected because the "mean free path of the gas particles in the mercury vapor in the range of the slit" decreases steadily. The maximum was 620 cm³/cm²s, only 5% of the theoretical maximum. Gaede calculated the solid curve for a "pumping probability" of 8%, the results confirmed his expectations exactly. They determined his attitude toward the design of diffusion pumps for a long time.

Leybold started the manufacture of Gaede's mercury diffusion pump in 1913. The first model, shown in Fig. 11, had a pumping speed of 0.08 l/s. After 60 pumps had been sold, a new version was marketed in 1917 of 0.25 l/s speed. This one was more successful.

Gaede wrote a summarizing report about diffusion pumps in 1923 (10). He mentioned therein a pump "built in 1914 made completely of steel for technical applications," but the development was only completed in 1923. It is not known whether these pumps, (there were two types, of 1.5 and 6 l/s pumping speed), were ever manu-

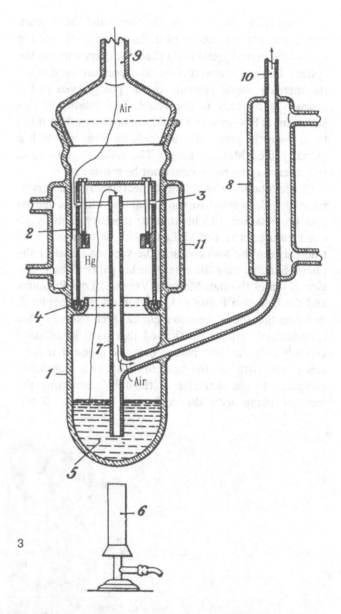

FIG. 9. Principle of the first mercury diffusion pump of Gaede, made of glass (German patent 286404, September 9, 1913). The mercury vapor streams through the vapor tube 2, made of steel, past the slit 3 and entrains the air entering through the inlet port 9. The air is moved to the forevacuum port 10 through the tube 7, while the mercury vapor is condensed in the cooler 8 and is returned to the boiler 5.

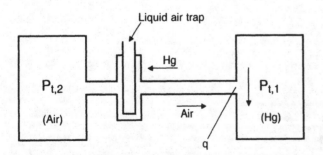

FIG. 8. Relationship between the partial pressures of air and mercury with full and empty coldtrap.

factured by Leybold. They did not differ from the performance of glass pumps at higher heater powers.

We shall not discuss here whether Langmuir was stimulated (38,38a) by the Leybold catalog (1913)* or Gaede's publication and his patent of 1915 (9), or who has the priority, nor whether the stimulus were the steam ejector pumps or whether there is a real difference between condensation pumps, Langmuir's definition, and diffusion pumps as visualized by Gaede. Gaede stated in

---

*On pp. 204–205, there is the first description of "The Principle of the Mercury Diffusion Pump," written by Gaede in July 1913.

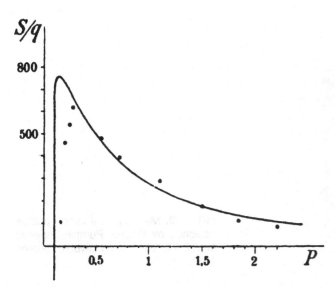

FIG. 10. A plot of the pumping speed per unit area (cm³/cm² sec) as a function of the mercury vapor pressure *P* in Torr, calculated from its temperature in the pump (Fig. 14 of ref. 9).

today's diffusion pumps. Their pumping speed of 3–4 l/s (more if needed) was substantially higher than that of all other high vacuum pumps at the time and largely independent of the heater power, contrary to Gaede's diffusion pumps.

It is not documented when Gaede learned of Langmuir's work. It must have been a great disappointment for him to accept the superiority of the latter's design, though he attempted (10) to prove his priority and the correctness of his modes. He had to realize that his physics, though correct in principle, had led him astray, as previously in the case of the molecular pump. However, this does not diminish the importance of his fundamental paper (9), which has not found the attention it deserves. Most texts on vacuum physics and technology do not mention the error caused when measurements are performed with a McLeod gauge located behind a cold trap (see ref. 6a).

Gaede continued to work on mercury diffusion pumps. He devotes only two pages to the description of his probably most successful creation (ref. 10, p. 366, Fig. 35 and ref. 10a): "Large size Gaede diffusion pump made of

ref. 10 "the air penetrates into the vapor stream by diffusion rather than mechanical mixing at low pressures." It remains a fact that the pumps shown in Figs. 3 and 8 of ref. 38a (1916) have the essential characteristics of

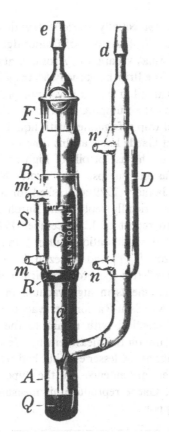

FIG. 11. Mercury diffusion pump made of glass designed by Gaede from Leybold Catalog Oct. 1913 p. 204.

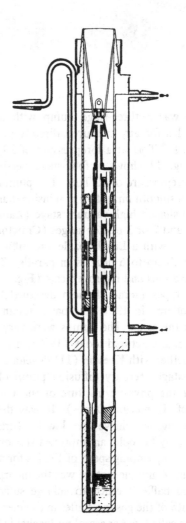

FIG. 12. Mercury diffusion pump with one diffusion and three ejector stages from Leybold Catalog 1927.

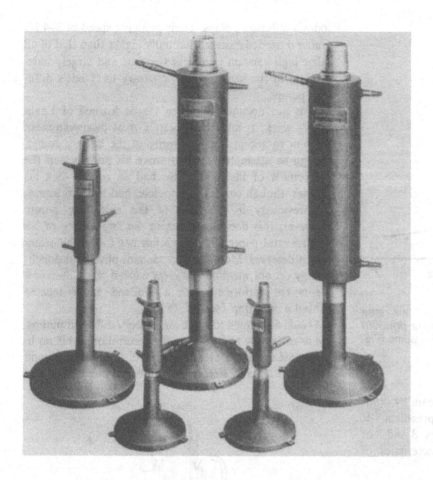

FIG. 13. Mercury diffusion pumps designed by Gaede. Pumping speeds from 2.5 to 20 l/s from Leybold Catalog 1937.

steel." This was a three-stage pump with a pumping speed of 60 l/s for air, measured directly at the pump inlet at 1×10^{-4} Torr at a forepressure of 20 Torr. The sketch of Fig. 12 shows a four-stage version with a maximum forepressure of 40 Torr. The pumping system is located in a smooth pipe of 40 mm inside diameter with a mushroom-shaped high vacuum stage (Langmuir, ref. 38a, Fig. 7) and 2 or 3 ejector stages (Crawford, ref. 39, Figs. 1 and 2) with a Laval nozzle and diffuser. Pumps were built with up to 6 nozzles in parallel. There were other modifications and improvements (Fig. 13); 5 metal and 5 glass types (some of them designed by Volmer [40]) were offered in 1927. The basic design principles were always the same. The pumps were very successful and are still manufactured today (1990), if requested.

Gaede together with Keesom (11) developed in 1928 a large single-stage mercury diffusion pump of 200 mm diameter for the physical institute of the University of Leiden (Prof. Kamerling-Onnes). It was designed for helium pumping, to achieve very low temperatures, and manufactured by Leybold; an improved version, designed by Gaede, had a pumping speed of 130 l/s for air and 422 l/s for helium. A unique feature was the incorporation of a refrigerated baffle "needed to enlarge sufficiently the mean free path of the gas molecules in the mercury vapor stream (later called 'vapor seam' by Jaeckel [41]) in view of the large pump dimensions."

Gaede was not equally successful with oil diffusion pumps. He met Prof. Aston (of Cambridge University) during a Christmas vacation in Arosa, Switzerland, who told him that "the British company Metropolitan-Vickers now produces an oil of such a low vapor pressure, that it can be used as the pumping fluid in a diffusion pump without a cold trap refrigerated with liquid nitrogen." It is unlikely that Gaede knew at this time Burch's publication (42) about the use of oils of high boiling point in mercury diffusion pumps. He wrote Metropolitan-Vickers on 9 January 1929 and asked for a sample. He received a very friendly reply from Burch a week later with a detailed report of 12 December 1928 about his results with "oil condensation pumps" and one sample each of Apiezon B Oil and Apiezon L Grease. Burch discussed in his letter the problematical nature of total pressure measurements in high vacuum in great detail. He liked the Knudsen-Radiometer vacuum gauge best, but was very sceptical with regard to the influence of radiation pressure on the measurement. He proposed the ionization gauge as the lesser evil, but had reservations on account of the dependence on the type of gas. He suggested that Gaede reproduce his measurements, and offered his support.*

---

*From this letter we know that Burch has used a Langmuir mercury condensation pump (see ref. 38a, Fig. 8 or ref. 44, p. 188, Fig. 12) for

Gaede put a larger boiler on a single-stage mercury pump and his assistant W. Molthan started an investigation in March of 1929. The results were obviously disappointing, for he writes in his report of 11 November 1929: "The results were unfortunately totally negative, it was not at all possible to reproduce your high vacua." (The cooperation was greatly hampered because Gaede knew hardly any English and all correspondence had to be translated.) It is not known whether he accepted the invitation "to see everything being done in the laboratory" when he went to England in the Spring of 1933 to receive the Duddel Medal of the Physical Society.

Gaede held the opinion that Apiezon oil was not suited for diffusion pumps, but designed in 1930 an "oil diffusion pump for machine oil with the forepump stage and the high vacuum stage arranged in two separate housings." The obvious purpose was to circumvent foreign patents and to be independent of special oils. Neither this pump nor another version with one vapor tube designed in August 1931 and reported in 1932 at a meeting in Karlsruhe (12) can match the specification of the Burch pump. It is interesting to note that the pump was developed at Metropolitan-Vickers for the same application as the Holweck pump, i.e., for demountable high-power transmitting tubes. A 500 kW tube was described in 1931 (43), with 13 oil diffusion pumps and one forepump.

Leybold maintained contact with Metropolitan-Vickers and in the spring of 1932 M. Dunkel, who had become the manager of Leybold in August 1931, entered a license agreement regarding the Burch pump. Undoubtedly it was an important factor that Gaede's patent of 1913 referred to diffusion pump operation with any kind of vapor. Gaede showed no further interest in oil-diffusion pumps, the Burch pump license did not affect him, for the sale of oil-diffusion pumps was only a small fraction of mercury-diffusion pump sales (3.7% in 1944).

MEASUREMENT OF VACUUM

Gaede was concerned with vacuum measurements from the very beginning. The Leybold catalog of 1908 already contains a McLeod gauge designed by Gaede. Later several other gauges were designed by Gaede, among them the "swivel gauge", which is still produced today.

Gaede wanted to know how low a pressure could be reached with oil-diffusion pumps; hence he needed a suitable total pressure gauge and started to develop one himself (13). First he built a hot cathode vacuum gauge. The electron path, and thereby the sensitivity, was increased by a magnetic field produced by a solenoid. The gas clean-up caused him to conclude "that the ionization gauge cannot be used for the measurement of low pressures, because it affects the gas balance, by clean-up,

degassing, and dissociation of gaseous compounds."

Next he considered the radiometer principle of Knudsen. He had quoted pressures measured with such gauges in his reports on oil-diffusion pumps without giving any details. Now he developed a gauge which he called the *Molvakumeter* (13a), which measured the thermal molecular pressure and, without heating, the gas friction (ref. 41, p. 61 and ref. 44, p. 309). The Molvakumeter was produced by Leybold and is described in the catalog of 1934 (Fig. 14). This gauge, like its predecessors and successors, was not a practical commercial product and it was discontinued.

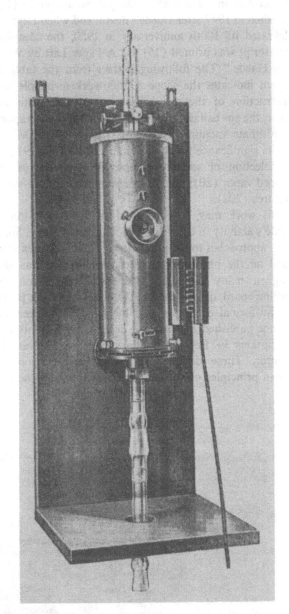

FIG. 14. Molvakummeter of Gaede from Leybold Catalog 1934.

his experiment. Gaede used, by recommendation of Burch, a Leybold type "C" diffusion pump, the smallest on Fig. 13.

GAEDE THE PUMP DESIGNER AND THE GAS BALLAST

Gaede's publication about low-pressure measurements (13) was the last report published during his life. There still exists a draft of a paper on "Thermopumps" (1941–42), where he proposed the use of thermal diffusion for pumping. This draft emerged after 20 years and was carefully examined and supplemented (also by experiments) by Albrand and Ebert, as shown in their paper (46) "The Applicability of Gas Flows in the Molecular Range." In March 1944, Gaede submitted a lengthy manuscript "Gas Ballast Pumps and the Vacuum Technique of Vapors" to the *Z. Techn. Phys.*, emphasizing applications, an attitude certainly influenced by Dunkel, the former industrial chemist. This paper could not be published at this time, mainly on account of its length. M. Dunkel had an excerpt about the gas ballast pump (14) published in 1947 after Gaede's death, and when Leybold celebrated its 100th anniversary in 1950, the complete manuscript was printed (15) as "A Paper Left by Wolfgang Gaede." The following extract from the table of content indicates the scope of this work: principle and construction of the gas ballast pump; the protective cooler; the gas ballast required for the pumping of vapors; the ultimate vacuum of gas ballast pumps; the gas veil and its significance for the vacuum distillation (19) and the selection of suitable pumps; the recovery of the pumped vapor (20); pressure measurements of gas-vapor mixtures (20a).

This work may be considered as the conclusion of Gaede's activity in the field of vacuum technology. Some of his approaches to the attainment of low pressures were based on the physics of gas behavior, but he has also designed many oil-sealed mechanical pumps on the drafting board, despite the fact that Gaede (ref. 23, p. 60) had no formal education as a mechanical engineer. He has not published his machine designs, but patents and descriptions in Leybold catalogs refer to him as the inventor. These designs are mainly modifications of known principles of vacuum pumps whose purpose was to circumvent foreign patents or to improve performance. Based on this activity, Leybold has offered since 1937 a complete line of single-stage, and some two-stage, rotary vacuum pumps with pumping speeds of 2 to 150 m^3/h.

The first "box" pump, a single-stage rotating vane pump (1.5 m^3/h), was made of bronze (45,16) and was produced in large quantities from 1908 to 1930. This pump was replaced by a new version made of steel (16a). This was a forepump for the rotating mercury pump and subsequent high vacuum pumps (Fig. 15). This pump differed from the first vane pump of Kaufmann of the Siemens-Schuckert Company, Berlin, which had been shown first at the same meeting where Gaede had presented his mercury pump, since it was quite handy and the mechanism not covered with oil. It could also produce compressed air at one atmosphere pressure, useful for blowing glass in the laboratory.

Oil-sealed piston pumps (e.g., the Fleuss type) were obviously still popular at the time for physical demonstrations. This explains why Gaede developed a pump of this type (Fig. 16), production of which began in 1913 (8). It is noteworthy that this pump was equipped with a "water separator." Leybold became interested in rotary, oil-covered vane pumps much later than the competitors. Hence Gaede started only in 1925 to develop such one- and two-stage pumps (17), which were the first ones "where the back-up of oil into the vacuum line with the pumps at rest was prevented" (Fig. 17). Aggregates of pumps, up to 12, driven by one motor and contained in one box filled with oil, became popular.

Starting in 1930, Gaede employed the principle of the rotary piston pump (18) to obtain greater pumping speeds. A 7.5 m^3/h rotary piston pump called *Wälzpumpe* has been manufactured since 1931, and another one of 150 m^3/h since 1937 (Fig. 18).

The introduction of the gas ballast principle, conceived after a discussion with Dunkel, was of major importance since it increased the water vapor tolerance of rotary oil pumps. The problem was that the water vapor, condensed during the compression, forms an oil emulsion instead of being discharged. Gaede found a simple solution (23):

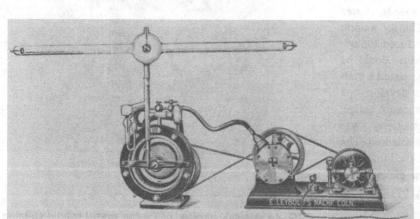

FIG. 15. Rotary "box" pump (*at right*), as forepump for a rotary mercury pump. Demonstration all modes of gas discharges (within one lesson) from Leybold Catalog 1908.

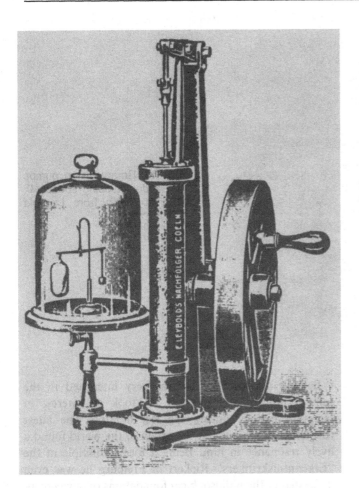

FIG. 16. Oil sealed, three stage, piston pump of Gaede from Leybold Catalog 1913.

"Air, the gas ballast, is bled into the pump shortly before the maximum compression, and the vapor is discharged" (Fig. 19). This method not only removes water vapor, but also other gases and vapors contained in the oil.

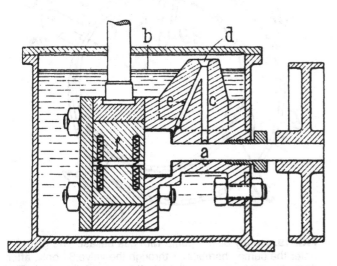

FIG. 17. Rotary, oil-covered vane pump of Gaede from Leybold Catalog 1927.

FINAL REMARKS

Gaede's extraordinary productivity until 1915 is very impressive. He finished high school very early and graduated from the university at 23 years of age. (He did not serve in the army, because a broken arm had healed improperly.) He had his own laboratory with a machine shop from 1907. This period ended when he submitted his publication about the diffusion of gases (9) on 19 November 1915. Therewith he had completed his most important scientific work when 36 years old. The ideas and inventions of the next 30 years were derived from the scientific knowledge he had acquired at that time.

Of major support were the contributions of the Leybold company and its workers. They created special manufacturing facilities and marketed several new pumps, some of them technically very demanding, within a few years without benefit of prior experience. This cooperation was economically very successful, and it was rendered possible by the balanced interests of the partners. Gaede had a very small or no income at the university until 1919 and could pursue his business and personal interests only if Leybold were successful, and Leybold needed new, successful products, for the competition was fierce.

It is also interesting to consider the consistency of the sequence of the events. The demand for vacuum pumps started in the 17th century, when physics laboratories became fashionable. The equipment included a vacuum pump, often as a formal present. It is reported that Otto von Guericke, while mayor of Magdeburg, presented one of his pumps to the city council of Cologne (47), though this cannot be documented. In the 19th century, a vacuum pump was part of the inventory of physical instrumentation of a university or a high school, especially for the demonstration of gas discharges. Therefore Leybold has listed vacuum pumps and demonstration equipment in their catalogs since 1878.

The industrial application of vacuum, particularly for the manufacture of incandescent lamps, began at that time. Leybold had difficulties in satisfying the needs of the new industry because it was a commercial establishment without its own manufacturing facility and its prices were too high. In addition, the potential customers often manufactured their own vacuum equipment. Leybold became successful only after they had started to manufacture the pumps themselves. Since Gaede's rotating mercury pump required a forepump, the manufacture of these pumps was also started, which was a risky undertaking at the time.

ACKNOWLEDGMENTS

The author wishes to thank H. Adam for his critical review of the original manuscript, G. Lewin for transla-

FIG. 18. Rotary piston pumps of Gaede; pumping speed 10, 30, 150 m³/h from Leybold Catalog 1937.

tion into English, K. Weedon for submitting the references 30 to 35, and A. Franz for providing essential documents. Many thanks go also to P. A. Redhead for his initiative in suggesting the preparation of this paper.

APPENDIX 1: IN MEMORIAM WOLFGANG GAEDE

Eulogy by Arnold Sommerfeld at the Memorial Service for W. Gaede in Munich, June 28, 1945*

Our friend, so suddenly taken from us, had an unusually rich harvest of his life. Each radiotube, each incandescent lamp bears witness to the success of his work. There is hardly any physical apparatus in the field of spectroscopy, even modern physics in general, which is not equipped with one or more of his masterly achievement, the diffusion pump.

He is the culmination of all the researchers, beginning with Otto v. Guericke, who labored in vacuum technique. Whoever worked in this field after him, stands on his shoulders. Certainly Gaede was a specialist but one of the highest class. He added to the foundations of his technical achievements, thermodynamics and kinetic theory of gases, his own original ideas. They deviated somewhat from conventional science, but lead him from success to success. When I asked him to write down his conception of the gas theory as a special contribution to one of my text books, he was very sceptical. He did not want to be a theoretician, he respected theory almost too much. His head was full of plans for new apparatus awaiting only the mechanic and the materials of the shop. Humanity has lost so much by his premature death.

*Z. Naturw. 1947; 2a:240.

Beyond his specialty he was very interested in the general problems of physics. He took an interest in quantum biology during the last months. The latest cosmological ideas about the origin of the world found a lively resonance in him. Being a former disciple of the "Erkenntniskritikers" Rickert of Freiburg he was even interested in the philosophical foundations of our perception of nature. He hardly showed these interests in public.

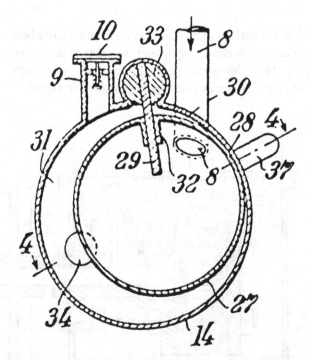

FIG. 19. Rotary piston pump with gas ballast. The gas can enter the pump chamber 31 through the valve 34 only, after the volume 31 has been closed off from the inlet port 8 (Fig. 3 from ref. 14a).

He expressed them only in the most intimate circle perhaps with a bottle of good wine.

He was very modest not only toward the outside, very remarkable indeed considering his technical successes. His contributions were widely recognized and rewarded as attested by his medals and diplomas, both domestic and foreign. He only found ingratitude at the place of his main activity. Incredible malicious accusations caused his dismissal from the Karlsruhe Institute of Technology. He barely escaped being arrested. Gaede suffered greatly by this ingratitude. He left his native state Baden and moved to Munich, where he was readily given a new laboratory. He became soon a highly valued member of our physics circle and a dear friend and colleague.

We shall honor the memory of this kind and extraordinary man.

APPENDIX 2: PUBLICATIONS ABOUT WOLFGANG GAEDE

Wolfgang Gaede on the occasion of his 60th birthday. Molthan W. *Z techn Phys* 1938; 19:153.

Wolfgang Gaede on the occasion of his 65th birthday. Justi E. *Elektrotechn Z* 1943; 64:285.

Wolfgang Gaede 65 years old. Mey K. *Z techn Phys*, Gaede issue, 1943; 24:65.

Address at the Funeral of Wolfgang Gaede, June 28, 1945. Sommerfeld A. *Z Naturw* 1947; 2a: 240.

Wolfgang Gaede. Wolf F. *Phys Blätter* 1947; 3:384.

Wolfgang Gaede, the Creator of the High Vacuum. Lecture given at the Technischen Hochschule Karlsruhe, May 9, 1947. Wolf F. Karlsruhe. C. F. Müller, 1947.

Wolfgang Gaede. Wolf F. Festschrift at the 125 Year Celebration of Technischen Hochschule Karlsruhe, S.46. Selbstverlag der Technischen Hochschule Karlsruhe 1950. Karlsruhe Institut, 1950; 46.

Wolfgang Gaede, the Creator of the High Vacuum. Gaede H. Karlsruhe: G. Braun, 1954; 127.

Wolfgang Gaede. Moebius W. ETZ-B 8, 8 (1956).

Wolfgang Gaede. An Appreciation of His Life on the Occasion of the 50th Anniversary of the Invention of the Diffusion Pump. Dunkel M. *Vakuum-Techn* 1963; 12:231./*Vacuum* 1962; 13:501.

Gaede's Influence on the Development of Mechanical Vacuum Pumps. Flecken FA. *Vakuumtechn* 1963; 12:249./*Vacuum* 1962; 13:583.

Memories of Wolfgang Gaede on the Occasion of the 100th anniversary of His Birth. Dunkel M. *Phys Blätter* 1978; 34:228./*Vakuumtechn* 1978; 27:99./*Vacuum* 1978; 29:3.

Wolfgang Gaede, a Pioneer of the Modern Vacuum Science and Technique. Rutscher A. *Phil-Hist '78* IV, 216 (Dresden 1979).

The Vacuum and Wolfgang Gaede. Auwärter M. *Vakuumtechnik* 1983; 32:235.

Gaede and his time. Reich G. *Vakuumtechn* 1986; 35:139.

Wolfgang Gaede and Johannes Stark (correspondence). Hoffmann D. *Wiss Z Techn Hochsch Magdeburg* 1987; 31:95.

Gaede Again. Adam H. *Vakuumtechn* 1988; 37:131.

Natural scientists and developers of technology. The Werner-von-Siemens-Ring award holder, (page 20–26: Wolfgang Gaede) D. Kind and W. Mühe, VDI Verlag (Düsseldorf 1990).

REFERENCES*

Publications by W. Gaede

Papers and Patents

DE-P = German Patent

1. Concerning the effect of temperature on the specific heat of metals. Inaugural Dissertation, University of Freiburg, 1902. *Phys Z* 1902; 3:105.
2. *Ann Phys* 1904; 14:641. US-P 852947.
3. *Verh dt phys Ges* 1905; 7: 287. *Phys Z* 1905; 6:758.
4. *Phys Z* 1907; 8:852. DE-P 202451 (14.9.1907).
5. *Verh dt phys Ges* 1912; 14:775. *Phys Z* 1912; 13:864. *The Electrician* (London) 1912; 70:48. DE-P 239213 (3.1.1909). US-P 1069408 (5.8.1913).
6. *Ann Phys* 1913; 41:289.
6a. The external friction of gases. Treatise submitted to the University of Freiburg in 1909. (C. H. Wagner, Freiburg 1910) and Berichte der Naturforschenden Ges.zu Freiburg 1910; 18:1. (Both identical papers were printed in 1911.)
7. *Ann Phys* 1913; 41:337.
8. *Phys Z* 1913; 14:1238. *Z phys u chem Unterricht* 1914; 27:92. DE-P 281977 (13.9.1913). DE-P 281595 (19.9.1913).
9. Leybold Catalog, Oct., 1913; 204–205. *Ann Phys* 1915; 46:357. DE-P 286404 (25.9.1913).
10. *Z techn Phys* 1923; 4:337.
10a. DE-P 401048 (1.10.1921). DE-P 416065 (1.10.1921). DE-P 419054 (1.10.1921). US-P 1490918 (22.8.1922) and 9 foreign appl. DE-P 436016 (22.7.1925).
11. Gaede W, Keesom WH. *Z Instrum* 1929; 49:298. CH-P 144942 (11.11.1929).
12. *Z techn Phys* 1932; 13:210. DE-GM 1237393 (8.12.1931), FR-P 730832 (1.2.1932), De-P 871495 (21.10.1942).
13. *Z techn Phys* 1934; 12:664. DE-P 634981 (4.6.1933). US-P 2081429 (1.6.1934).
13a. DE-P 648380 (13.9.1934).
13b. *Z VDI* 1935; 79:795.
14. *Z Naturforsch* 1947; 2a:233. Submitted posthumously by M. Dunkel.
14a. DE-P 702480 (22.12.1935). US-P 2191345 (21.12.1936) and foreign appl.
15. A paper left by Gaede after his death, first submitted to *Z techn Phys* March 1944. *Eine Schrift aus dem Nachlass*. Munich: R. Oldenburg, 1950.
16. DE-P 214504 (29.3.1908), DE-P 225286 (29.3.1908).
16a. DE-P 569212 (12.1.1933).
17. DE-P 442185 (29.11.1925/471827 (4.12.1926)/678962 (15.1.1937).
18. DE-P 596978 (6.9.1932)/602226 (1.1.1933)/661031 (27.3.1936)/ 687216 (20.3.1937).
19. DE-P 748618 (2.11.1939).
20. DE-P 840745 (23.7.1944).
20a. DE-P 842858 (17.6.1944).

Chapters in Handbooks

Auerbach-Hort, *Handbuch der physikalischen und technischen Mechanik* 1927; 6:90–121. Vacuumpumps and Vacuumtechnique.

W. Wien u. F. Harms, *Handbuch der Experimentalphysik*, 1930; 4 T. 3, 413–461 (Leipzig), Vacuumpumps.

21. *Handwörterbuch der Naturwissenschaften* (G. Fischer, Jena) 1912; 6:499; and 1931; 6: 587, Vacuumpumps.

*All papers, patents, and other documents quoted are sampled and stored at the "Gaede archives" within the "Gaede Foundation." (Enquiries to the author.)

Papers by Other Authors

22. Schmidt A. History of the Co. E. Leybold's Nachfolger 1850–1925. Cologne: Paul Gelhy, 1926.

22a. Gaede H. Wolfgang Gaede. The Creator of the High Vacuum. Karlsruhe: G. Braun, 1954.

23. Dunkel M. History of the Co. E. Leybold's Nachfolger 1850 to 1966. Cologne: Published by the author, private edition, 1973.

24. Knudsen M. *Ann Phys* 1909; 28:75.

25. Smoluchowski Mv. *Ann Phys* 1910; 33:1559.

26. Clausing P. *Ann Phys* 1930; 4:533.

27. Comsa G, David R. *Surf Sci Rep* 1985; 5:145.

28. Holweck F. FR-P 536278 (1921). *C R Acad Sci* 1923; 177:43.

29. Goes K. *Phys Z* 1913; 14:170.

30. Coolidge WD. *Phys Rev* 1913; 2:409, 415.

30a. Dushman S. *Phys Rev* 1915; 5:212, 224.

31. Llewellyn FB. *Radio Television News* 1957; 57:43. Tyne GFJ. *Saga of Vacuum Tubes*. Indianapolis: Howard W. Sams, 1957; 88.

32. Rukop H. *25 Jahre Telefunken*, Berlin 1928, p. 141.

33. Dejussieu-Pontcarral P. *L'Epopée du Tube Electronique*. Paris:, 1961.

34. Holweck F. *C R Acad Sci* 1923; 177:164. 1924; 178:1803. 1931; 193:151.

34a. Elwell CF. *J Inst Elec Eng* 1927; 65:784.

35. Holweck F. *L'Onde Electrique* 1923; 21:497.

36. Becker W. *Vakuum Techn* 1966; 15:211.

37. *Müller-Pouilets Lehrbuch der Physik*, ed. L. Pfaundler, Vieweg Braunschweig 1906; 1:505.

38. Langmuir I. *Phys Rev* 1916; 8:48.

38a. Langmuir I. *J Franklin Inst* 1916; 182:719.

39. Crawford WW. *Phys Rev* 1917; 10:557.

40. Volmer M. *Z angew Chem* 1921; 34:149.

41. Jaeckel R. *Kleinste Drucke, ihre Messung und Erzeugung*. Berlin: Springer and Bergmann, 1950; 149.

42. Burch CR. *Nature* 1928; 122:729.

43. Metropolitan-Vickers Leaflet No. 900/1-1, 1931.

44. Dushman S. *Scientific foundations of vacuum technique*, New York: John Wiley 1949; 309.

45. Meyer G. *Verh phys Ges* 1908; 10: 753./*Phys Z* 1908; 9:780.

46. Albrand KR, Ebert H. *Vakuum Technik* 1963; 12:237. *Vacuum* 1963; 13:563.

47. Berthold G. *Wied Ann Phys* 1883; 20:345.

48. Reich G. Early development of the molecular drag pump. *History of Vacuum Science and Technology*, 1993, Volume II, page 114.

Fernand Holweck *(1890–1941)*

P. S. Choumoff

Holweck mounting a 10 kW triode.

INTRODUCTION

On the centenary of Fernand Holweck's birth, it is appropriate for the French Vacuum Society to honor the memory of this scientist, engineering physicist, and pioneer of vacuum technology. We should remember that the creation in 1945 of the *Société Française des Ingénieurs et Techniciens du Vide*, the first name of the Society, was inspired by one of his own projects before World War II. The *SFITV* was placed under the posthumous patronage of Holweck, since he was the first French scientist to be the victim of his patriotism, having died as a result of torture by the Gestapo. In the first issue of the *Le Vide* in January 1946, later reproduced in *Le Vide* on the 40th anniversary of the Society, Professor G. Ribaud paid tribute to Holweck, praising his career as "that of a

life of labor dedicated entirely to science, that of a fierce and serene independence continued until death" (1).

The first college-level course on the applications of physics to the vacuum and electronics industries in France was created under his aegis, as attested by the first, very moving, lecture given by Prof. G. A. Boutry at the Conservatoire National des Arts et Métiers in October 1944. He didn't hesitate to compare Holweck to Irving Langmuir in the United States, for the industrial importance of his inventions, emphasizing that every one of his works had an original note, that each step of his thinking added something useful to the areas covered by the course.

HOLWECK'S CAREER

Fernand Holweck was born on 21 July 1890 into a family of Alsatian ancestry who opted to remain French in 1870.

Editor's note: This paper was first published in French in *Le Vide* 45 Suppl. No. 252 (1990). It was translated by N. Daigle.

His father, Louis Holweck, was close to the sculptor Bartholdi and collaborated on the latter's imposing work, the Statue of Liberty in New York.

At the age of 20, Holweck, an engineering physicist who finished first in his class at l'Ecole de Physique et Chimie Industrielle de Paris, became personal assistant to Madame Pierre Curie at the Institut du Radium of the University of Paris. In 1912, during his military service, he was called as an engineer to the radio station at the Eiffel Tower by General Ferrié, pioneer of wireless telegraphy. Being inspired by Fleming's thermionic tubes, he produced his own models; two of these served in 1913 and 1914 to determine by radiotelegraphy the difference of longitude between Washington and Paris. It was probably this work that directed him toward problems in pure science as well as applications related to vacuum technology. These detectors were the theme of his first publication in 1913 (2), the year he obtained his *licence ès sciences*, and he presented his first patent application in July 1914.

The war lead to his mobilization in August 1914 in a regiment of engineers, and it was in the trenches that in 1916 he took part in the realization of the first revolving direction finder intended to pick up the enemy's radio transmissions (Fig. 1). In 1917 he was recalled from the front to be assigned to the navy's study center in Toulon where, under the direction of Prof. Langevin, he devoted his time to the applications of ultrasonic sounding.

He had to wait until July 1919 to be demobilized and to resume his work, started at the Institut du Radium, in 1914, necessary for his thesis on the continuity of the electromagnetic spectrum between x rays and the visible, to find soft x rays, then unknown, in the band from 136 to 12 angstroms. He used electrons accelerated from 90 to 1200 volts for the production of x rays in an enclosure under a sufficient vacuum, then studied the radiation in a detection cell after it had passed through absorption chambers. The wavelengths were estimated by the energy of the electrons and the nature of the corresponding radiation demonstrated by showing a purely atomic absorption. In February 1922 he became *Docteur ès Sciences*, discovering one year later the total reflection of soft x rays. In the meantime, he had been nominated *chef de travaux* at the Science Faculty of the University of Paris. He became Director of Research at the CNRS in 1938.

The work necessary for his thesis revealed that

FIG. 1. Holweck near the revolving direction finder he used in 1916 to pick up the enemy's radio transmissions beyond the trenches of the War of 1914–1918.

Holweck, man of science, was also gifted with an exceptional manual skill, as illustrated in Fig. 2. Besides conducting experiments, he undertook himself the design and fabrication of the different devices and apparatus that he developed, whether it was mechanical parts, glassblowing, electronic circuits, or vacuum technology. Some essential elements had to be entirely designed, such as a Knudsen gauge. The use of the ionization gauge was already well established through the use of the anode as an ion collector in the glass triode used in military radiotelegraphy—the famous "TM" tube, with which Holweck was very familiar.

At first, the vacuum in all the equipment was obtained by means of Gaede's molecular pump, whose defects brought Holweck to design and create as early as 1922 the molecular pump bearing his name (3); it was developed by the Beaudoin company with the help of one of his fellow classmates, H. Gondet. A modified version of this pump is still manufactured today. It was Holweck who had the idea of maintaining the single rotor in an airtight envelope without any axial passage. The pump could operate unattended for many days, giving a pumping speed of 5 l/s at a pressure of 10^{-2} Pa.

At first, the equipment was essentially made of glass pipes connected by a conical ground joint to the pump's inlet. To avoid the propagation of vibrations, even as low as they were, it was necessary to fix the large connecting glass tube to a separate table with two heavy lead ingots. The difficulty of dismantling prevented the easy replacement of filaments or defective parts. This led Holweck to look for other solutions, noting as early as 1921 that "contrary to general belief, it was possible to attain and maintain a good vacuum in complicated metallic devices." Thus demountable apparatus was born. If it seems evident today that all installations are *a priori* demountable, we will better understand the consequences

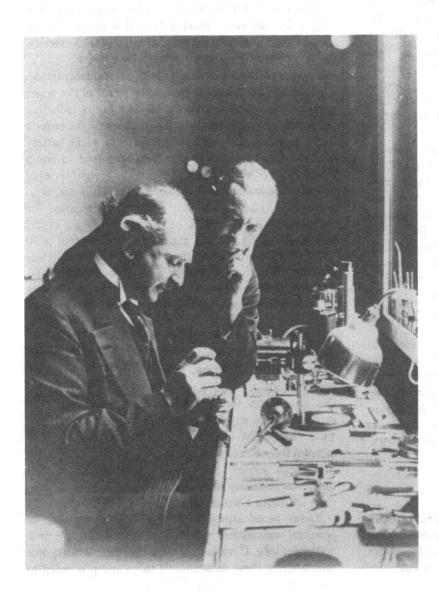

FIG. 2. Man of science, Holweck was also gifted with an exceptional manual skill.

of this observation for the realization of thermionic devices by continuously pumping instead of being sealed-off, e.g., as power tubes, x-ray tubes, or high-power transmitting tubes for the new radiotelephony.

Before discussing these issues, let us briefly survey the variety of areas that interested him and often led to the creation of original apparatus.

In 1920 Holweck's work with radium led to a device that allowed hearing the arrival of individual α and β particles to be heard on a loudspeaker, and later, to a device for the extraction of radium emanations that for years ensured the daily emanation readings at the Curie and Pasteur laboratories.

Attracted by the possibilities of transmitting animated images, he proceeded to do successful experiments in television with E. Belin, as reported by a note (4) in February 1927. However, these trials remained unexploited due to apparent lack of interest from the industry.

In 1929 Holweck and Professor Lejay published a note (5) on a quartz tuning fork in high vacuum that led them to develop an inverted pendulum made of Invar to within 10^{-6} of the exact value. This allowed France to be equipped with a precise gravimetric network suited to the search for oil and mineral deposits. This gravimeter earned him the Albert I of Monaco Prize in 1936 awarded by the *Académie des Sciences*. Only his patience and skill allowed him to assemble the gravimeter, despite the fact that he didn't want it to take up too much of his time. The manufacture of gravimeters was stopped in 1940, and none have been made since.

During the same period, other research interests attracted Holweck to a new area, biology, and as early as 1928 he proceeded to do trials on the action of soft x rays on microbes (6). In collaboration with Dr. A. Lacassagne, he began the systematic study (at the microscopic scale) of the biological action of various radiations on microorganisms and bacteriophages (7); this was the subject of his last paper in 1940 (8). He even measured the elementary dimensions of viruses by a method of statistical ultramicrometry. This led Holweck to consider the principle of an electron microscope and in 1940 he discussed his first trials with P. Grivet.

He was also fascinated by astronomy.

Some 60 papers, a book titled *De la lumière aux rayons X*, and many patents are proof of his tireless activity, which was rewarded by many prizes, the first in 1923 by the *Société Française de Physique*, in which he was an assiduous member, in 1923 and 1927 by the *Académie des Sciences* and in 1932 by the *Société des Electriciens* (9).

DEMOUNTABLE TUBES

The importance of the concept of a demountable tube can only be understood in the technical and chronological context of the development of the first tubes intended for long-wave radiotelegraphic transmitters, and later, for short-wave broadcasting. Developed for small transmitters from the TM tube used since 1919, the increase in power was limited to a maximum of 2 kW per tube because the envelope was made entirely of glass. A decisive change did not occur until 1924–25 when, after the invention of copper-glass seals, the first tubes with envelopes partially constituted by the metal anode became industrially available, allowing much higher thermal dissipation.

As early as 1921, Holweck had foreseen the means of significantly increasing the available power by using a demountable tube continuously evacuated by his molecular pump. The problems of degassing during the treatment of the electrodes at temperatures higher than those during operation were therefore eliminated, whereas the replacement of defective elements, among others the filament, was made possible. From 1922, a tube of 10 kW nominal power was in service at the Eiffel Tower station, as indicated by the first paper (10) presented to *l'Académie des Sciences* on 16 July 1923, by General Ferrié. This tube mounted on a molecular pump is shown in Fig. 3. This model was perfected the same year, since on 6 May 1924, Holweck reported (11) that many of these triodes were in service on a wavelength of 2600 meters, and sealed with greased, conical, ground joints.

Figure 4 shows the elements of the demountable triode. There is no doubt that the easy replacement of the upper part containing the filament, in 10–15 minutes including the pumping and formation time, was one of the criteria for its adoption. It successfully replaced a whole battery of TM tubes mounted in parallel and helped to equip France with powerful transmitters. At least eight series of ten of these triodes were ordered from Holweck, who personally took responsibility for their fabrication for many years. Also in 1923, he produced a larger model with an anode 64 cm long, allowing a power output of 30 kW at 7000 volts; when the grid was removed it was possible to use it as a rectifier with a power of 70 kW. These models were produced directly by the Laboratoire de la Marine Nationale. The figure on the first page of this article shows Holweck mounting a 10 kW triode.

The continuous progress on sealed-off tubes led to output powers of 25 kW in 1930 and 35 kW at the beginning of 1932. Holweck, assisted by P. Chevallier, described in July 1931 a new tube of 150 kW, mounted this time under the molecular pump, which was thus protected from any falling metallic particles (12). This triode had eight filaments (Fig. 5). It took only 50 minutes from atmospheric pressure to get this triode into continuous service. Such power, unparalleled at that time, was necessary for contact with submerged submarines requiring a wavelength of 10,000 meters (13).

Radio broadcasting needed a limited number of powerful transmitters requiring more than 120 kW peak power per tube. It was natural that the radio industry was interested by these demountable tubes and proceeded to compare the overall possibilities with those of sealed-off tubes. In 1934 and 1935 Maurice Ponte reported in two articles on sealed-off tubes of 130 kW useful power and, among others, a Holweck system tube of 100 kW (14).

Previously, the *Société Française Radioélectrique (SFR)*, had been forced to substitute an oil-diffusion pump for the molecular pump because the pumping speed of the latter was too limited for such power. Considering the constraints of permanent pumping, the necessity of having replacement tubes, and taking into account the longevity in thousands of hours of sealed-off tubes, the conclusion was reached that the niche for demountable tubes was the production of very short wavelengths with high power.

But the Compagnie Générale de Radiologie (CGR), of the Thomson-Houston group, then a competitor of SFR, had already delivered in August 1933 five short-wave sets to the Croix d'Hins station near Bordeaux, and then more to stations in Bamako, Brazzaville, Dakar, and Fort de France. They were operating 24 hours a day with a useful

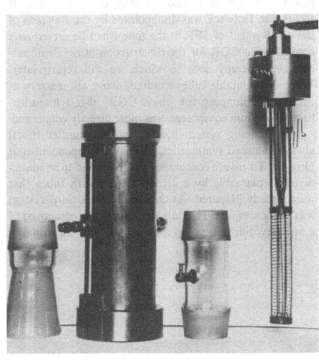

FIG. 4. Demountable elements of the 10 kW triode.

FIG. 3. Holweck's demountable triode of 10 kW, 40 cm high, mounted on a molecular pump, in service since 1923 at the military radiotelegraphic station of the Eiffel Tower.

FIG. 5. Holweck triode of 150 kW used for radiotelegraphy (1931).

power of 60 kW at an anode voltage of 12,000 volts. It seems that Holweck was disappointed by the slowness of the involvement of SFR at the time when he served as a consultant to CGR for the construction of his demountable radiotherapy sets, to which we will return later. These demountable tubes required, above all, mastery of continuous pumping for which CGR, also a manufacturer of vacuum equipment, was more directly concerned.

The pumping system holding two transmitter tubes, always mounted symmetrically, was an independent unit (Fig. 6). To ensure continuous service, it had to be immediately replaceable by a second system with tubes that were already prepared. At the same time, a third system was being serviced. The pumping group was composed of a mercury diffusion pump ensuring the easy starting of a low-vapor-pressure oil vapor pump. It must be emphasized that, due to the presence of a reservoir forming a vacuum regulator between these pumps and the mechanical pump, the latter functioned only intermittently under the action of a manometric relay. This practice seems to have been forgotten since. Among the advantages of these tubes we should mention those emphasized in a paper by M. Matricon (who became, at the end of 1940, responsible for the radio transmitter activity at Thomson), the possibility of adopting forms favorable to the production of short waves, the use of quartz as an insulator, the easy replacement of electrodes and the very large overload capacity without any destructive vapor release (15).

Figure 7 shows a cross-sectional diagram of such tubes with copper anodes cooled by water circulation. The

FIG. 6. System for a set of two CGR triodes of 100 kW, with continuous pumping, in service since 1939 at the short wave broadcasting transmitter of Allouis.

quartz cylinder has a height of 35 cm with a diameter of 25/28 cm. Before having appropriate elastomer seals as we have today, the assemble of such different materials involved a delicate heating and cooling procedure for the achievement of wax joints, melting at 80 °C, their surface being protected by a carefully applied layer of vacuum plasticine. The search for leaks was done with a discharge tube. In service, the pressure was never more than 7×10^{-4} Pa.

At the beginning of the Second World War, the Allouis short-wave station was unique among its kind on 16-, 19-, 25-, and 31-meter wavelengths, RF power circuits being composed of triodes with even more performance allowing a normal carrier power of 104 kw which, later, in the 49 meter band were increased to 145 kW. With the tubes limited to the use of pure tungsten cathodes in order to last as long as those in sealed-off tubes (16), the twelve systems had a life of approximately 25 years, except for the period of destruction during the Liberation of France, which made it possible to wait until technical progress in sealed-off tubes became significant and permitted their replacement in 1965 by vapotrons of 300 and even 500 kW.

X-ray tubes played an important role in the scientific

activity of Holweck and as early as the nineteen thirties, he started the construction of a demountable x-ray tube of 250 kV, 10 mA, for radiotherapy, which was assembled at the Institut du Radium (Fig. 8 shows Holweck near this tube). The CGR developed until 1939, four installations of 600 kV, opening a large field in medical treatment at the Institut du Cancer in Villejuif and the Centre Anticancéreux in Bordeaux. These units weighed two tons, were hung from the ceiling, and were composed of a horizontal x-ray tube and its pumping group placed above it (Fig. 9). The anode measured 1.40 meters and weighed 40 kg. The industrial applications of radiography at 15 cm thicknesses of steel thus became possible. To judge these characteristics in the scope of prewar technological possibilities, a comparison can be provided by the demountable x-ray tube of similar construction, fitted with a revolving anode, but with the tube resting on the diffusion pump, built as a unique specimen in 1937 at the California Institute of Technology and operating at 300 kV and 100 mA (17).

Broadcasting tubes, x-ray tubes and even neutron tubes

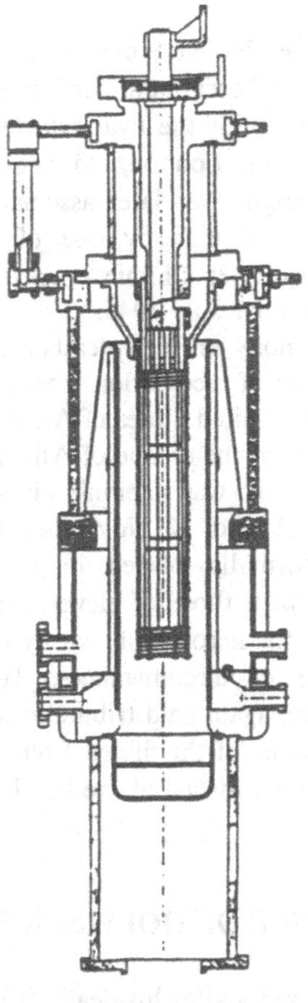

FIG. 7. Schematic diagram of a 100 kW CGR triode, 1.50 m high, quartz cylinders ensuring the insulation of the copper electrodes.

FIG. 8. Holweck near his 250 kV demountable x-ray tube.

(generated by the nuclear reaction on lithium of accelerated deuterons at 600 kV) were also produced at the CGR in 1939 (6), all demountable, show the validity of Holweck's ideas from the time of his thesis work. It contributed significantly to the progress of vacuum technology in France as well as abroad—for example, in the metal tetrodes with mechanically adjustable resonant cavities called resnatrons which were developed by Westinghouse during the war for jamming of German transmissions by the Allies.

THE PATRIOT AND THE POST-WAR TRIBUTES

During the war Holweck became advisor to Paul Reynaud, the prime minister. Antoine de Saint-Exupéry also asked him to resolve a problem encountered by fighter pilots, e.g., the freezing of machine guns at high altitudes (8). Later he stopped all activity that might be used by the German occupying forces, such as gravimetry. In fact, even if he appeared to take no political stand, he was deeply saddened by the defeat and always carried a picture of General de Gaulle in his wallet.

With the help of his colleagues, he started the fabrica-

FIG. 9. CGR radiotherapy installation with a 600 kV demountable tube.

tion of false papers, and took other actions of resistance such as organizing the escape of Allied airmen for which he made contact, too confidently, with a so-called British agent. The latter came in person with the Gestapo to arrest Holweck on 11 December 1941. We now know that, because he was the cause of numerous denouncements, this British deserter was finally shot in January 1945. A recent English book was written about him, presenting him as "the worst traitor of the war" (17). It is also said that it was his statements on secret-weapon research, apparently undertaken by Holweck to be transmitted to the British, that were the cause of the horrible tortures Holweck endured which led to his quick death.

The announcement on 12 December 1941 of the arrest and deportation of 1000 Jewish "notables" and the execution of hostages three days later (93 were executed at Mont Valérien) is an example of the sinister context of the period of Holweck's arrest.

An important text was dedicated to Holweck by Charles Fabry as early as June 1942, in the *Cahiers de Physique* published in the unoccupied zone of France. This exceptional fact followed the tribute given at his burial by A. Debierne on behalf of his friends and colleagues, who were shocked by his dramatic end. The latter was responsible for the campaign launched in February 1946 for the funding of an F. Holweck memorial award, since presented every three years by *l'Académie des Sciences*.

A conference in Holweck's memory, given in New York in 1943 by S. Rosenblum, confirmed his worldwide reputation, as attested by the award started in 1945 by the Physical Society in London, to which the *Société Française de Physique* was later associated; it has since been awarded every year by one of these societies. L'Ecole de Physique et Chimie opened an auditorium bearing his name on 7 July 1945.

During a ceremony on 6 September 1946, his family received a diploma of recognition from the governments of Britain and the United States of America, for the help given by Holweck in the escape of Allied airmen.

The nation's tribute was solemnly given by the French government on 11 and 12 November 1946, when his remains were formally placed in the crypt of the Sorbonne along with those of eleven other scholars, to testify forever to the heroism shown by the university in the most tragic of circumstances. Ten years later, Professor R. Heim again paid tribute to him on behalf of the university as one of the fifteen French prize-winners of the Institut who perished at the hands of the German authorities.

PERMANENCE OF HOLWECK'S MEMORY

More than forty years after his death, it is comforting to see how the memory of Holweck has lasted: by dedicating a street bearing his name in Paris on 22 October 1986, due to the efforts of R. Latarjet; by the choice of his name on 19 May 1988 for a technical *lycée* in Paris that is also trying to devote a small museum to him, being already in possession of a molecular pump; by the existence of a fund in his name at the *Fondation Curie*, at the Conservatoire National des Arts et Métiers, in which the museum has some of his devices. A special issue of the journal of the *Amis de l'Histoire des PTT d'Alsace*, dedicated in part to Holweck and abundantly illustrated (19), was published in 1984.

I would like to thank all those who have given me their recollection of Fernand Holweck—in particular his son Jacques Holweck, his friend Prof. R. Latarjet, and his colleague Lucien Desgranges. I will end with the words delivered by André Debierne over Holweck's grave on 3 January 1942: "With his great stature, his open face, his artist's hands so strong and so skilled, his mind full of prolific ideas, his tender and compassionate heart and that vitality full of self-confidence and audacity, he leaves us the image of a strong man, provided with the best gifts, and of a scholar who accomplished high-quality work."

REFERENCES

1. Ribaud G. *Le Vide*, 1985; Suppl. No. 228.
2. Holweck F. Etudes des detecteurs thermioniques. *Revue de TSF* 1913.
3. Holweck F. *Revue d'optique* 1922; 1: 274; *CR Acad Sci* 1923; 177:43; *L'onde Electrique* 1923; 21:457.
4. Holweck F. *CR Acad Sci* 1927; 184:518.
5. Holweck F, Lejay P. *CR Acad Sci* 1929; 188:1541.
6. Holweck F. *CR Acad Sci* 1928; 186:1318.
7. Holweck F, Lacassagne MA. *CR Soc de Biol* 1930; C3:60.
8. Holweck F, Luria S, Wollman E. *CR Acad Sci* 1940; 210:639.
9. Documents of the Curie Foundation.
10. Holweck F. *CR Acad Sci* 1923; 177:164.
11. Holweck F. *CR Acad Sci* 1924; 178:1803.
12. Holweck F, Chevallier MP. *CR Acad Sci* 1931; 193:151.
13. Brachet C. *Science et Vie* 1931; 40:466.
14. Ponte M. *La Nature* 1934; 210. *Bull Soc Franc de Radio* 1935; 71.
15. Matricon M. *Bull Soc Franc des Elect* 1939; 9:511.
16. Gilloux H. *Rev Techn CFTH* 1945; 2:33.
17. DuMond JNM, Youtz JP. *Rev Sci Inst* 1937; 8:291.
18. Murphy B. *Turncoat*. London: MacDonald; 1988.
19. Petitjean G. F. Holweck. *Diligence d'Alsace* 1984; 30:38.

Rudolf Jaeckel *(1907–1963)*

H. G. Nöller, H. L. Eschbach, and H. Pauly

Rudolf Jaeckel

Rudolf Jaeckel was born on 14 December 1907 in Elberfeld, Germany. Like many other vacuum specialists he came from a totally different field of activity. Jaeckel was closely associated with Gustav Hertz at the Technical University of Berlin, and under the latter's guidance he passed his main diploma examination in 1932 with a thesis on "The Separation of two Gases in the Positive Column of a Glow Discharge." Then he became an assistant to Lise Meitner at the Institute of Chemistry of the *Kaiser Wilhelm Gesellschaft* in Berlin-Dahlem. Under his tutor, Gustav Hertz, he also gained his doctorate in 1934 with his first research work at the Dahlem Institute. His thesis was "Experiments with Aluminum and Berylium Neutrons" (1). Four more publications followed at short intervals, among which were the remarkable measurements of the resonance capture of slow neutrons by tungsten, iridium, and rhodium. Until 1938 he worked in the Kaiser Wilhelm Institute, where at the end of 1938 Otto Hahn and Fritz Strassmann discovered the fission of

uranium. Jaeckel's last publication from the Dahlem Institute was submitted on 5 July 1938 (2).

Like Lise Meitner and other scientists Jaeckel was forced by the Nazi regime to break off his career with the Kaiser Wilhelm Institute in 1938. In that year the Nuremberg racial laws came into force, which concerned Jaeckel, who had a Jewish grandmother and who was thus considered to be a "Quarter Jew." Not only his professional career but also his personal fate was in danger from these inhuman and preposterous laws. His fiancée, the physicist Barbara Fuchs, was also a "Quarter Jew," and had the engaged couple not married shortly before the introduction of the laws, their marriage would have been forbidden.

Nevertheless Jaeckel wanted to stay in Germany. There were persons more in danger and more urgently needing a refuge abroad. In 1938 Jaeckel decided to venture a new start in industry with the Leybold company, certainly not an easy decision in the prevailing political situation in Germany. Dr. Manfred Dunkel, head of Leybold's recollects the episode as follows:

It was in March 1938 that a young physical scientist presented himself at Leybold's in answer to an advertisement. He had a very fine training behind him. This young man was Rudolf Jaeckel. In answer to my question as to his experience in vacuum science he mentioned his diploma thesis with Gustav Hertz in which he had treated the subject of gas discharge in vacuum tubes. In the meantime, however, he had become a true nuclear scientist and in applying for a post in industry had turned his back on a sphere of work which he had grown to love and in which he had achieved notable success. He had already told me his reason for leaving the Kaiser Wilhelm Institute, first in writing before applying for the post, and then again at the interview before he even sat down: he had left because of the ever-increasing oppression of the Hitler regime. It was typical of the man that he nevertheless decided to remain in Germany and seek an opening in industry and we can today see how fortunate it was that through the Leybold Company he became associated with Gaede and vacuum technology. As a physical scientist of outstanding ability he was soon at home in his subject and developed a very happy personal relationship with Wolfgang Gaede who, during his remaining years, groomed him as a successor (40).

Rudolf Jaeckel studied his new field of activity systematically, and he got a comprehensive survey of the state of the art. For his own daily use he collected working documents that resulted from a critical screening of publications and of his own experience. Later this material formed the basis of his well-known monograph "Very Low Pressures," which was ready for print in 1945 but due to post-war difficulties could be published only in 1950 (8). Until the late 70's this monograph served as an indispensable aid to scientists and engineers who had to deal with vacuum. Jaeckel contributed another survey of vacuum science and technology to the Encyclopedia of Physics in 1958 (22).

A key problem of vacuum physics in those days was the development of large diffusion pumps and the theory of their operation. In 1943 Jaeckel and Schröder had reported on a large diffusion pump with a pumping speed of more than 2000 l/s (5). Jaeckel very soon spotted the shortcomings of the existing theoretical approaches. In Gaede's first diffusion pump the gas to be exhausted entered through a narrow gap into a relatively low-velocity vapor stream, and Gaede stipulated that the width of the gap should not be much larger than the mean free path in the vapor. In the meantime Langmuir and Copley and others (also Gaede himself) had constructed far more effective diffusion pumps with relatively fast vapor jets and with much larger gap widths than required by Gaede's specification. Nobody really knew how this worked; from a physical point of view such a diffusion pump was an extremely complicated thing. Jaeckel considered a simple model of a pump, which he calculated in two approximations. From these, the influence of important technical parameters of the pump on the

pumping speed and the critical forepressure could be described correctly. From Jaeckel's theory, Gaede's gap condition results in the limit of a vapor jet of negligible low velocity. In a second approximation, the model was improved by considering a "vapor border," which takes into account the fact that at the edge of the vapor jet the component of the velocity in the pumping direction is reversed. The influence of this vapor border increases with the vapor pressure and therefore with the heater power of the pump. This results in a reduction of the pumping speed, a phenomenon well known to every designer of diffusion pumps. At that time, Jaeckel's theory represented a significant improvement and made the development of many diffusion pumps feasible. The simple model, however, did not give a complete description, and therefore the diffusion pump engaged Jaeckel's attention for many more years.

Early in 1947 Prof. Kopfermann of the Göttingen Physical Institute entrusted one of us (Nöller) with the task, as a diploma thesis, to produce a hydrogen molecular beam as intense as possible and to measure its intensity by a sensitive electrical balance. Jaeckel's large diffusion pump, which was to enable this, was in the high vacuum laboratory of Leybold's (at that time still in Andreasberg in the Harz mountains, where it had been evacuated during the war because of the bombing of Cologne). Even two years after the war it was not possible to transport the pump to nearby Göttingen, and therefore the experimenter had to go to the pump. Jaeckel initiated him in vacuum technique, the theory and practice of diffusion pumps, and many little tricks of the trade. The young novice got to know Jaeckel's great helpfulness, as well as his abundance of ideas and experience that helped to overcome many obstacles in the unfamiliar field of vacuum technology. The diffusion pump had to be operated at slightly higher inlet pressures than foreseen, and it turned out that the theory was not adequate for this. Nevertheless, with some working hypotheses and experiments, the goal was finally achieved (7). With his theory of the diffusion pump, Jaeckel had qualified in 1947 as a lecturer at Bonn University, and he made plans for further projects. From the work with a molecular beam two themes were derived for the following years:

1. The investigation of gas flow in the transition region (i.e., in the region between molecular and viscous flow) especially with application to diffusion and vapor ejector pumps.

2. The investigation of elementary collision processes by means of the molecular beam technique, particularly with two crossed molecular beams.

The economic recovery after the monetary reform in 1948 enabled Jaeckel to tackle both themes. After 1954 an additional theme was added:

3. The study of the interaction of gases with solids and

liquids in the bulk, especially at their surfaces, in connection with gas release in vacuum devices.

Before treating the three themes in more detail, Jaeckel's strong influence on the advancement of the industrial application of vacuum technology should be emphasized. Typical of his way of working was his combination of scientifically oriented projects and technical and commercial projects that were directly connected with his responsibilities as an industrial physicist. Under his management at Leybold's, the following items were developed and produced: rotary piston pumps (Fig. 1), vacuum valves, the system of couplings with small flanges, Roots pumps, diffusion and vapor ejector pumps. The large-scale vacuum degassing of steel was made feasible by employing large rotary piston pumps. Large pumping systems were developed for molecular distillation; for the drying of insulating oil, condensers, transformers, and cables; and for the evaporation of thin films, always in close collaboration with the corresponding industries. A sensitive method for the measurement of vapor pressures below 10^{-2} Torr was developed

(43) on the basis of the electrical microbalance used with the work on molecular beams (7), it was also applied to the study of liquids with high boiling points (19). Both these measurements and the determination of solubilities of gases in organic liquids were of great importance for technical applications. Many conference papers (12) review articles in scientific journals (9,11,13) and contributions to books (23,29) promoted these new vacuum applications.

The book in ref. 24 was written for those working with vacuum equipment. Users and researchers will find relevant tables and diagrams in the *Leybold Vacuum Pocket Book* (38). Of equal usefulness for the expert were the literature surveys on "Vacuum Technique and Molecular Distillation" which appeared every other year in *Fortschritte der Verfahrenstechnik* (35).

During the first two or three decades after the war, efficient diffusion and vapor jet pumps were a prerequisite for the fast-growing application of vacuum procedures in industry and science. Jaeckel realized that only a deeper understanding of the physical processes in these pumps

FIG. 1. Large rotary piston pump (700 m³/h), developed by Jaeckel in 1945.

could lead to their optimization and to a more methodical and less empirical approach in their development. The research that he performed to this end with Nöller and Kutscher (14,26,41) between 1948 and 1953 in Cologne, had as its first objective to study the vapor jet (mercury, oil). Experimentally the vapor jet was made visible by means of a glow discharge; in addition, some probe measurements were carried out. Finally it became apparent that the flow behavior which had been made visible could be described over a large range by well-known gas-dynamic methods. This is not true for the border region of a jet expanding into the high vacuum, there the vapor molecules leave the jet without further collisions.

Not only a free-expanding vapor jet was investigated, but also the vapor flow in real pumps. In addition, an experimental and theoretical study was made of the pumped gas diffusing into the vapor jet. This finally led to a thorough understanding of diffusion and vapor jet pumps (Fig. 2). With a diffusion pump operating in high vacuum and requiring a low vapor pressure, the gas penetrates deeply into the vapor jet. There is virtually a complete mixing of gas and vapor. Therefore the pumping speed of diffusion pumps is relatively high, and in the high vacuum region independent of the inlet pressure. This is different with vapor jet pumps, which operate at higher pressures and therefore require a higher vapor pressure, thus (a) the gas diffuses only into a thin border region of the vapor jet, and (b) the expansion of the vapor jet is dependent on pressure. From this results the well-known maximum of the pumping speed with pressure. Later an approximation was found to describe the border region of the vapor jet at the transition from viscous to molecular flow; this led to scaling rules for the change of scale of geometrically similar pumps (42).

These investigations contributed substantially to the fact that pumps of virtually any size can be planned, developed, and manufactured efficiently. The great era of diffusion pumps is long since gone. In many applications, diffusion pumps have been replaced by molecular, cryogenic, or getter pumps because these avoid the cardinal drawback of diffusion pumps, the possibility of contaminating the vacuum by the vapor. It is Jaeckel's great merit that he promoted the development of the diffusion pump at a time when there was no alternative and that he thereby contributed essentially to the development and use of vacuum technology.

In 1948 Jaeckel started, together with his doctoral student W. Jawtusch, to construct an apparatus for the study of the scattering of two crossed molecular beams with high angular resolution. At that time, scattering experiments with molecular beams were performed nowhere in the world and the experimental techniques had to be established, based upon earlier work by Otto Stern's school in Hamburg during the twenties and early thirties. The measuring device was conceived for the detection of differential cross-sections, but at first the easier problem, the measurement of total cross-sections, was tackled. Potassium atoms were chosen for the primary beam and mercury atoms and heavy molecules (pumping oils) served as the secondary beam. Potassium can be detected simply and efficiently by surface ionization and, due to the high atomic or molecular weight of the secondary beams, the influence of target motion on the cross-section is much reduced. The use of the results for the quantitative evaluation of the processes in fluid

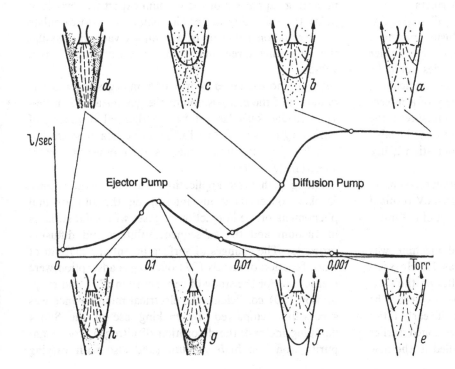

FIG. 2. Vapor stream and gas density (schematic) at several points on the pumping speed vs. pressure curves of diffusion and ejector pumps.

entrainment pumps also played a role in choosing the first target particles. After the graduation of W. Jawtusch in 1950, and partially with the collaboration of G. Schuster, the measurements were extended to other collision partners (K-Zn, K-Cd) (15). At that time the scattering of identical particles (K-K) and the investigation of a reactive collision pair were studied, although no quantitative results could be achieved because of experimental difficulties.

The vapor pressure measurements already mentioned using an electrical microbalance were also of special importance for the study of collision processes, since the balance made it possible to determine the intensity of the target beam and thus to perform an assessment of the absolute scattering cross-section.

When Jaeckel took over the direction of the Department of Applied Physics in the Physical Institute of the University of Bonn in 1955, the scattering experiments and other new activities were set on a broader basis. Now the collaboration with W. Paul's molecular beam group that had moved to Bonn two years earlier had its effect; in particular, the experimental technique was rapidly developed and refined. For his doctoral thesis, H. Pauly constructed a new apparatus with which he studied differential cross-section at very small angles, a region where quantum effects become noticeable, and (for the first time) the velocity dependence of total cross-sections. A little later (1956), the equipment of W. Jawtusch was rebuilt by E. Gersing to measure differential cross-sections at large angles. The first measurements of large-angle scattering were performed with the system K-Hg. These investigations were continued by E. Hundhausen. By comparison with classical calculations it was possible for the first time to make quantitative statements on the interaction potential of these scattering partners.

The investigation of the scattering at small angles was extended by R. Helbing to further collision partners, from which precise assertions on the long-range van der Waals forces between the scattering particles could be derived. These measurements had a practical application, because it could be shown that the intensity of a molecular beam is a precise measure of relative pressure in the high vacuum region. This was used to show the bistable behavior of the sensitivity factor of an ionization gauge with glass walls (33).

Further scattering experiments with atomic beams in the energy from a few eV to several hundred eV resulted in valuable information on the repulsive part of the interaction potential.

In 1961 the planning of an universal machine was started to perform investigations such as He-He scattering and the influence of particle statistics on the scattering process. Here for the first time jet beams were to be used that had been thoroughly studied in the early fifties by Jaeckel, Nöller, and Kutscher, as mentioned earlier (14). This new machine was to be installed in the new

Institute of Applied Physics that was under construction. Jaeckel did not live to see its completion.

From the middle fifties onward, research with molecular beams increased all over the world. Numerous research groups were formed that were concerned with the physics and chemistry of molecular impact processes and with gas-solid interaction. In the subsequent twenty years, extensive knowledge would be gained on intermolecular potentials, inelastic and reactive collision processes, and gas-solid interactions.

Rudolf Jaeckel was the first to resume this field after the Second World War, and his work showed the many possibilities offered by this experimental technique. He has inspired many a young scientist to work in this domain. It is Jaeckel's great merit that German scientists had an excellent starting position for the international competition in the sixties and seventies and that they could make important contributions.

In the third theme mentioned above, Jaeckel was intensely concerned with the physical processes and material properties that play a role in the production and measurement of low pressures. Already in his first publications on vacuum work (1942) he had investigated the usefulness of buna rubber as sealing component for high-vacuum equipment (3) and organic oils as pump fluids for diffusion pumps (4). After Jaeckel left Leybold in 1955, he devoted himself in his institute at Bonn, with a growing group of students, to the study of the interaction of gases with solids. In the following period great effort was applied to measure systematically the degassing rate of vacuum materials, first at room temperature and later at elevated temperatures and after different pretreatments (20). The main interest was focused on materials that were used for vacuum containers, sealing elements, and mounting supports. For the vacuum expert who wants to judge the suitability of certain materials or to calculate the evacuation behavior of a complex vacuum installation, these measurements of degassing rates are of great value.

To get more detailed information on degassing, a determination of the composition of the gas released is necessary. In Jaeckel's laboratory the thermal degassing of refractory metals (W,Mo,Ta,Ti) and of stainless steel at very low pressures was studied using an omegatron mass spectrometer (39).

Along with these application oriented investigations, Jaeckel concentrated his interest on the fundamental phenomena of the interaction of gases with solid surfaces in vacuum and the subsequent solution and diffusion processes. The changes of surface potential (25) and of work function of metals (37) due to gas adsorption were measured. For the study of condensation (27) and evaporation (30) coefficients the electrical microbalance was successfully employed. By making use of the Snoek damping method, the desorption of nitrogen from high-purity iron into high vacuum (and also with varying hydrogen pressures) was investigated (21).

Soon after the publication of the fundamental paper of D. Alpert and co-workers, Jaeckel directed his interest to the physics of ultrahigh vacuum. Jaeckel realized immediately, and he emphasized in his lectures, the great importance of the new technique for basic research on clean solid surfaces. However, experimental difficulties encountered with the production and measurement of ultrahigh vacuum in small glass vessels had first to be surmounted. This resulted in detailed measurements of diffusion, permeability, and solubility of helium, neon, and hydrogen in glasses of different compositions (Eschbach, Müller) and in laminate materials (28). The flow of helium and hydrogen through glass capillaries was studied in order to investigate adsorption processes (34). Data on the mean sojourn time of helium atoms on the wall could be obtained from the observation of the initial flow. The measurements were performed at very low pressures, i.e., with very low concentrations of adsorbed molecules and in the temperature range between 13.8 K and 20.4 K (Müller).

Applying the technique that was developed for investigations with glass membranes, the diffusion of hydrogen in low-carbon steel and chromium-nickel steels was determined. Large differences dependent upon structure and composition of the diffusion membrane were observed. These measurements were of particular interest in view of the degassing of large metal chambers being developed for ultrahigh vacuum research.

Electrical clean-up of gases, a process that is of special importance for the production and measurement of ultrahigh vacuum, was studied (32). It was shown that the clean-up of helium and argon was produced by ionization, whereas for nitrogen and carbon monoxide, dissociation and chemisorption played a role.

Another approach for the study of gas-solid interaction was made by combining the above-mentioned technique of molecular beams with ultrahigh vacuum procedures. First measurements were made of the angular distribution of atoms rebounding from a clean solid surface. In order to facilitate the detection, an atomic beam of a radioactive isotope was used.

The foregoing brief survey shows how diverse was Jaeckel's research. This could amply be shown by pointing to the materials research, the vapor pressure measurements and the study of gas release and diffusion and their practical application in vacuum technology. Further research in the Bonn laboratory showed good promise; however, in the last years of his life Jaeckel had neither the time nor the strength to conclude these projects.

In addition to his research Jaeckel committed himself to teaching. He took a particularly keen interest in the training of candidates for the teaching profession, more especially in those destined to teach physics. He insisted that the candidates should learn to use simple and orig-

inal means to make physical relationships clear.

The ever-growing importance of vacuum technology led to the formation of professional associations and other organizations. In Germany the *Arbeitsgemeinschaft Vakuum* was formed, which consisted of the technical committees of the Physical Society, the VDI, and the DECHEMA. The chairman of the committees of these three organizations was Rudolf Jaeckel. Later he was appointed president of the German National Committee following its establishment; he was also a member of many foreign vacuum societies.

Jaeckel was a passionate democrat with a marked sense of democratic freedom. He appreciated candid discussion and he stimulated discussions beyond the range of the natural sciences. There were many occasions for this at the numerous lectures for advanced students and during the many cheerful Institute parties. Shortly after the war, students were particularly responsive to such talks, and often things were pretty lively. From these discussions regular advanced political courses evolved to which competent speakers from other departments were invited.

Rudolf Jaeckel integrated a large number of students and collaborators in his institute. With his frank and open nature he created a lively and personal atmosphere for a growing community of colleagues and friends that still exists today. In his daily contacts with his young co-workers, Jaeckel was their teacher and their fatherly adviser who made every effort to create the best basis for their scientific work. He often secretly supported indigent collaborators with material aid. He maintained close contacts with his students after they had left his institute.

Jaeckel's wife, Barbara, took an active part in creating a lively and personal atmosphere in his institute and she often supported him in his activities.

Jaeckel's many scientific projects and his intensive preparations for a new building for his institute were abruptly terminated by his death on 17 January 1963.

REFERENCES

Publications by R. Jaeckel

The complete list of publications by R. Jaeckel comprises more than 70 titles. In the following only the papers mentioned in the text are listed.

1. Jaeckel R. Experiments with aluminum and berylium neutrons. *Z Phys* 1934; 91:493.
2. Jaeckel R. The neutron resonance levels of iridium and rhodium and the overlap of their respective resonance zones. *Z Phys* 1938; 110:330.
3. Jaeckel R, Kammerer E. On the suitability of buna rubber for high vacuum seals. *Z Techn Phys* 1942; 23:85.
4. Jaeckel R. The properties of organic fuels for diffusion pumps in operational use and a new oil diffusion pump shape. *Z techn Phys* 1942; 23:177.
5. Jaeckel R, Schröder H. An oil diffusion pump having high pumping speed and large working range. *Z techn Phys* 1943; 24:69.

6. Jaeckel R. On the theory of the diffusion pump. *Z Naturforsch* 1947; 2a:666.

7. Jaeckel R, Nöller HG, Oetjen GW. Generation and measurement of an intense molecular beam of hydrogen. *Z Naturforsch* 1949; 4a:2, 101.

8. Jaeckel R. *Very low pressures, their generation and measurement.* Berlin: Springer-Verlag, 1950.

9. Jaeckel R. High and ultra-high vacuum techniques in chemistry. *Dechema-Monographien* 1950; 16:79.

10. Jaeckel R, Jawtusch W. Scattering measurements of atoms and molecules with crossed molecular beams. *Phys Verhandlungen* 1950; 6.

11. Jaeckel R. The use of high vacuum technique in chemistry. *Chimia* 1951; 5:129.

12. Jaeckel R. Measuring methods and instruments in the high and ultra-high vacuum regions, Trans. of Instruments and Measurements, Conf., Stockholm, 1952; 227.

13. Jaeckel R. Vacuum technique in the pressure range below 1 torr. *Techn Mitt* 1953; 46:304.

14. Jaeckel R, Nöller HG, Kutscher H. The physical processes in diffusion and vapor jet pumps. *Vakuumtechn* 1954; 3:1.

15. Jaeckel R, Jawtusch W, Schuster G. The effective cross-sections for K-Zn and K-Hg collisions with scattering measurements of thick and thin atomic beams. *Z Naturforsch* 1954; 9a:905.

16. Jaeckel R. The present state of vacuum technique. *Berg und Hüttenmänn Monatshefte* 1955; 100: Heft 7/8. Wien: Springer-Verlag, 1955.

17. Jaeckel R, Jawtusch W, Schuster G. Investigations into collisional behavior between neutral atoms and molecules. *Forschungsberichte des Wirtschafts- und Verkehrsministeriums* NRW, Nr. 157. Köln/Opladen: Westdeutscher Verlag, 1955.

18. Jaeckel R. New research results in vacuum techniques. *Chemie-Ingenier-Technik* 1956; 201.

19. Jaeckel R, Reich G. Measurement of vapor pressures in the range below 10^{-2} torr. *Forschungsberichte des Wirtschafts- und Verkehrsministeriums* NRW, Nr332. Köln/Opladen: Westdeutscher Verlag, 1956.

20. Jaeckel R, Schittko FJ. Gas evolution from materials into vacuum. *Forschungsberichte des Wirtschafts- und Verkehrsministeriums* NRW, Nr369. Köln/Opladen: Westdeutscher Verlag, 1957.

21. Jaeckel R, Junge H. Measurement of the release to vacuum of nitrogen dissolved in high purity iron. *Ann Phys* 1957; 20:331.

22. Jaeckel R. General vacuum physics. In: Flügge S, ed. *Handbuch der Physik XII.* Berlin: Springer Verlag, 1958; 515.

23. Jaeckel R. Distillation and sublimation in high vacuum. In: *Houben-Weyl Methoden der organischen Chemie Bd.I, Allgem. Laboratoriumspraxis.* Stuttgart: Georg Thieme Verlag, 1958; 901.

24. Jaeckel R, Diels K. Evacuation, processes in vacuum technology. In: *Angerer-Ebert, Technische Kunstgriffe bei physikalischen Untersuchungen.* Braunschweig: Vieweg Verlag, 1959.

25. Jaeckel R, Eberhagen A, Strier F. The measurement of changes in contact potential during adsorption on metal surfaces. *Z angew Phys* 1959; 11:131.

26. Jaeckel R, Kutscher H. The behaviour of supersonic flow at pressures below 1 torr. *Forschungsberichte des Wirtschafts- und Verkehrsministeriums* NRW, Nr 683. Köln/Opladen: Westdeutscher Verlag, 1959.

27. Jaeckel R, Peperle W. On the relation of the coefficient of condensation to the partial pressure above the volatilizing crystal surface. *Z Naturforsch* 1960; 15a:171.

28. Jaeckel R, Eschbach HL. Laminate walls for ultra-high vacuum containers. *Z Naturforsch* 1960; 15a:268.

29. Jaeckel R. The creation of vacuum and work with low pressures. In: *Houben-Weyl Methoden der Organischen Chemie, Bd 1/2 Allgemeine Laboratoriumspraxis* II. Stuttgart: Georg Thieme Verlag, 1960; 577.

30. Jaeckel R, Peperle W. On the relation of the coefficient of vaporization to the partial pressure over the volatilizing crystal surface. *Z Phys Chemie* 1961; 217:321.

31. Jaeckel R. Degassing. In: Trans. 8th National Vacuum Symposium, AVS. 1961; 17.

32. Jaeckel R, Teloy E. Electrical gas clean-up by the excitation of metastable states. In: Trans. 8th National Vacuum Symposium, AVS. 1961; 406.

33. Jaeckel R, Helbing R, Pauly H. An accurate relative method for the determination of pressure in a high vacuum. In: Trans. 8th National Vacuum Symposium, AVS. 1961; 525.

34. Jaeckel R, Eschbach HL, Müller D. Molecular flows at very low pressures. In: Trans 8th National Vacuum Symposium, AVS. 1961; 1110.

35. Jaeckel R, Eschbach HL, Müschenborn G. Vacuum technique and molecular distillation. *Fortschritte der Verfahrenstechnik 1960/61,* Bd 5. Weinheim: Verlag Chemie, 1961.

36. Jaeckel R, Eschbach HL, Müller D. Diffusion of neon through glass. *Z Naturforsch* 1963; 18a:434.

37. Jaeckel R, Wagner B. Photo-electric measurement of the work function of metals and its variation due to gas adsorption. *Vak techn* 1963; 12:306.

38. Jaeckel R, Diels K. *Leybold vacuum pocketbook,* 2nd ed. Berlin: Springer Verlag, 1963.

39. Jaeckel R, Müschenborn G. Investigation of thermal degassing of metals in ultrahigh vacuum using an omegatron partial pressure gauge. *Forschungsberichte des Wirtschafts- und Verkehrsministeriums* NRW, Nr. 1452. Köln/Opladen: Westdeutscher Verlag, 1965.

Other Publications Quoted

40. Dunkel M. The late Rudolf Jaeckel and the significance of his scientific work. *Vacuum* 1963; 13:505.

41. Nöller HG, Kutscher H. Physical processes in diffusion and ejector pumps. *Z angew Phys* 1955; 7:218.

42. Nöller HG. The significance of Knudsen numbers and laws of similarity in diffusion and ejector pumps. *Vak Techn* 1977; 26:72.

43. Herlet A, Reich G. A device for the measurement of vapor pressures below 10^{-2} Torr. *Z angew Phys* 1957; 9:14.

Martin Knudsen *(1871–1949)*

H. Adam and W. Steckelmacher

Martin Knudsen

Martin Hans Christian Knudsen was born 15 February 1871 in Hasmark, Fyn, Denmark. His parents owned a small estate and he enjoyed a simple country life working as a shepherd boy in the summer. His mental faculties were soon discovered, when at the age of 13, he entered the Cathedral School in Odense. After his final examination he entered the University of Copenhagen in 1890 to study physics, mathematics, astronomy, and chemistry. He was granted a master's degree in science with physics as his main subject in 1896, and published a paper on the production of X-rays (1). In 1892 he became a private assistant of C. Christiansen, professor of physics at the University. Christiansen very soon discovered Knudsen's unusual skills as an experimental physicist. In 1895 he received the University's gold medal award by answering one of the prize questions on electric sparks. Following this, in 1896, he became an assistant lecturer at the Poly-technic Institute. Following the establishment of the posi-

tion of docent in 1901, he taught physics to medical students.

Knudsen's comprehensive and pioneering scientific activities covered two fields of sciences, hydrography and kinetic theory of gases at low pressures. In addition, due to his outstanding talent for organization, he became a well-known figure in national and international scientific institutions.

His close contact with hydrography was established with his participation in the Ingolf expeditions in the summer of 1895 and 1896. Knudsen's skill in the construction of scientific apparatus and complicated equipment soon gave him a leading position in this field. He developed methods to define the various properties of sea water, and was very active as an administrator. From 1899 until 1902 he was the Danish delegate to international meetings for investigations of sea water. It was mainly through his initiative that the Central Committee for Oceanic Research of the International Council for

75

Exploration of the Sea was based in Copenhagen. He was the Danish delegate to the Council from 1902 to 1947, the last fourteen years serving as vice president. Related to this, his hydrographical constants formed the basis of the well-known "Hydrological Tables, (Copenhagen-London, 1901)." He edited the *Bulletin Hydrografique* from 1908 to 1948; in 1930 he became president of the International Association for Physical Hydrography. He published some 51 publications in this field from 1896 until 1940.

Not less significant are Knudsen's achievements in physics, in particular his research activities involving aspects of vacuum technology and the properties of gases at low pressures. Apart from several physics textbooks and exercise books, the list of his important publications in this field includes some 38 papers from 1894 to 1934. To every physicist, Knudsen is seen as a very important scientist. Whenever you read a basic textbook or physics encyclopedia you are bound to come across such entries as Knudsen gas (molecular flow), Knudsen number (*Kn*), Knudsen cell, Knudsen gauge, Knudsen effect, etc.

We should note that the kinetic theory of gases "invented" by Clausius, Maxwell, and Boltzmann around 1860 was virtually dormant for several decades, so that Knudsen may be regarded as having revived it, i.e., he "excavated its hidden treasures," leading to far-reaching and important practical results that form the basis of vacuum science and technology. To start with, in 1909 there were three important publications: on molecular flow through tubes (2); on molecular flow through apertures and effusion (3); and on experimental determination of saturated vapor pressure of mercury at 0 °C (4). He was able to confirm the correctness and predictions of the kinetic theory of gases in four further publications (5,6,7,8). Thus he was able to confirm, in his experiments on the flow of gases through small apertures and on molecular effusion, the calculations based on Maxwell's law. These studies also indicated that they could be applied to molecular diffusion, from which he could determine the vapor pressure of mercury (4). The molecular flow through small apertures also led to the concept of the *Knudsen cell*, as discussed in more detail later, in connection with his studies in 1915. He examined the behavior of gas flow under conditions of temperature gradients (5), i.e., thermal molecular flow, and thermal molecular pressure (6), and hence, the possibility of an absolute manometer (7). The development of such instruments became known as *Knudsen gauges*, for a review of these and a typical design of a Knudsen gauge (some 40 years ago), see ref. 9.

The theoretical analysis by Knudsen of rarefied gas flow was discussed by Smoluchowski (10) for flow in tubes of circular and rectangular cross section, and with regard to the theory of the Knudsen gauge (11), to which Knudsen (12) immediately replied. The problem of molecular heat conduction and the concept of accommo-

dation coefficient (introduced by him) were studied by Knudsen (13), leading to the analysis of a hot wire gauge (14). This again led to comments by Smoluchowski (15) on the theory of heat conduction in rarefied gases, and an immediate reply by Knudsen (16).

This hot wire gauge was, of course, the one invented by Pirani 1906. In time Knudsen became aware of certain shortcomings when handling this gauge (e.g., due to vibrations), which he analyzed in a later publication (17) on the hot wire manometer (then the usual name for the Pirani gauge).

Following the retirement of Professor C. Christiansen in 1912, Knudsen was appointed to succeed him as professor of physics. He held this position until his retirement in 1941, and also taught at the Polytechnic Institute. To start with, those appointments kept him rather busy. In his next publication on rarefied gases Knudsen (18) discussed the determination of molecular weights, followed by the interaction of rarefied gas with moving plates (19) related to the viscosity vacuum gauge.

He used an interesting application of diffusion to describe the vapor pressure of mercury at low temperature (20). By introducing an evaporation coefficient α, strongly dependent on the condition of the mercury surface (typically due to surface contamination), the evaporation rate equation could be adjusted, resulting in what is sometimes referred to as a Hertz-Knudsen equation. He found that by using carefully purified mercury, evaporating from a series of droplets falling from a pipette, this could yield the maximum evaporation rate. It had in fact already been shown by Langmuir in 1913 (21) that the same equation also applied to evaporation from a free solid surface, e.g., tungsten from filaments in an evacuated glass bulb, referred to as "Langmuir free evaporation."

In an alternative technique, already established by Knudsen (4), evaporation occurs as effusion from an isothermal enclosure with a small orifice, generally referred to as the *Knudsen cell*, widely employed since then in many experimental procedures to determine the vapor pressure of materials, as well as heats of vaporization. In further studies (22), of evaporation (under high vacuum conditions), the cosine law in relation to the kinetic theory of gases was investigated, including the uniform deposition inside a sphere, which is fundamental to modern thin-film deposition techniques (see also, e.g., the review by Holland and Steckelmacher (23). The evaporation, condensation, and reflection of molecules and the mechanism of adsorption was also discussed by Langmuir in 1916 (24). This also includes comments on the cosine law and results of experiments by Wood (25,26) and Dunoyer (27). This stimulated further contributions (28) on the condensation of metal vapors on cooled surfaces, and on condensation, evaporation, and adsorption by Langmuir in 1917 (24a); and evaporation from crystal surfaces (29).

The thermal molecular pressure in tubes was studied by Knudsen (30); later Knudsen (31) reported on investigations of the radiometer effect, i.e., radiometer pressure and coefficient of accommodation. This showed the complete radiometer effect up to higher pressures, and also the effect of gas flow along heated surfaces.

In the autumn of 1933 Knudsen gave three lectures at the University of London on the "Kinetic Theory of Gases," the manuscript of these was the basis of the booklet (32), that gives a good summary of his work on this topic.

Knudsen's way of working was influenced by his comprehensive technical experience and clear vision of what could be realized in practice. He rarely asked for assistance from his co-workers, preferring to do everything himself. His experiments, to prove the gas laws at low pressures, benefited greatly by the availability of Gaede's mercury pump and later, the diffusion pump, both facilitating enormously the achievement of the required low gas pressures.

This review must not end without mention of Knudsen's activities in the promotion of natural science, in which his Norwegian wife was of considerable help. When still fairly young, he showed great interest in teaching physics at a higher level, particularly with respect to the development of experimental methods. At the proposal of his "boss," Prof. Christiansen, Knudsen became at the age of 30 the chairman of the Society for the Promotion of Natural Sciences (founded by Oersted [1777–1851]), and he retained this chairmanship for 40 years. On the occasion of the 100th anniversary of Oersted's discovery of electro-magnetism, Knudsen suggested the introduction of the Oersted Medal, and at the same time he founded the Oersted museum to display Oersted's original instruments and certain of his family belongings. In 1916 Knudsen himself was awarded the Oersted Medal in appreciation of his scientific achievements. Prof. P. K. Prytz, in his speech on this occasion, included the following noteworthy sentence: "When, at a meeting of the Physical Society you reported for the first time your initial results in the field of kinetic theory of gases, I found it necessary to praise you, because you did find gold in an unploughed field. Since then you dug up a great amount of that precious metal."

With his fundamental investigations of the behavior of gases at low pressures, Knudsen became a well-known figure in a large circle of scientists at an international level. Thus, apart from his many Danish commitments, he became a member of the International Solvay Institute of Physics in 1911 and was for two years secretary of the Danish Solvay Committee. In 1923 the International Union of Physicists appointed him their vice-president, a post he held for several years. He was awarded the honorary Doctor's degree of the University in Lund (Sweden) in 1918 and also became a member of the Academy of Göttingen (Germany) and the Royal Insti-

tution in London; in addition, he was a member of the Russian Academy for Oceanography and honorary member of the Royal Danish Geographical Society as well as the Danish Physical Society. He was decorated with three Danish medals of high order. When he retired in 1941, a commemorative volume was issued in his honor. This list, far from being complete, is evidence of Knudsen's most extraordinary personality, both as a scientist and human being. Knudsen died in Copenhagen on 27 May 1949.

REFERENCES

1. Knudsen M. On the production of X-rays. (in Danish). *Overs Danske Vid Selsk Ferh* 1896; 150.
2. Knudsen M. The laws of molecular and viscous flow through tubes. (in German). *Ann d Phys* 1909; 28:75. English translation: AEC-tr-3303.
3. Knudsen M. Molecular flow of gases through orifices (and effusion). (in German). *Ann d Phys* 1909; 28:999. English translation: AEC-tr-3715.
4. Knudsen M. Experimental determination of saturated mercury vapour pressure at 0 °C and higher temperatures. (in German). *Ann d Phys* 1909; 29:179.
5. Knudsen M. A revision of the conditions of equilibrium for gases; the thermal molecular flow. (in German). *Ann d Phys* 1910; 31:05.
6. Knudsen M. Thermal molecular pressure of gases in tubes and porous bodies. (in German). *Ann d Phys* 1910; 31:633.
7. Knudsen M. An absolute manometer. (in German). *Ann d Phys* 1910; 32:809.
8. Knudsen M. Thermal molecular pressure of gases in tubes. (in German). *Ann d Phys* 1910; 33:1435.
9. Steckelmacher W. Knudsen gauges. *Vacuum* 1951; 1:266.
10. Smoluchowski Mv. The kinetic theory of transpiration and diffusion in rarified gases (Derives equations for flow in tubes of circular and rectangular cross-section). (in German). *Ann d Phys* 1910; 33:1559.
11. Smoluchowski Mv. Remarks on the theory of the absolute manometer of Knudsen. (in German). *Ann d Phys* 1911; 34:182–184.
12. Knudsen M. Reply to Mv. Smoluchowski (1910). On the theory of transpiration and diffusion in rarefield gases. (in German). *Ann d Phys* 1911; 34:823.
13. Knudsen M. Molecular thermal conductivity of gases and the accommodation coefficient. (in German). *Ann d Phys* 1911; 34:593.
14. Knudsen M. Molecular flow of hydrogen and the hot wire gauge. (in German). *Ann d Phys* 1911; 35:389.
15. Smoluchowski Mv. On the theory of heat conduction in rarefied gases and the resulting pressure. (in German). *Ann d Phys* 1911; 35:983.
16. Knudsen M. Reply to M. Smoluchowski (1911). On the theory of heat conduction in rarefield gases and the resulting pressures. (in German). *Ann d Phys* 1911; 36:871.
17. Knudsen M. The hot-wire manometer. (in German). *Ann d Phys* 1927; 83:385. English translation in: *Viden Selsk Matem fysiske Meddelser* 7 (15):18.
18. Knudsen M. A method for the determination of the molecular weight of very small amounts of gas or vapours. (in German). *Ann d Phys* 1914; 44:525.
19. Knudsen M. The molecular gas resistance in opposition to a moving plate. (in German). *Ann d Phys* 1915; 46:641.
20. Knudsen M. The maximum evaporation rate of mercury. (in German). *Ann d Phys* 1915; 47:697.
21. Langmuir I. The vapour pressure of metallic tungsten. *Phys Rev* 1913; 2:329.
22. Knudsen M. The cosine law in the kinetic theory of gases. (in German). *Ann d Phys* 1916; 48:1113.
23. Holland L, Steckelmacher W. The distribution of thin films condensed on surfaces by the vacuum evaporation method. *Vacuum* 1952; 2:346.
24. Langmuir I. The evaporation, condensation and reflection of molecules and the mechanism of adsorption. *Phys Rev* 1916; 8:149.
24a.Langmuir I. *The Condensation and Evaporation of Gas Molecules.* Proc Nat Acad Sci 1917; 3:141.

25. Wood RW. Experimental determination of the law of reflection of gas molecules. *Phil Mag* 1915; 30:300.

26. Wood RW. The condensation and reflexion of gas molecules. *Phil Mag* 1916; 32:364.

27. Dunoyer L. The realization of material radiations of purely thermal origin. *Radium* 1911; 8:142; see also *Comptes Rendus Acad Sci* (Paris) 1911; 152:592.

28. Knudsen M. The condensation of metal vapours on cooled surfaces. (in German). *Ann d Phys* 1916; 50:472.

29. Knudsen M. The evaporation from crystal surfaces. (in German). *Ann d Phys* 1917; 52:105.

30. Knudsen M. Thermal molecular pressure in tubes. (in German). *Ann d Phys* 1927; 83:797. English translation in: *Viden Selsk Matem fysiske Meddelser* 8 (3):35.

31. Knudsen M. Radiometer pressure and accommodation coefficient. (in German). *Ann d Phys* 1930; 6:129. English translation in: *Kgl Dansk Vidensk Selsk* 1930-2; 11 (1):75.

32. Knudsen M. *The kinetic theory of gases: Some modern aspects.* London: Methuen; 1934.

Suggested Readings, Biographical and Historical Reviews

For a complete bibliography (of all Knudsen's 106 publications) see: *Fysisk Tidsskrift* 1949; 47:159. *Note*: References 16–54 refer to physics, kinetic theory, etc.; Refs. 55–106 cover hydrography and related topics.

Bohr N, Rasmussen REH. An appreciation of M. Knudsen, (Two obituaries). *Fysisk Tidsskrift* 1949; 47:145.

Hansen HM. Knudsen. *Dansk Biografisk Leksikon*, Copenhagen 1937; 12:615; and Copenhagen 1981; 12:83.

Hertz H. Concerning the evaporation of liquids, particularly mercury into a vacuum. (in German). *Ann d Phys* 1882; 17:177.

Jacobsen IC, Bohr N. Two lectures on Martin Knudsen. Danish Science Society, Dec. 9, 1949. (Reprints issued by the Society.)

Knudsen M, Fischer WJ. The molecular and frictional flow of gases in tubes. *Phys Rev* 1910; 31:586.

Knudsen M, Weber S. Atmospheric resistance to the slow motion of small spherical balls. (in German). *Ann d Phys* 1911; 36:981.

"Mogens Pihl, Knudsen, Martin Hans Christian." *Dictionary of Scientific Biography*, Vol. 7. New York: Charles Scribner, 1974;416.

Steckelmacher W. Knudsen flow 75 years on: The current state of the art for flow of rarefied gases in tubes and systems. *Rep Prog Phys* 1986; 49:1083.

Irving Langmuir *(1881–1957)*

George Wise

Langmuir with film balance and trough for surface chemistry research, 1932.

More than any other individual, Irving Langmuir demonstrated that support of scientific research by industry could be a good bargain, both for industry and for science. Trained in physical chemistry, he contributed in major ways to surface chemistry, electron and plasma physics, and vacuum science and technology. He made significant inventions that helped improve the light bulb, usher in the first (vacuum tube) electronics revolution, and apply atomic hydrogen to manufacturing in the form of a new welding torch (Fig. 1). His model of the atom, based on earlier work by Gilbert N. Lewis, though now obsolete was an important step toward a theoretical foundation for the use of the valence bond concept in chemistry.

These diverse achievements had a common denominator. They all resulted from subjecting astutely chosen problems to conceptually simple experiments, and modeling those experiments with relatively simple mathematics in terms of objects visualized on the atomic and molecular level.

Irving Langmuir was born in Brooklyn, N.Y., on 31 January 1881, the son of Charles Langmuir, an insurance executive, and Sadie Comings Langmuir. He was one of four brothers, all of whom were very successful in their chosen fields. He attended private schools in Paris (where his father served on European assignment) and Philadelphia. His independence, conscientiousness, and interests were encouraged by the provision of scientific equipment, the encouragement to keep regular daily records, and permission to engage independently in a widening round of outdoor activities, such as hiking, mountain climbing, bicycling, and skating. He attended Columbia University School of Mines, graduating with a degree in metallurgical engineering in 1903.

He then earned his Ph.D. in chemistry in 1906 at the University of Gottingen, Germany, working with Nobel Laureate Walther Nernst. Nernst was consulting on the area of nitrogen fixation, a key to the production of fertilizer and explosives, and had earlier successfully patented an incandescent lamp. Nernst suggested that Langmuir study whether the high temperature in the vicinity of the lamp's light-producing element could sufficiently speed the reactions that oxidized the nitrogen in the air to suggest commercially promising possibilities. Langmuir's experiments in this field of "light bulb chemistry" provided him with a thesis topic and a basis for his most important scientific work. He did not, however, find Nernst a helpful adviser, but benefited greatly from the teaching of Felix Klein, an applied mathematician. Simple yet powerful mathematical models would become another hallmark of the Langmuir style.

In 1907 he joined Stevens Institute of Technology as an instructor, ready to combine teaching and research. However, in the next two years he found himself overloaded with teaching responsibilities, and short of time, equipment, and assistance. In 1909 he learned from a Columbia classmate about a summer job with the General Electric Research Laboratory in Schenectady, New York. This was the first laboratory in U.S. industry created to combine fundamental and applied science. He got the summer post, was offered at its conclusion a full-time job, and accepted it with no intention of permanence. "While at Schenectady," he wrote his mother, "I will be looking around for a really good job in a university."

Instead, Langmuir was still working at GE at the time of his death nearly 50 years later. Why? Much of the credit goes to the man who hired him—Willis R. Whitney, GE's director of research. He saw a scientist unusually talented and versatile, yet thin-skinned. He quickly made Langmuir a kind of general consultant and

sage of the laboratory, permitted him to pick his problems, and equipped him with a highly skilled technician, Samuel Sweetser.

Langmuir in his turn selected problems related to improving the light bulb. (Had he chosen otherwise, who knows how long Whitney's generosity would have lasted?) He chose to study why continued improvement in the vacuum of an incandescent lamp did not lead to a corresponding improvement in its life.

He and Sweetser built a vacuum system and developed measuring methods that enabled them to determine precisely the fate of small amounts of gases introduced into a light bulb, including nitrogen, hydrogen, oxygen, mercury vapor, and water vapor. He found that the blackening of the bulb that shortened its useful life was due to a "ferrying" action by water vapor, carrying tungsten from filament to glass, and then returning for more.

Meanwhile, he traveled weekly to Pittsfield, Mass., on another GE assignment: he was helping a prominent inventor, William Stanley, with heat transfer problems on stoves and thermos bottles. This seemingly mundane effort provided the key to Langmuir's most valuable invention. He combined his knowledge of the behavior of gases in light bulbs with the heat transfer studies to invent the gas-filled incandescent lamp 25% more efficient than the vacuum lamp then in use.

All this would have marked Langmuir as a useful inventor but not a major scientific figure. However, his growing knowledge of chemical reactions at low pressure and on surfaces meshed in a timely way with some critical problems then at the frontier of chemistry.

Many scientists in the generation of his teachers believed atoms and electrons were only metaphors. Langmuir moved to the forefront of world science as a person

FIG. 1. (l. to r.) W. R. Whitney, Langmuir, and G. Marconi view Langmuir's electronics apparatus, ca. 1920.

whose work demonstrated that accepting atoms, molecules, and electrons as real objects similar to macroscopic ones, only smaller, provided a very fruitful way of understanding nature. When the quantum revolution later made those entities again strange and unlike the objects of the everyday world, the value of Langmuir's style receded.

Langmuir's attitude can best be illustrated with a story from his youth. He had weak eyes, and finally was fitted with spectacles. To his amazement, the fuzzy masses of trees suddenly resolved into individual leaves. Similarly, his light-bulb studies provided a way to deal with atoms at such low pressures that the fuzzy concepts of physical chemistry resolved into the behavior of individual atoms and molecules. His work served as a set of atomic spectacles, resolving problems of electrons, atoms, and molecules on the atomic scale.

Focusing those spectacles on electronics in near-vacua, he cleared up mysteries of thermionic emission and the space charge effect and developed the high vacuum tube, a key component in the first electronics revolution.

Focusing on chemistry, he developed the theory of heterogeneous catalysis, based on a picture of single layers of atoms occupying spaces on a kind of atomic checkerboard at surfaces, that explained the rates of reactions on surfaces. He described heterogeneous catalysis as something that occured in a single layer of gas molecules held on a solid surface by the same forces that held the atoms of the solid together. Other scientists have since shown Langmuir's model to be an oversimplification. But it proved a powerful oversimplification. It revitalized the field, shaped research and theory, and remains a practical tool today.

Focusing on atomic physics, he built on the work of Lewis to provide a fruitful (but eventually discarded) explanation for valence bonding in terms of stationary shells of electrons around nuclei. Focusing on surface phenomena, he developed a theory of floatation that explained a process long important in the mining industry.

Focusing again on surfaces, he built on the work of Agnes Pockels and Lord Rayleigh to develop methods of measuring the size of molecules by spreading layers of oil on water until they covered the surface with a uniform thickness of one molecule. His methods provide the foundations for modern biochemical and biophysical studies of cell membranes.

And, in the course of all this work, he invented important new apparatus. His "Langmuir balance and trough" became a standard tool of surface chemistry. In 1913 he developed two new vacuum gauges, both more sensitive than any previously in use.

In 1915 he invented the mercury condensation vacuum pump. Gaede had invented a pump in which a blast of mercury vapor through a nozzle, its pressure low due to hydrodynamical principles, draws into itself the gas to be exhausted out of the vessel and carries it away. Langmuir used pumps of this type and wished to speed up their operation. He noticed that the limitation on the speed of Gaede's pump was in the condensation and subsequent evaporation of mercury from the walls of the nozzle. This resulted in a vapor through which the gas being removed diffused slowly. So he cooled the walls of the pump near the nozzle. This condensed the mercury vapor stream onto the wall surface and prevented it from leaving the surface again by re-evaporation. As a result, the gas to be pumped out diffused much faster into the jet of mercury vapor, which meant greatly increased pumping speed for the final stages of evacuation. The resulting Langmuir condensation pump remains a major tool of vacuum research.

The years from 1912 to 1920 were Langmuir's peak period of productivity. But he maintained for the next three decades a record any scientist would envy, averaging more than ten papers a year (Fig. 2).

He did not do this merely by sticking to his earlier established specialties. He helped develop methods for the sonic detection of submarines in World War I. He branched out into biological studies and plasma physics (a field he named, based on a somewhat obscure analogy between electrically neutral low-pressure discharges and blood plasma). Building on observations he made in his thesis research, and the work of Robert Wood, he

FIG. 2. Langmuir with thyratron electronic tube, 1927.

invented the atomic hydrogen welding torch. He entered his final field of concentration, atmospheric physics, due to some important and militarily useful work undertaken during World War II on smoke screens and the prevention of aircraft icing. In 1946, his associate Vincent Schaefer discovered the way to seed clouds with dry ice to cause rain or snow. In this field, in the early 1950s, Langmuir would gain his greatest public notoriety with claims that weather modification experiments carried out in the Far West were influencing the weather in the Ohio Valley. The claims have not been confirmed. But the work of Langmuir and his team on "Project Cirrus" became the basis of modern efforts in weather modification and cloud physics.

The full details of his work can be followed in the 12 volumes of his collected works. He received numerous honors for these discoveries, most notably the 1932 Nobel Prize for Chemistry, honoring his contributions to surface chemistry.

As the rainmaking episode shows, he did not shirk from controversy or responsible speculation. His address "Pathological Science" (posthumously published) is a devastating critique of scientists who go public with half-baked results, and has been widely referenced in the context of the recent cold fusion furor. A less cited but equally insightful paper, "Science as a Guide to Life," discusses differences between what he terms convergent and divergent phenomena that foreshadow issues now being dealt with in the mathematical study of chaos.

Langmuir married Marian Mersereau in 1912. They adopted two children. His concentration could result in legendary levels of absentmindedness. One story depicts him leaving a tip on the plate for his wife after a breakfast in the family kitchen. Another depicts a GE employee collapsing on the stairs in front of him, and Langmuir calmly stepping over the prone victim and walking on. If true, these stories illustrate not an uncaring nature, but absorption in his work.

He could also be effective outside the laboratory. He organized Schenectady's first Boy Scout troop, ran for city council on a reform ticket, contributed to a book of essays advocating international control of nuclear power, and helped to protect New York State's Adirondack Park from assault by developers. He remained a vigorous outdoorsman, walking over 50 miles in a single day, swimming in the icy waters of Lake George from the dock of his camp, learning to fly a plane, and exploring the Adirondacks.

He died on 16 August 1957 (Fig. 3). His legacy is not only his own work, but the example he set. Langmuir attributed his scientific success to the chance that placed him in an industrial laboratory, where he got the challenges and the freedom that brought out the best of his abilities. He described research as unplannable, and offered as the most promising strategy that of letting the researcher choose the problems, based on a thorough understanding of the goals and needs of the organization paying the bills. Alongside such other peak achievers in industrial laboratories as Wallace Carothers of DuPont and John Bardeen of Bell, he demonstrated in his own career how well that strategy could work when the right times met the right person.

SUGGESTED READINGS

The best source of information on Langmuir is *The Collected Works of Irving Langmuir*, C. Guy Suits ed., 12 vol. (N.Y., Pergamon, 1962). Volume XII is a biography by Albert Rosenfeld, giving an excellent account of the personal events that accompanied Langmuir's scientific productivity. Some of his more accessible papers were also published in *Phenomena, Atoms and Molecules* (1950). The place of Langmuir's work in the understanding of surface phenomena is treated superbly in Charles Tanford, *Ben Franklin Stilled the Waves* (Durham, NC: Duke, 1989). Langmuir's personal papers are in the U.S. Library of Congress, Washington, DC. The two quotes by him in this article are in that collection.

FIG. 3. Irving Langmuir, 1957.

Marcello Pirani *(1880–1968)*

H. Adam and W. Steckelmacher

Pirani on his 85th birthday.

Marcello Stefano Pirani was born in Berlin on 1 July 1880 (his parents were of Italian descent), where he spent the most prolific years of his life. After his grammar school years he studied at the Technical University in Berlin-Charlottenburg and soon joined the Berlin section of the German Physical Society, headed at the time by such famous scientists as Max Planck among others. He completed his studies with a Ph.D. and in October 1904 joined the Siemens & Halske incandescent lamp factory (Glühlampenwerk) in Berlin. Although already much concerned with the day to day problems arising from the mass production of incandescent lamps, Pirani continued his studies in physics at an advanced level and in 1911 became a *Privatdozent* (i.e., unpaid lecturer) at the Technical University in Berlin, and held a professorship there from 1918 to 1936.

As he pointed out much later (1), the interest in high vacuum measurement is the outcome of modern technical development with its need for speedy and, if possible,

automatic operation. Evidently the start was made in the vacuum lamp industry with which he was associated at Siemens & Halske, particularly in the manufacture (in 1905) of tantalum lamps, which required a higher vacuum than the previously produced carbon filament lamp (2,3).

Pirani is generally best known for his thermal conductivity vacuum gauge, the "Pirani gauge," ever since his publication that described it in 1906 with the title "*Selbstzeigendes Vakuum-Messinstrument*" ("directly indicating vacuum gauge") (4). The importance of this development was apparently not properly appreciated at that time, when he was a young man with many other interests, particularly in sources of light (5,6). This invention came about because Pirani had a lot of trouble with some 50 Mcleod gauges used for low-pressure measurements in the factory for incandescent lamps. The Mcleod gauges were of course made of glass and each gauge was filled with about 2 kg of mercury. Broken gauges were quite

83

common, so that poisonous mercury spilled all over the floor. In order to avoid such disasters, the Mcleod gauge had to be replaced by something else, but what? Pirani's diary of 8 June 1906, prior to his lecture given on the 14 December 1906, reads:

> The present attempt to use heat conductivity for the measurement of low pressures does not present a new concept, but seems anyway a promising practical application of the observations of Warburg and others.

Very much later in one of his lectures on high vacuum gauges (delivered to the Electronics Group of the Institute of Physics on 10 June 1944 [1]), Pirani mentions those earlier publications by Kundt & Warburg (7) and Smoluchowski (8), as well as the above quotation from his 1906 diary. This confirms his general attitude in all his activities. In his classical paper (4) he described the pros and cons of the three modes of operation: constant voltage, constant resistance (i.e., constant temperature, exploited again much later in wide-range commercial instruments) and constant current. Pirani realized of course that the readings of his gauge depended on the nature of the gas, so that separate calibrations for each individual gas were required. The most appropriate electric circuit appeared to Pirani to be the Wheatstone bridge circuit (Fig. 1), where the Pirani gauge (W_4) is one branch of the four-element bridge circuit. Toward the end of his paper, Pirani also mentions the thermocouple gauge, first described in a paper by Voege (9), published just a few months prior to his own article. In the last sentence of Pirani's paper (4), he states: "The method, along with a few special arrangements, was subject to a patent application by Siemens & Halske." However, it is not clear if in fact a patent was granted and, at a later date, he even denied that such an application had been made. As Yarwood (6) noted:

> Though Pirani is thought of and remembered for this gauge he also did other pioneer work in vacuum technology proper (e.g., his work in 1905 on the sorption of gases by tantalum, a topic on which research papers are

still published today), his main activities during his life were not concerned primarily with vacuum, but with illumination and, in particular, sources of light. Indeed, he often said whimsically (though certainly not sardonically or in any critical sense; malice was completely foreign to him) that he considered his outstanding work was the development of the sodium lamp as a source of light. In Britain, at least, his name is not usually coupled with this work, yet there is little doubt that the present widespread use of discharge lighting is probably more due to the pioneer work of Pirani than any other single person.

His activities in industry were primarily concerned with illumination, particularly with sources of light. When the OSRAM company was founded in 1919* Pirani became head of the OSRAM Research Laboratories. In this capacity he initiated the *Research & Development OSRAM Reports*, also called by his colleagues "Pirani's Annals." From 1928 until 1936 he was director of the *Studiengesellschaft für elektrische Beleuchtung* (Study group for electrical illumination) at the *Technische Hochschule Berlin*. The transition of the development and research from incandescent lamps to gas-discharge lamps essentially took place during this period under his guidance. It was during this period that he had the most fruitful and finest success of his professional career.

At the *Physiker-Tagung* (Congress of Physicists) in Königsberg in September 1930, Pirani announced in a lecture that the OSRAM concern had developed a light source with an electrical-to-light energy conversion efficiency of 70%. This lecture caused a great sensation in scientific, technical, and commercial circles. As early as 1931, further development was leading to the production of the sodium vapor lamp on a commercial scale at OSRAM (6). Further developments in the field of discharge lamps, leading to the fluorescent lamps, widely used today, were very much in the hands of Pirani. Apparently Pirani was responsible for more than 100 patents in various branches of science and technology (6); he said he never earned any money from any of them in the form of royalties. Apart from his many patents, he was author or co-author of some 300 papers on science and technology, many of which are constantly referred to in standard treatises, and of three books.

When Pirani was in his seventies he met Yarwood who noted (6) how lively and interesting a personality he was:

> . . .willing to discuss anything from vacuum, education (in which he was profoundly interested) lighting and diet (about which he had original ideas) to ancient Chinese philosophy (for which he had a profound respect) communism (which he distrusted as practiced and which he regarded as severely limited as a philosophy and creed, particularly from the point of view of the individual) and the problems of the undeveloped countries (for which he had a deep sympathy, which did not stop at mere senti-

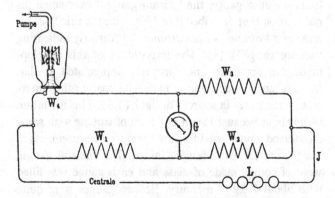

FIG. 1. Pirani's thermal conductivity gauge (W_4) for the measurement of low pressures in a Wheatstone bridge circuit. Courtesy of *Verh d D Phys Ges*, 1906; 8: 686.

---

*An amalgamation of the three companies: Siemens and Halske, AEG (the German equivalent of GEC), and Auergesellschaft.

ment but was translated into material assistance through his Quaker activities).

In 1936 Pirani left Germany to come to England, being highly dissatisfied with the prevailing policy there. For five years, he worked as a physicist at the Research Laboratory of the General Electric Co. (GEC) in Wembley, engaged with problems of discharge tubes, special glasses, quartz, and other high temperature resistant materials. From 1941 until 1947 he took on the task of scientific consultant and head of a department of the British Coal Utilization Research Association, at Leatherhead (Surrey) and in London. His field of activities were mainly related to new developments, such as heating contrivances without grates, high-temperature resistant materials, coke from non-coking coal, and the utilization of flue dust. However, he still retained his basic interest in vacuum science and technology. Thus he was invited by the Electronics Group of the Institute of Physics, London on 6 June 1944 to give a lecture on high vacuum gauges (1), at which he also presented a most interesting, brief historical introduction. It was only in 1947, when Pirani became scientific consultant at the British-American Research Laboratories in Glasgow and London, that he came again in close contact with vacuum technology, in particular with industrial applications. In 1950 he published another interesting historical review on the "Development and Applications of High Vacuum Technique" (10).

In 1953 Pirani returned to Munich, Germany, and in 1954 he became a scientific consultant to the OSRAM company and moved to Berlin in 1955. There he continued with his activities without any sign of retirement. For OSRAM he wrote several reports on many aspects of their scientific work.

Pirani was a deeply religious man. Although his contributions to science and technology were outstanding and have had an influence all over the world, particularly in topics such as illumination, material science (e.g., high-temperature metallurgy), and many aspects of vacuum technology, yet he believed the spiritual life to be more important (6). He was very close to the Quaker movement. When he died in Berlin on 11 January 1968, his urn was buried in the old Quaker cemetery in Spa Pyrmont (Germany). Pirani was honored by many scientific societies, mainly in Germany and Great Britain, and he was also an honorary member of IUVSTA (nominated at the Namur Conference 1958).

We acknowledge our appreciation of the many useful and informative comments in the Obituary by Professor

J. Yarwood (6), with whom in his later life (after they first met in 1950) he continued to be in close contact, leading to the publication of their very extensive joint book (11).

REFERENCES

1. Pirani M, Neumann R. High vacuum gauges (based on paper read by M. Pirani: June 6, 1944, Electronics Group Inst. Phys., London). *Electronic Engineering* 1944; Dec: 277; 1945; Jan: 322; 1945; Feb: 367; 1945; Mar: 422.
2. Pirani M. Tantalum and hydrogen. (in German). *Z Elektrochemie* 1905; 2:555.
3. Pirani M. An early method of vacuum melting tantalum. *Vacuum* 1952; 2:159.
4. Pirani M. Continuously indicating vacuum gauge. (in German). *Verh d D Phys Ges* 1906; 8:686.
5. Barth JA. Licht-technik *1*. (in German). Die Physik in regelmässigen Berichten. 1934; 2:127.
6. Yarwood J. Obituary: Dr. Marcello Stefano Pirani. *Vacuum* 1968; 18:233.
7. Kundt A, Warburg E. Friction and thermal conductivity of gases. (in German). *Poggend Ann d Phys* 1875; 156:177.
8. Smoluchowski M. On thermal conductivity in rarefied gases. (in German). *Wiedem Ann d Phys* 1898; 64:101.
9. Voege W. New type of vacuum gauge. (in German). *Phys Z* 1906; 7:498.
10. Pirani M. Development and application of high vacuum technique. *Research* 1950; 3:540.
11. Pirani M, Yarwood J. *Principles of vacuum engineering*. London: Chapman & Hall; 1961.

Additional Papers on Vacuum Technology by Pirani

1. Becker K, Pirani M. The bending of bi-metallic foils. (in German). *Z Tech Phys* 1932; 13:216.
2. Lax E, Pirani M. Tungsten. (in German). *Lehrbuch der Techn Physik* (Liepzig, 1929); 3:317.
3. Lax E, Pirani M. Technology of illumination, p. 351; Pyrometer lamps, p. 400; Life times of W-filaments, p. 440. In: *Handbuch der Physik*, vols 19 and 21. Berlin: Springer; 1929.
4. Pirani M. Water adsorbed on glass under the influence of electrons. (in German). *Z Phys* 1922; 9:327.
5. Pirani M. *Elektrothermie* (in German). 2nd ed. Berlin: Springer; 1960.
6. Pirani M, Fischer J. *Graphische Darstellung in Wissenschaft und Technik*. (in German). Berlin: De Gruyter; 1957.
7. Pirani M, Lax E. Electrolytic migration of sodium through glass. (in German). *Z Techn Phys* 1922; 6:232.
8. Pirani M, Wangenheim G. Thermo-element made of W-WMo. (in German). *Z Techn Phys* 1925; 9:358.

Other References

Herrmann O. Obituary: Prof. M. Pirani. OSRAM GmbH, Berlin-Munich, 22 Jan, p. 168 (private communication).
Persönliches (in German). Personal note. *Vakuum-Technik* 1968; 17: 25.

John Yarwood *(1913–1987)*

K. J. Close

John Yarwood in 1977.

John Yarwood was an outstanding experimental physicist whose research effort was centered around the high vacuum laboratory. One of his early textbooks, *High Vacuum Technique*, first published in 1943, was recognized internationally for its clarity in presenting the essential theory and practice of vacuum to students of science and technology and was revised and rewritten many times over a period of 25 years. His ability to present scientific ideas in such a lucid manner, a style immediately identified with John Yarwood, was appreciated by undergraduates and research workers using his 14 textbooks and some 40 research papers.

He was born on 5 October 1913 and chose a scientific career because the prospects of employment seemed brighter in the economic climate of the 1930s. He gained his first industrial experience as a research assistant in 1933 in the vacuum laboratories of EMI, involved with the early development of television. Further vacuum experience was acquired with Hilger & Watts Ltd. while he continued his part-time studies at Sir John Cass College, London, until he graduated in 1940 with a second-class honors degree in physics from London University. In 1941 he moved from industry into education, serving as a civilian instructor in radio communication with the RAF at Sir John Cass College. In 1944 he was awarded a Master's degree by London University for a study in electron optics. After a short period as lecturer in physics at North-East London Polytechnic, he moved to the Polytechnic of Central London in 1948 as a senior lecturer, and in 1953 was appointed head of the department of mathematics and physics. He very much enjoyed academic life and was a first-class lecturer, his enthusiasm for his subject and a lively sense of humor making him a popular figure with students. Expansion dictated that mathematics and physics should exist as separate departments in 1972 and he remained head of the physics department with the title of professor; his was one of the first professorial appointments made in the polytechnic sector.

In 1961 he published, with Marcello Pirani, *Principles of Vacuum Engineering*, a text that for the first time described the many applications of vacuum engineering. Without doubt he could foresee the impact of vacuum in high technology, and he used vacuum physics to provide his undergraduate and postgraduate students with rewarding experimental problems. Research funds were difficult to obtain for a relatively small educational establishment, but John saw vacuum physics as providing the opportunity to work, with a limited budget, in a controlled environment with problems in residual gas analysis, instrumentation, ion implantation, and surface studies. These fields he exploited, and the work is contained in his research publications, generally written in conjunction with a research student working under his supervision. Very often he would supervise work carried out in an industrial research laboratory for submission for a higher degree. He valued these links and his advice was constantly sought by industrial research laboratories. In 1974 he was elected dean of the faculty of engineering and science and served in this capacity for three years, returning to head the physics department until 1978, when he retired.

His enthusiasm for the formation of an international body, the International Union for Vacuum Science, Technique, and Application (IUVSTA), was important in gaining the full support and active participation of vacuum scientists and engineers in the United Kingdom. He accomplished much in open committee and worked tirelessly behind the scenes. At the national level he was very active in the formation of the Joint British Committee for Vacuum Science and Technology and in working for its development and evolution into the British Vacuum Council.

John played a leading role in establishing the journal *Vacuum* as it evolved from a house journal of Edwards High Vacuum. When the British Vacuum Council took responsibility for the editorial policy he was appointed editor-in-chief and remained in that post from 1971 to 1980, injecting life into the journal and soon establishing its scientific integrity.

His physical and mental stature could have made him a remote figure, but his gentle manner and natural courtesy with students and colleagues created an atmosphere of friendliness. The British Vacuum Council has established a John Yarwood medal, awarded each year for an outstanding contribution to the subject of vacuum physics and the Polytechnic of Central London (now the University of Westminster) has established a John Yarwood medal awarded annually for a first-class performance in physical science. When John Yarwood retired in 1978 he was awarded the title of emeritus professor and continued to lecture to postgraduate students for a number of years. A motor neurone disease caused a deterioration in his health over the final three years of his life and he died on 27 October 1987.

SECTION II
Some Major Advances

The development of three types of vacuum pumps that occurred in the first part of the 20th century are described in this section: diffusion pumps (Dayton), molecular-drag pumps (Reich), and turbopumps (Hablanian). Brief descriptions of the invention of important vacuum equipment prepared by the inventors themselves have been assembled by Lafferty. Ultrahigh vacuum is the subject of the last two papers. In the first Redhead examines the 40-year search for methods of producing and *measuring* very low pressures, culminating in 1950 with the development of the Bayard-Alpert gauge, and in the second Alpert comments on the subsequent development of UHV at Westinghouse Research Laboratories.

History of Vacuum Science

A Visual Aids Project

J. M. Lafferty

This Visual Aids Project on the history of vacuum science was undertaken by the Vacuum Science Division, in cooperation with the Education Committee of the International Union for Vacuum Science, Techniques, and Applications (IUVSTA), and completed in 1989. It is used primarily as an aid in teaching vacuum science and technology. It consists of a series of 56 overhead viewgraphs accompanied by a brief written description giving pertinent historical information and dates on various topics related to vacuum. The subjects covered date from the time of Torricelli and von Guericke to the present and are written by experts in the field. In addition, it was possible to obtain historical accounts directly from some of those people responsible for setting up recent milestones in the development of vacuum science. Those articles were written in the first person over the signatures of the authors.

It seems appropriate to reproduce these modern historical accounts in this anniversary volume of the American Vacuum Society, since they have nearly all occurred within the past 40 years. The topics covered are:

Oil-free reciprocating piston fore-vacuum pump (1974), p. 93

Becker's turbomolecular pump (1958), p. 95

Getter-ion pump (1953), p. 97

Single-cell ionic pump (1954), p. 99

Multi-cell sputter-ion pump (1958), p. 101

Bayard-Alpert ionization gauge (1950), p. 103

Helium mass spectrometer leak detector (1942), p. 105

While these accounts illustrate some of the events in modern history of vacuum science, older historical accounts are also of considerable interest and are covered in the Visual Aids Project (Part 2) on the History of Vacuum Science.

Ancient scholars, believing that "nature abhors a vacuum," denied the concept of a vacuum and the possibility of a void. The belief that vacuum was impossible existed well into the 17th century, but skeptics were beginning to question the ancient "fear of vacuum" early in that century. This was motivated by failure of siphons and the inability of pumps to "suck" water above heights of about 10 meters. By the mid–17th century, piston vacuum pumps were in use and vacuum science was on its way. It is not surprising that these early pumps resembled the water pumps that preceded them (just as the first automobiles resembled horsedrawn carriages).

The next 200 years were spent in improving these pumps and in measuring and characterizing vacuum. Progress was very slow by today's standards. However, at that time there was not a crying need for vacuum. A few scientists began to use vacuum as a tool for their research. The study of electrical discharges at low pressures was an early and prominent example all through the evolution of vacuum science and technology. By the latter part of the 19th century, vacuum technology had advanced to the point where pressures of 10^{-3} Torr, or better, were possible. This environment made possible the discovery of the electron, thermionic emission, x rays, and the incandescent lamp.

The commercial implications of these discoveries created an unprecedented need for the production of vacuum on an industrial scale. Consequently vacuum technology advanced at an almost explosive rate by the turn of the 20th century, followed by many new innovations in the production and measurement of vacuum for industrial needs. Vacuum science and technology continued to serve industry well through the period of World War II. At that time pressures of 10^{-6} Torr were routinely produced and measured.

Immediately after World War II there were isolated cases indicating that lower pressures had been produced, but they could not be readily measured by vacuum pressure gauges available at that time. The two decades (1945–65) following the war saw a flurry of activity in fundamental studies of the physics and chemistry of vacuum technique. Out of this research came a better understanding of the sorption and migration of gases and vapors on surfaces and the permeation of gases through solids. This was made possible by the development of new vacuum equipment with the ability to produce and measure pressures of 10^{-9} Torr and below. Notable developments were improved ionization gauges, ion and cryogenic pumps, partial-pressure analyzers, and demountable high-temperature seals for use at ultrahigh vacua.

By 1960 commercially available vacuum systems routinely attained pressure of at least 10^{-9} Torr in just a few hours, starting from atmospheric pressure. This was adequate for practically all industrial applications and most scientific requirements of that period. Pressures of 10^{-10} Torr gave the surface physicist many hours to perform experiments on atomically clean surfaces without contamination by residual gases.

Since 1965 fundamental measurements of sputtering phenomena, surface sorption, and diffusion have continued to give a better understanding of the principles on which vacuum technology is founded, but they have not led to any fundamentally new concepts for vacuum equipment. The information acquired has been used to refine and optimize existing equipment. Most improvements in recent years have been directed to making vacuum equipment more convenient and durable for specific applications and to the incorporation of modern solid-state electronics and displays.

In conclusion, history shows that vacuum science and technology is not only responsible for some of the world's most important discoveries, but is also responsible for their economic development and production. Shakespeare's title *Much Ado About Nothing* is most appropriate for the field of vacuum. In hardly more than 300 years, vacuum, which was at first thought not to exist, has been harnessed to become the workhorse for many modern industries.

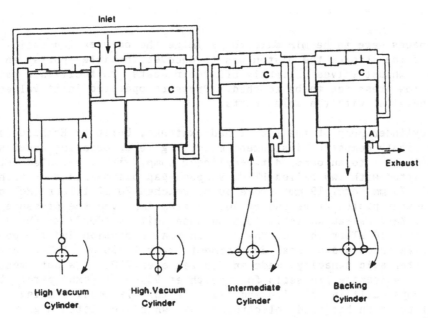

C — (Cylindrical Working Space)

A — (Annular Working Space)

OIL-FREE RECIPROCATING PISTON FORE-VACUUM PUMP (1974)

The oil-free reciprocating piston fore-vacuum pumps Varian makes under licence from the Commonwealth Scientific and Industrial Research Organization (CSIRO) were invented in Melbourne at the Electron Microscopy Section of CSIRO's Div. of Chemical Physics. CSIRO imported Australia's first electron microscope, a RCA EMU-1, from Radio Corp. of America in 1945 and appointed me to head the E.M. Group to introduce e.m. here. In '47 RCA's James Hillier announced exciting improvements in e.m. resolving power. Hillier helped me adopt his methods and generously judged my micrographs to be the highest resolution shown at the 1950 Internat. E.M. Congress. Hillier's infectious enthusiasm for higher resolution alerted me to the need for oil-free pumps.

During my '49 visit, Hillier described evidence that pump oil was the major source of the carbonaceous contaminating material deposited on microscope specimens and objective lens aperture diaphragms by electron irradiation. Contamination rapidly obscured specimen details and charge on the diaphragm deposit impaired optical performance. To me oil-free fore-pumps were needed but no materials then available seemed suitable. In 1968 I read that wear-resistant, low-friction materials were being made from polyfluoroethylene (PTFE) filled with finely divided graphite, bronze, etc. By then oil-free pumps were needed for many other types of electron and ion beam scientific and industrial equipment as well as for ultra-high-resolution microscopes, so in '70 I returned to the tantalizing challenges of oil-free pump design.

Seals for rotating shafts needed oil to transfer heat to the shaft and dry rotary pistons posed formidable sealing problems. Years earlier I noticed that air under pressure leaked extremely slowly through the long narrow gap between the piston and barrel of dry glass hypodermic syringes so I chose a reciprocating piston design. I drew a steam-engine-like pump with a long filled-PTFE-sleeved piston and the cylindrical and annular working spaces in series so the mean pressure difference across the piston would be small.

Unswept spaces were to be minimal to maximize the compression ratio. Near the ends of the strokes the piston was to uncover the inlet ports and also raise the exhaust valves slightly off their seats to facilitate gas flow and allow low pressures to be reached. Pressure operated inlet valves were shown in parallel with the inlet ports.

A single cylinder pump of the type I had sketched, built by Eckhard Bez, EM Section's highly competent instrument maker, gave encouraging results. The design evolved into an opposed two-cylinder pump, with air-cooled stepped pistons fitted with two filled-PTFE, stepped-gap piston rings on each diameter. This 70 mm bore, 19 mm stroke pump reached 70 milliTorr (McLeod) in '74 (1). Better piston rings and exhaust system improved the pressure to 15 mTorr. With Karl Balkau assisting Bez we then built a 100x27 mm (bxs) model which attained 10 mTorr in '76. The high thermal expansion and low conductivity of PTFE made piston seal development very tedious. Balkau and Bez, with characteristic tenacity, made and tested over 200 seals but eventually we evolved low-leakage lip seals for the three 90x45 mm (bxs) parallel twin cylinder pumps built in '77, '78 and '79. For the '79 model I tried Al cylinders plated with hardened electroless Ni which permitted higher piston speeds. In January '80, we added an intermediate third cylinder to increase the pumping speed at pressures below 1 Torr. It was effective so we began a 90 vee-four higher capacity pump with 2 high vacuum cyls., an intermediate and a backing cylinder; all 90x45 mm. Our inner piston seals had worn too rapidly so we discarded them and reverted to filled-PTFE sleeves. We fitted thin sleeves to both diameters of each piston, leaving a clearance of about 0,025 mm. The atmospheric outer seals which wore very slowly were retained. At 776 rpm the pump attained 310 l/m and 1.2 mTorr (McLeod).

In 1978 I visited Varian, Leybold-Heraeus, etc. to assess demand. Leybold's Hansen Pfaff promptly came to Repco Ltd, CSIRO's licensee, and negotiated a marketing franchise but Repco, disappointed by Leybold's sales forecasts, withdrew in late 1979. Other Australian manufacturers lacked the requisite resources so I again visited Varian which arranged for me to demonstrate our 4 cyl. pump in California for 90 days in 1982. Varian then negotiated a licence under which I returned to USA with Bez in April 1983 to help design a direct drive 108x25.4 mm opposed 4 cyl. pre-production pump. At CSIRO we then built two 90x25.4 mm 90 vee-four, 1750 rpm pumps with internal ducts replacing external manifolds. In '85 CSIRO seconded Bez to Varian as Senior Project Engineer. Varian's Dry Vacuum Pump appeared in '86. Ducts like ours are used but the filled-PTFE and the cylinder wall coating are more durable than those available in Australia. DVP-500 pumps reach 420 l/m and a total pressure of 10-20 mTorr (2). Wear is small even after 10,000 h operation.

Several senior Chem. Phys. men, notably former Chief, Lloyd Rees, Alexander Mathieson and Clive Coogan enthusiastically supported us. Richard Scholl was our first ardent advocate at Varian and many others there are involved now. Marsbed Hablanian has energetically publicised and promoted the pumps.

John Farrant.

John L. Farrant
(1) E. Bez & J. L. Farrant, E. Micros. (Austra. Acad. Sci.) 1 , 192 (1974).
(2) M. Hablanian, E. Bez & J. Farrant, J. Vac. Sci. Tech. A5 , 2612 (1987).

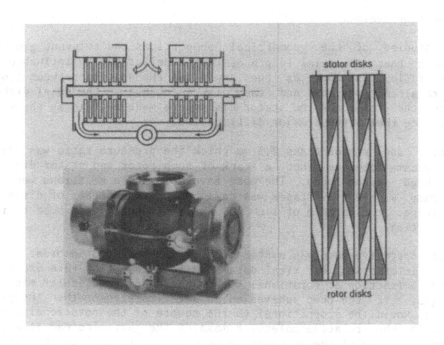

BECKER'S TURBOMOLECULAR PUMP (1958)

When in charge of the vacuum technology R & D laboratory of Messrs. "Arthur Pfeiffer Hochvakuumtechnik GmbH" in Wetzlar, West Germany, I became interested, as far back as 1945, in the possibility of improving the performance of high vacuum pumps, in particular diffusion pumps and Gaede's molecular pump and its various descendants.

The considerable backstreaming of oil vapor in diffusion pumps and the very narrow gaps between the moving and stationary parts in molecular pumps motivated me to investigate the theoretical background of both phenomena.

In the course of these investigations I had an inspiration for a novel type of baffle, but I couldn't pursue the new idea for quite a while because of lots of other work. So it was only in 1955 that I started to construct a baffle for diffusion pumps which consisted of a fast rotating propeller, the blades of which were so shaped as to make the propeller opaque in the axial direction. With the correct sense of rotation and corresponding high speed of the propeller the gas molecules to be pumped would pass uninhibited downwards whilst the backstreaming oil vapor molecules moving in the opposite direction would be caught by the propeller blades.

It soon became evident that the rotating propeller produced a measurable pressure ratio in the molecular flow regime, easily interpreted as a pumping effect. This aroused my greatest interest and accelerated further developments. As the design of the propeller resembled very much that of a turbine, we fitted mirror-inverted guide blades on both sides of the propeller blades. Bearing in mind the hazard of small gaps in molecular pumps we made sure that all gaps were not less than one millimeter. The guide blades increased the pumping effect to a pressure ratio of 3:1 in the molecular flow regime.

Further studies of the geometrical proportions and relevant experiments have shown that the axial length of the propeller had no influence on the pressure ratio, as long as the angle and the spacing between slots and blades remained constant and each disk is opaque in the axial direction. Therefore the rotor and the stator disks were made as thin as the workshop could produce them without major difficulties.

With rotor and stator disks 2.5 mm thick the pressure ratio was the same. Now it became obvious that a stack of such rotor and stator disks would make a high vacuum pump. The most favorable angle relations were chosen and a pump with 2 x 19 stages was designed where one pair of disks counts as one stage. A clearance of 1 mm was strictly observed in both radial and axial direction.

The first operative pump was systematically investigated in order to obtain firsthand information on vital data such as volume flow rate and pressure ratio as a function of rotational speed and relative molecular mass of the pumped gas. If the pump behaved as an axial compressor then the pressure difference would be proportional to the square of the rotational speed and linear with the relative molecular mass of the gas. Instead it was found that the logarithm of the pressure ratio was proportional to the rotational speed and the root of the relative molecular mass of the gas. These relations are identical to what Gaede has found for his molecular pump. Hence I had the proof that the new pump was a molecular pump but without the hazards of very narrow gaps. My efforts to find a novel and more effective type of baffle for diffusion pumps led to the invention of a novel type of molecular pump! What a thrill!

The first models with two 19-stage stacks in parallel gave a pressure ratio of 50,000,000 : 1 for air, derived from the average pressure ratio of 2.5 : 1 per stage. After thorough tests and a running time of 10,000 hours the "New Molecular Pump" - as designated in the beginning - was first presented to the public on June 12, 1958 on the occasion of the First International Congress on Vacuum Technology in Namur, Belgium, and published soon afterwards (1),(2).

Somewhat later the theory of the now called "Turbo-Molecular Pump" was published (3) relating to Gaede's fundamental equations and including also the hitherto accumulated practical experience with the new pump. Since then, operational reliability, efficiency and versatility of the turbo-molecular pump in numerous applications has been generally acknowledged.

W. Becker

(1) W. Becker, Vak. Techn. 7, 140 (1958).
(2) W. Becker, Advances in Vacuum Science and Technology, E. Thomas, Editor, (Pergamon Press, New York, 1960), Vol. 1, p. 173.
(3) W. Becker, Vak. Techn. 10, 199 (1961).

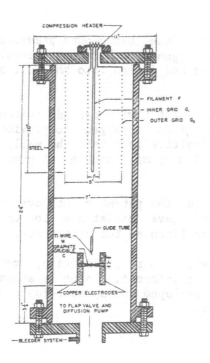

GETTER-ION PUMP (1953)

In the late 1940's I had a development program underway aimed toward pushing the voltage achieved by electrostatic accelerators to 5 MV and beyond. Experimental work was done by undergraduate hourly workers and by beginning graduate students. The most difficult component in the multi Mev machines was in all cases the subdivided accelerating tube consisting of large numbers of glass or ceramic cylinders separated by metal electrodes with seals formed by a wax, rubber gaskets or organic cements. Vapor from these materials, I thought, could play a role in the discharges that limited accelerating tube voltage holding capabilities.

Elimination of accelerating tube vapors required two developments: one, development of metal to ceramic bonding so as to provide an all metal and ceramic, bakeable, accelerating tube, and two, development of a vapor free substitute for the diffusion pump. We solved the first problem with two different methods. Both led later to wide industrial applications.

For the second problem I asked an undergraduate student, David Saxon, to explore pumping (getting) affects of various metals when they are evaporated. Results were uninteresting until he tried titanium, then available only in chip form utilizing the iodide process. He was astonished. The vacuum gauge reading appeared to drop to zero. David Saxon left his hourly work because of study requirements and two beginning graduate students, Robert Davis and Ajay Divatia carried on. We wanted titanium wire to be fed continuously into an evaporator. I soon learned that titanium wire had never been made but I then heard of the formation of a new company, "Titanium Metals Corporation." I wrote and was pleased to get a quote from the general manager for one pound of 0.020 inch diameter wire. I believe it was the world's first titanium wire. It was black coated but ductile. It gave fast continuous pumping of nitrogen, oxygen and hydrogen, but the background pressure slowly rose. This, I suspected, was due to argon or other nongetterable gases which could be trapped only if ionized and driven into the wall onto which titanium was condensing.

We installed a central tungsten wire cathode surrounded by two concentric cylindrical wire anode grids arranged to provide radial oscillation of electrons and to drive positive ions into the wall. It worked and was the first practical getter-ion pump capable of pumping large quantities of gases.

We published (1), turned the patent rights over to the Wisconsin Alumni Research Foundation and gave assistance to the Consolidated Vacuum Corporation which had been licensed for production.

Russel Varian wrote after our publication, much interested in the development. I kept him informed. Later he stopped writing. I learned then that Lew Hall had developed the sputter-ion pump. It was simpler and soon dominated the market.

Raymond C. Herb

(1) R. Herb, R. Davis, A. Divatia, and D. Saxon, Phys. Rev. 89 , 897 (1953).

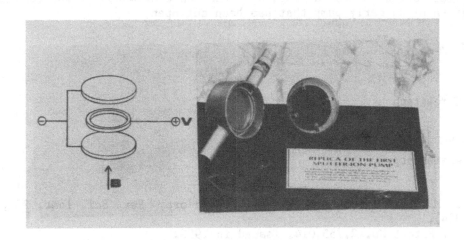

SINGLE CELL-IONIC PUMP (1954)

The ionic pump (1) was developed for removing various gases including noble gases from a sealed-off device used in experimental work in our laboratory. Because of the presence of noble gases, chemical getters could not be used. It occurred to us that it might be possible to solve this problem by ionizing the gas to be removed by sufficiently energetic electrons and then driving the ions by means of electric fields into suitable surfaces where they would be trapped indefinitely.

It was clear that a pump based on the above idea should, among other things, provide streams of electrons moving in very long trajectories through the gas before collection in order to enhance ion production. We experimented with two distinctly different pump designs. One resembled a magnetron structure where a radial electric field was perpendicular to an axial magnetic field. The second design resembled a Penning vacuum gauge. Here the electric and magnetic fields were essentially parallel.

In both structures the electrons would move in very long orbits. Preliminary experiments with both designs proved very encouraging: both pumped various gases including noble gases. Because the Penning type design appeared simpler, we continued our work with it.

We discovered that the driving of ions into surfaces was not the only pumping mechanism. Ions impinging upon the cathode surfaces produced sputtering and much of the pumped gas was found to be trapped not in the cathodes bombarded by the ions, but in areas where the sputtered material was deposited.

We experimented with various materials for the cathodes and found that titanium provided a particularly effective pumping action. We also found that in general auxiliary electron sources in the pump could be dispensed with provided that the voltage used was raised above a certain limit and the magnetic field was increased appropriately.

The illustration shows a sketch of the basic elements of the pump and a photograph of an early pump that has been cut open.

A. M. Gurewitsch (signature)

A. M. Gurewitsch

Willem F. Westendorp (signature)

W. F. Westendorp

(1) A. M. Gurewitsch and W. F. Westendorp, Rev. Sci. Inst. 25 , 389 (1954).
U. S. Patent No. 2,755,014, issued in 1956.
U. S. Patent No. 2,858,972, issued in 1958.
U. S. Patent No. 2,925,214, issued in 1960.

IUVSTA 07.029

MULTI-CELL SPUTTER-ION PUMP (1958)

Getter-ion pumping had its origins in the laboratories of four pioneers - F. M. Penning in the Netherlands, R. C. Herb and R. J. Connor in the U.S., and R. S. Barton in England. In the early 1930's, Penning introduced the techique of using a gas discharge in a magnetic field to provide a new kind of vacuum gauge. He also used this effect to produce pumping by trapping gas molecules in sputtered metal films. In the early 1950's, Herb demonstrated the ability of evaporated titanium films to pump various gases. In the middle 1950's, Connor developed the first practical sublimation pump and used it to maintain a vacuum in a small particle accelerator. At about the same time, Barton invented a similar device for the United Kingdom Atomic Energy Authority.

The original motivation for developing the commercially practical sputter-ion pump was to provide a high-vacuum pumping techinque for microwave tubes which would not introduce long-chain hydrocarbon molecules into the vacuum system. These molecules were known to "poison" tube cathodes and impair electron emission. I had been working, with little success, on a study of cathode materials at Varian Associates. It was suggested that a change of direction might be advisable, and I began to look into pumping techiques which might replace the oil-diffusion pump. It came to my attention that Prof. Simon Sonkin on Stanford University had obtained a device from Connor which looked interesting. It consisted of a chamber about 10 cm diameter and 15 cm high, with a Penning gauge connected to the side. The chamber contained a replaceable filament composed of tungsten wire wrapped with smaller diameter titanium wire. It had been used to maintain a high vacuum in a small sealed accelerator tube. It was believed by Connor that the Penning gauge generated a certain flux of metastable atoms and molecules which were more easily trapped by the sublimed titanium film than were their stable counterparts.

One or two of the Connor pumps were built and tested in my laboratory. They had some promising aspects, but filament burnout proved to be a very serious problem. In thinking about this limitation, it occurred to me that a Penning gauge with titanium cathodes might provide an alternative solution.   Enough titanium might be sputtered to provide a net trapping of gas atoms and molecules, thus eliminating the need for a sublimation filament.   It seemed logical to provide increased pumping by subdividing a large anode, so an experimental pump was fabricated, containing a 4-cell "egg-crate" anode and two titanium cathodes (1). The pump body was a section of copper waveguide. The magnetic field was provided by a large magnetron magnet.   The power supply consisted of an autotransformer and a high-impedance high-voltage transformer, with no current limitation.

A small vacuum system was assembled, consisting of the pump, a mechanical roughing pump, and the usual vacuum plumbing. Rus and Sig Varian, the Company's founders, had been told of the impending test, so they appeared and waited with interest. I turned on the roughing pump and pumped the system down to approximately 0.001 Torr, then turned on the ion pump. Instead of decreasing, the pressure rose rapidly. In addition, the pump and plumbing began to heat up considerably and the pressure stayed high. The pump became hotter and hotter, and so did the high-voltage transformer, but there was no observable pumping. After what seemed like a long time, but was probably only 15 or 20 minutes, Rus and Sig lost interest and left. Finally, however, just before the circuit breaker was ready to trip, the pressure began to fall. I was able to valve off the roughing pump and let the ion pump operate independently.

We immediately began to think about ways to fabricate larger pumps. The first such design used an electromagnet, but the magnet power supply proved unreliable. Bill Lloyd and Glen Huffman then designed a pump housing with many separate modules, using permanent magnets. This was what was needed for future growth, and the sputter-ion pump was on its way.

Lew Hall

Lewis D. Hall

(1) L. D. Hall, Rev. Sci. Instrum. 29 , 367 (1958).

Note: The editor regrets to inform the readers that Dr. Lewis Hall died on June 9, 1985, shortly after writing this account of his invention of the "egg-crate" sputter-ion pump.

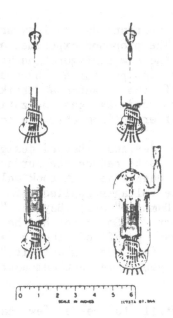

BAYARD–ALPERT IONIZATION GAUGE (1950)

The work that led to the Bayard–Alpert gauge was stimulated by the need for gases of extreme purity in other researches (on entrapment of resonance radiation) in which I was engaged in the late 1940's as a research physicist at the Westinghouse Research Laboratories. Achieving high gas purity, of course, demanded an "empty" container in which to carry out the experiments, and I had for some time been developing new vacuum systems and gas-handling techniques to operate at base pressures below 10 exp(-9) Torr.

The specific stimulus for the work was the recognition by Nottingham and others that ionization gauges then in use did not read pressures below 10 exp(-8) Torr, suggesting the existence of a residual current to the ion collector which is independent of pressure. At the MIT Physical Electronics Conference in 1947, Nottingham proposed that this residual current was due to soft x-rays which release photoelectrons from the ion collector, the x-rays being created at the grid by the incidence of the thermionic electrons.

In 1949, I suggested the study of the residual effect as a thesis problem for Robert Bayard, then a research physicist at Westinghouse Research Laboratories and a graduate student in the Physics Department at the University of Pittsburgh, where I held an appointment as an adjunct professor. The task: develop an experimental method for determining whether the residual current was indeed due to an x-ray effect or to some other physical process; we were not averse to finding an alternative explanation for the residual effect.

Bayard considered various designs of apparatus aimed at reducing the residual currents by reducing the surface area of the collector, thus to differentiate between the current due to ions arriving at the collector and electrons leaving it. The proposed experimental procedure was then to compare the behavior of the new configuration with that of the standard cylindrical gauge at very low pressures - by measuring the current as a function of the energy of the electrons striking the grid. If the collector current were produced by gas ionization, the measured energy dependence would be quite different from that due to x-rays.

After several unsuccessful designs, Bayard designed and constructed the "inverted" geometry, which was to reduce the surface area of the collector by at least two orders of magnitude. It took only a few hours after the system was baked out and the voltages applied before Bayard rushed into my office with the data. "Dammit," cried Bayard, "I'm afraid Nottingham's x-ray hypothesis is correct." So intent had he become to disprove it that this, the first clear proof of the x-ray hypothesis, left him actually dismayed. "Don't feel bad about it," I replied, "either the results are in error or you have just demonstrated the best ion gauge known to man."

Inevitably, there were still to be a few bad moments before the "Bayard-Alpert" gauge could be considered a laboratory instrument. The most serious was occasioned by a gradual deterioration of the new "gauge" over the next few days - an irreversible increase in the magnitude of the measured residual current. This effect was soon correlated with the gradual evaporation of a thin tungsten film over the glass envelope: the "ion" collector was then measuring x-ray photoelectrons emitted from the metal-covered glass surface. The effect was removed by adding a glass "skirt" - as shown in the figure - to break the connection between ion collector and tungsten film.

Within a few months of the publication of the Bayard-Alpert paper (1) in April, 1950, two other scientists (2,3) published results which similarly verified the x-ray hypothesis and offered alternative designs for ionization gauges. We later showed (4), by processing the data of an early experiment, that pressures below 10 exp(-10) Torr had been achieved and the effect could have been detected as early as 1931.

Dan Alpert

Daniel Alpert

(1) R. T. Bayard and D. Alpert, Review of Scientific Instruments 21 , 571 (1950).
(2) J. J. Lander, Review of Scientific Instruments 21 , 672 (1950).
(3) G. H. Metson, British Journal of Applied Physics 2 , 46 (1951).
(4) D. Alpert, Journal of Applied Physics 24 , 860 (1953).

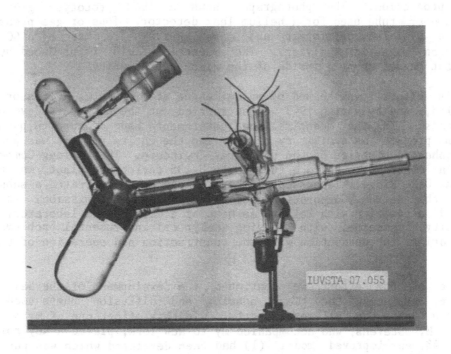

HELIUM MASS SPECTROMETER LEAK DETECTOR (1942)

During World War II the Manhattan Project in the United States was given the task of separating the uranium isotope U-235 from U-238. One of the promising enrichment methods under development at that time was gaseous diffusion. Basic studies were made by Professor John Dunning and colleagues at the University of Columbia and the M. W. Kellogg Company made pilot plant studies. A practical plant would require, in addition to diffuser units the size of railroad tank cars, many gas compressors, valves and other plumbing, all of which had to be vacuum tight - far beyond state-of-the-art industrial practice at that time.

By coincidence two of my close friends from Harvard postdoctoral days, Manson Benedict and Robert Jacobs, were employed by the Kellogg Company, so during 1942 the Columbia group, and my Kellogg friends and I had many discussions relating to the leak testing of components as well as of a completed plant itself. It was decided that a mass spectrometer, tuned to helium, would be the most sensitive and selective leak detector one might employ. An object to be tested would be vacuum pumped and the gas stream entering the pump sampled with the mass spectrometer. Since helium is a very minor constituent of the atmosphere, the helium signal would normally be very low, even in a leaky vessel. However, if helium were sprayed on a suspected leak, of say a weld, there would be an immediate and positive response. The method looked so powerful and promising that our contract with the Office of Scientific Research and Development which supported our mass spectrometer uranium isotope analysis work at the University of Minnesota was expanded to include the development of four helium leak detectors to be used for feasibility studies and as prototypes for eventual mass production. The photograph shows a 1942 prototype glass mass spectrometer tube used for a helium leak detector. Ions of gas present are produced by electron impact and mass analysis is made by a 60 degree deflection magnetic analyzer. The magnetic field is produced by small permanent magnet mounted inside of the glass tube housing.

In this effort I was helped by my colleague at Minnesota, Professor Andrew Hustrulid, and by spring 1943 we had carried out the necessary developments and tests. In the meantime the government had decided to pursue the uranium project as a military program and the entire project was put under the Manhattan District of the U.S. Army Engineers. The Kellogg Company was given a contract to build a full scale U-235 enrichment plant, known as K-25, in Oak Ridge, Tennessee. The Kellogg Company created a subsidiary called the Kellex Corporation for carrying out the mission. In 1943 I accepted a position with Kellex, as head of a development laboratory in New York City, concerned with solving analytical instrumental problems which might arise in connection with the construction and operation of the K-25 plant.

As part of the effort we continued the development of the helium leak detector, replacing the glass housing and diffusion pumps used in the prototype units by all-metal parts. In this effort one of my students, Charles M. Stevens, who accompanied my to New York, played a key role. By late 1943 an improved model (1) had been developed which was put in the hands of the General Electric Company for manufacture. We cooperated actively with General Electric as they solved the many problems encountered in producing a new product on a tight time schedule. Hundreds of leak detectors were built by G.E. and used by vendors of components for the K-25 plant and by the crews asesmbling by plant itself.

The leak detection program (2) at Kellex was under the direction of Robert Jacobs. It is interesting to note that two of his assistants, Albert Nerken and Frank Raible, formed the Veeco company after the war ended and transplanted their Manhattan Project experience in leak testing to the larger civilian sector.

The helium leak detector is by far the most sensitive device of its kind. In 1945, its sensitivity was in the neighborhood of 10 exp(-7) standard cubic centimeters per second. This was a hundred times more sensitive than an ionization gauge, the next most sensitive device. The present day mass spectrometer leak detectors can detect flows of 10 exp(-11) standard cubic centimeters per second, i.e., 10,000 times smaller leaks than the original models.

Alfred O. C. Nier

Alfred O. C. Nier

(1) A. O. Nier, C. M. Stevens, A. Hustrulid, and T. A. Abbott, J. Appl. Phys. 18, 30 (1947)

(2) R. B. Jacobs and H. F. Zuhr, J. Appl. Phys. 18, 34 (1947)

History of the Development of Diffusion Pumps

B. B. Dayton

INTRODUCTION

During the period from 1917 to 1960 the production of high vacuum, in systems requiring pumping speeds greater than could be attained with trapped mechanical pumps, was achieved with mercury and oil vapor diffusion pumps. Vapor diffusion pumps, as the name implies, depend for their pumping action on the diffusion of gases into fast-moving streams of vapor molecules which carry the gas toward a forepump and then condense on cooled walls. The vapor generated in a boiler rises through conduits to exit through nozzles forming a high-speed expanding jet of vapor. The use of diffusion pumps has declined somewhat since 1958, when W. Becker in Germany developed a practical turbomolecular pump which led to the development of modern high-speed turbomolecular pumps and when L. Hall introduced the sputter-ion pump which was modified and improved until large high-speed ion pumps became widely available, and since 1960 when W. Gifford and H. McMahon developed an efficient mechanical refrigerator such as used in modern cryopumps. However, because of their relatively low cost compared to other types of high vacuum pumps, diffusion pumps are still widely used and are of more than historical interest.

EARLY DEVELOPMENT

The invention of the diffusion pump arose out of studies made by W. Gaede in Germany in 1913 on the counterflow of mercury vapor and air in a vacuum system pumped by his rotary, mercury-sealed mechanical pump with a cold trap to condense mercury vapor. During this study to determine the lowest partial pressure of gas that could be produced beyond the trap, he discovered the effect of the counter-diffusion of air and mercury vapor

on the reading of a McCleod gauge and conceived of the pumping effect that might be obtained by allowing gas to diffuse through a narrow slit into a rapidly moving stream of mercury vapor. Fig. 1 shows his first embodiment of this concept. Note that the slit region is surrounded by a water cooling jacket to condense mercury vapor escaping through the slit.

In 1916 Irving Langmuir at the General Electric Company showed that the slit or diffusion diaphragm through which the gas was pumped in the Gaede pump (German patent No. 286,404) could be widened if the mercury vapor was directed away from the slit by a suitable nozzle as shown in Figs. 2 and 3. He emphasized the importance of adequate condensation of the mercury vapor and referred to his pumps as condensation pumps. The pumping speed was thereby greatly increased, and Langmuir was granted U.S. Patent 1,393,550 in 1921. In 1916 Langmuir also applied for a patent on the inverted or "mushroom cap" nozzle design, which issued in 1919 as U.S. Patent 1,320,874, as shown in Fig. 4. H. Williams in 1916 (1) published a few months prior to the announcement of the Langmuir pump a description of a wide slit mercury condensation pump in which the mercury could be purified by distillation during operation of the pump. In 1917 W. Crawford improved the Langmuir pump by using an expanding conical nozzle based on turbine theory and a well-designed condenser and entrance chamber (Fig. 5). He also developed a multiple-stage, inverted-nozzle mercury pump and obtained U.S. Patent 1,367,865 in 1921. In 1918 O. E. Buckley applied for a patent on a concentric nozzle multi-stage mercury pump—U.S. 1,307,999 issued in 1919.

In 1922 Gaede in Germany developed a highly effective four-stage steel mercury pump and applied for a U.S. patent which was granted in 1924 (U.S. 1,490,918). This pump was marketed worldwide by the E. Leybolds Nachfolger Company of Germany. In 1923 Gaede published an article (2) on the origin and development of mercury diffusion pumps, in which he defended his invention against the claims of Langmuir, Crawford and others in the U.S. Many designs of multiple-nozzle, mercury vapor diffusion pumps have been published since 1922. T. Ho in 1932 (3) and M. Copley and coworkers in 1935 (4) studied the relation between nozzle design and entrance chamber to improve the speed of mercury vapor pumps.

Mercury vapor pumps became widely used in the electronic tube industry in the U.S. from 1920 to 1940, but they required the use of refrigerated traps to keep the vapor out of the tubes (except for mercury rectifier tubes). In 1928, in connection with experiments on the vacuum impregnation of pressboard with transformer oils to improve the dielectric strength of insulators, C. R. Burch (5,6) at Metropolitan-Vickers in England became interested in high vacuum distillation at moderate

temperatures, which was known as "molecular distillation." He succeeded in obtaining some very low vapor pressure fractions from the petroleum oil used in rotary mechanical pumps. It then ocurred to him to try these oil fractions as the pump fluid in a diffusion pump in place of mercury. He was able to obtain a pressure reading on an ionization gauge of 10^{-6} Torr without a cold trap, and thus began a revolution in the application of diffusion pumps (British patent 303,078). The oils were later marketed by Shell Chemical Co. Ltd. under the trade name Apiezon. Metropolitan-Vickers began the development of diffusion pumps for use with these oils, and in 1931 British patent 346,293 on an oil-diffusion pump was granted to Burch and Bancroft. W. Gaede in 1932 (7) then developed an oil-diffusion pump with three water cooling jackets and a trap in the foreline for volatiles, so that fractional distillation of the oil could be carried out in the pump itself. Gaede's pump was marketed by the E. Leybolds Nachfolger Co. of Cologne, Germany.

Dr. Kenneth Mees, research director of the Eastman Kodak Laboratories, after visiting England, brought the

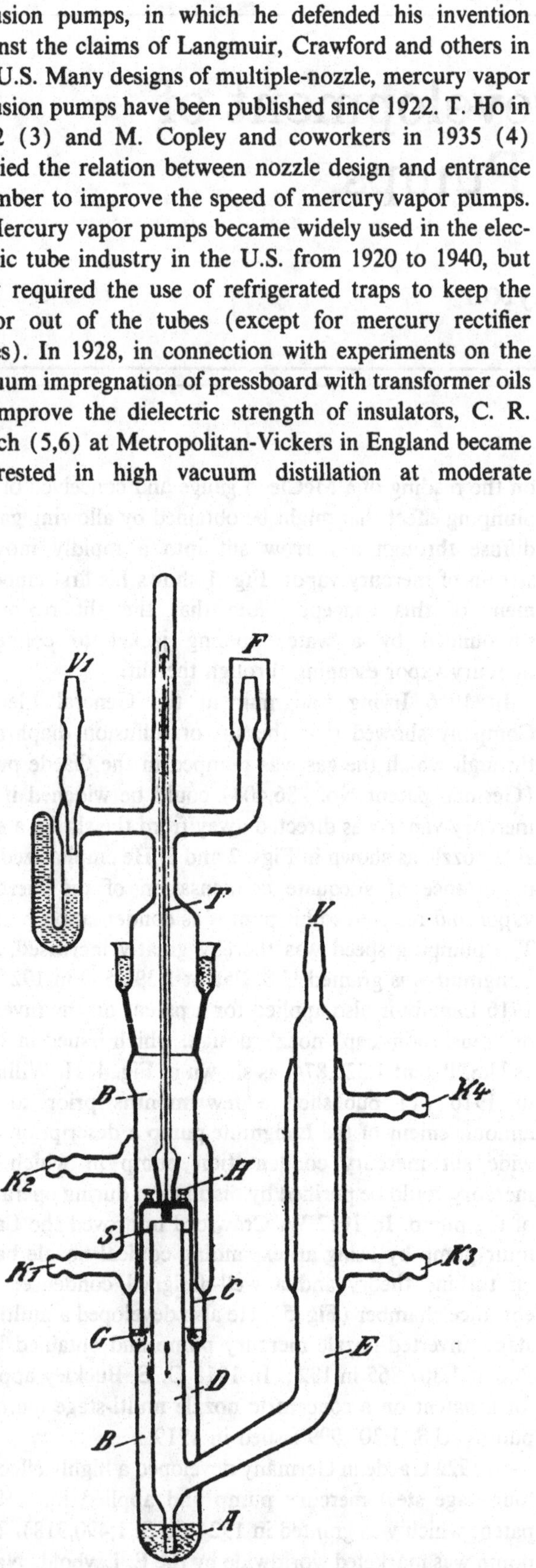

FIG. 1. Gaede's mercury diffusion pump with adjustable gas intake slit at *S* between steel cylinders *C*. Boiler *A* with mercury heated by Bunsen burner. Thermometer *T* for determining vapor pressure. Water cooling at *K*.

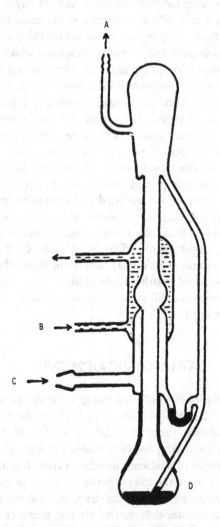

FIG. 2. Early, glass model of Langmuir's condensation pump. *A* connection to forepump, *B* water inlet, *C* high vacuum connection, *D* mercury boiler.

work of Burch to the attention of Dr. K. C. D. Hickman, who had been working with low-vapor-pressure synthetic organic esters in place of mercury in special manometers for measuring the pressure in apparatus for the drying of photographic film where mercury vapor was harmful. In 1929 Dr. Hickman found that these esters gave good results when used as the pump fluid in small glass diffusion pumps of simplified design, which he constructed himself (8). He applied for a patent on the use of butyl phthalate and similar esters in place of mercury in these pumps (U.S. 1,857,506 issued in 1932). Many articles were published in the period from 1932 to 1940 on the conversion of mercury vapor pumps to oil-diffusion pumps and the use of baffles and charcoal traps to reduce oil vapor contamination (9).

During the period from 1929 to 1937 the factors involved in designing diffusion pumps for use with the new synthetic oils were studied by Dr. Hickman and his co-workers in the research laboratories of the Eastman Kodak Company, resulting in the development of multi-boiler, "self-fractionating" oil-diffusion pumps, which extended the lowest pressure attainable without cold traps from 10^{-5} Torr to about 5×10^{-8} Torr. Fig. 6 shows an experimental fractionating pump in which the condensate from each pumping stage could be returned to either one of the two boilers by magnetically controlled ball valves. The lowest ultimate pressure was achieved with the ball seated in the valve opening b' on the right, so that condensate from the high vacuum stage flows first

into the boiler on the left where volatile impurities are purged from the oil before it returns to the boiler on the right feeding the high vacuum stage. Hickman applied for a patent on the fractionating oil diffusion pump in 1935 (U.S. 2,080,421 issued in 1937). Fig. 7 shows a three-stage fractionating oil diffusion pump made from glass, as designed by Dr. Hickman, which uses heater coils of nichrome wire immersed in the oil in the boilers, as first suggested in 1935 by J. Bearden (10).

DESIGN DEVELOPMENTS IN THE MODERN ERA

In 1937 Dr. N. Embree, a coworker of Dr. Hickman, applied for a patent on the streamlined inverted nozzle design (U.S. patent 2,150,676 issued in 1939), which

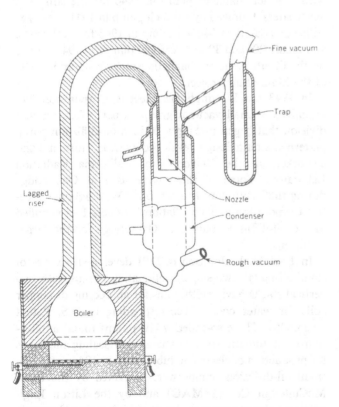

FIG. 3. Glass form of Langmuir's condensation pump, improved model.

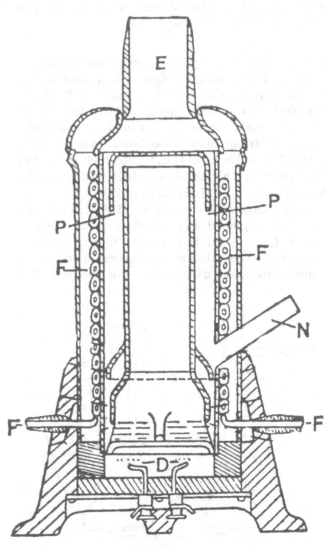

FIG. 4. Langmuir's all-metal, mercury vapor condensation pump as used by General Electric in the manufacture of lamps. *E* gas intake connection, *N* outlet to forepump, *P* gas intake region between "mushroom cap" and cooled walls *F*, *D* electric heater.

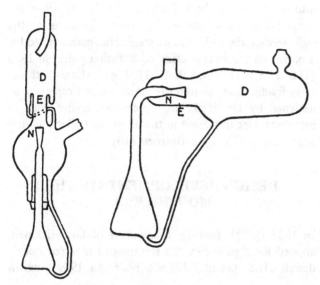

FIG. 5. Two versions of Crawford's mercury vapor diffusion pump with expanding conical nozzle *N* and air cooling on walls of condensing chambers *E* and *D*.

greatly improved the pumping speed and became the prototype for subsequent designs of high-speed vertical pumps (Fig. 8). Also in 1937, L. Malter of the Radio Corporation of America applied for a patent on the concentric vapor conduit tube system for vertical multistage, self-fractionating oil-diffusion pumps (U.S. 2,112,037 issued in 1938) (Fig. 9), and Hickman applied for a patent on a horizontal metal fractionating oil diffusion pump (U.S. 2,153,189 issued in 1939).

In 1938 Distillation Products Inc. (DPI) in Rochester, N.Y., was organized as a subsidiary of the Eastman Kodak Co. and jointly owned by General Mills to produce oil-soluble vitamin concentrates. In addition to applying molecular distillation to fish liver oils, DPI

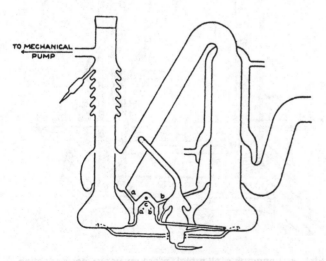

FIG. 6. Hickman's two-stage, self-fractionating, oil vapor diffusion pump. Experimental model with steel ball at *a,b* to control flow of returning condensate to first- or second-stage boiler.

began to market diffusion pumps in 1939. In 1940 Dr. Richard Morse, who had worked with Dr. Hickman in developing vacuum products, left DPI and founded the National Research Corporation in Cambridge, Mass. NRC soon began the development of a line of metal diffusion pumps which found use on lens coaters and vacuum furnaces for the production of magnesium.

In 1941, in response to a problem in handling the high gas load from 32-inch-diameter centrifugal molecular stills used at Distillation Products Inc. to produce vitamin oil concentrates, the gap in the range between diffusion pumps and steam ejectors was filled by Hickman's invention of an oil-vapor ejector pump. The oil-vapor ejector pump operates on a principle similar to that of the diffusion pump, but the vapor jet issues from an expanding conical nozzle at considerably higher vapor pressures and flows through a Venturi-tube diffuser with most of the condensation of the vapor occurring beyond the throat of the diffuser. DPI developed a series of oil-vapor ejector pumps (KB, or kerosene booster, pumps) that found application in production of wartime magnesium by the Pidgeon Process and in the dehydration of penicillin and blood plasma and foods. Diffusion-ejector pumps were developed by the Edwards High Vacuum Company in England by adding an ejector stage to a vertical metal diffusion pump design.

A series of 3-stage, vertical, all-metal oil-diffusion pumps with Embree streamlined top nozzle was developed by DPI in the period from 1940 to 1943, beginning with a 4-inch-diameter pump in 1940 for the stills and lens coaters, followed by a 6-inch pump in 1941 for larger stills and coaters, a 14-inch pump in 1943 for cyclotrons, and finally 20- and 30-inch vertical pumps in 1943 for use on the Calutron electromagnetic isotope separation units of the Manhattan Project.

In 1943 the Westinghouse Electric Corporation had organized a small vacuum group as part of its rectifier division that began making a 20-inch oil-diffusion pump patterned after a design developed by C. E. Normand and coworkers at the University of California Radiation Laboratory. Some of these were used at Oak Ridge during the Manhattan Project. Both Westinghouse Electric Corporation and Distillation Products Inc. supplied 32-inc oil-diffusion pumps to Oak Ridge for the larger Calutrons.

In 1944 M. E. Johnson of DPI developed a series of small all-metal, two-stage fractionating pumps with two vertical stacks having either fins for air cooling or copper coils for water cooling (corresponding to U.S. Patent 2,435,686). These so-called VMF pumps found application on automatic rotary exhaust and sealing machines for production of electronic tubes. Also in 1944 and 1945, small oil-diffusion pumps were marketed by the Eitel-McCullough Co. (EIMAC) and by the Litton Engineering Laboratories for exhaust of high-power electronic tubes. In the years that followed several other companies in the U.S. began to manufacture and sell diffusion pumps. Similarly, from 1930 to 1960, many firms in Europe developed their own line of diffusion pumps.

Although Hickman could not obtain a patent on the basic idea of an oil vapor ejector because of a prior invention by W. K. Lewis of MIT, U.S. Patent 2,379,436 was issued in 1945 to DPI for an improved design of oil ejector (KB pump) developed by Dr. Hickman and G. Kuipers.

In 1945 DPI began to manufacture vertical metal fractionating pumps after obtaining rights under the Malter patent owned by RCA. A 10-inch model, the MCF-1400, was their first pump in 1945, and was followed by a 6-inch pump, the MCF-700, in 1946. Several of the neon sign manufacturers made small glass oil-diffusion pumps for their own use, and some offered these for sale.

By 1947 the National Research Corporation was marketing four sizes of vertical metal 3-stage oil diffusion pumps, the H-2, H-6, H-10, and H-16, with inlet diameters ranging from 2 to 16 inches. NRC also introduced in 1947 the B-6 booster type diffusion pump. A summary of the vacuum pumping equipment and systems commercially available from U.S. manufacturers was published in the January 1948 issue of *The Review of Scientific Instruments* by H. M. Sullivan.

As a result of the success of the Soviets in developing and detonating an atomic bomb in 1949, the decision was made and announced by President Truman on 31 January 1950 that the U.S. would develop a thermonuclear weapon, the hydrogen bomb. In 1950 projects were begun at the Savannah River Plant of the Atomic Energy Comission (AEC) to produce large quantities of the hydrogen isotope tritium, which would be required for the construction of the "super" as originally designed. It was necessary to extract the tritium under vacuum from uranium in the reactors. However, the tritium would interact with the hydrogen in the organic pump fluids used in oil-diffusion pumps. At the request of the AEC a small, stainless-steel casing, mercury diffusion–ejector pump, the MHGS-20, was developed by the Vacuum Equipment Division of DPI. This pump could extract the tritium and compress it to pressures above 10 Torr.

Dr. Ernest O. Lawrence, director of the University of California Radiation Laboratory, was able to obtain funding for the construction of a very large linear accelerator at Livermore, California, to explore methods of producing materials for a hydrogen bomb. This materials-testing accelerator, or MTA project, required forty-eight 32-inch diffusion pumps to pump a vacuum chamber of about 220,000 cubic feet volume. At first it was assumed that oil-diffusion pumps with Freon-cooled baffles would give the required vacuum without contamination. Forty-eight surplus Westinghouse 32-inch pump casings were brought from Oak Ridge and Distillation Products Industries was asked to supply forty-eight fractionating-type jet assemblies to fit in the Westinghouse casings. However, after a short period of operation, it was decided in 1951 that oil vapor contamination was causing leakage paths on the insulators in the accelerator and contributing to high-voltage breakdown. DPI was then requested to convert all the 32-inch pumps from the use of hydrocarbon oil as the pump fluid to the use of mercury. This involved reducing the area of the nozzle throat openings and adding internal water cooling coils to help condense the mercury vapor. The Freon baffles were replaced by low-temperature, refrigerated, two-zone baffles designed by Hugh R. Smith in 1956 (11) of the University of California Radiation Laboratory.

Dr. Lawrence, anticipating a need for many very large diffusion pumps for production of materials for a hydrogen bomb if his MTA project proved successful, visited Distillation Products Industries early in 1950 and requested the development of the largest feasible oil-diffusion pump. It was decided that a 48-inch-diameter diffusion pump would be practical, and by December 1958 DPI had completed the development of a 48-inch

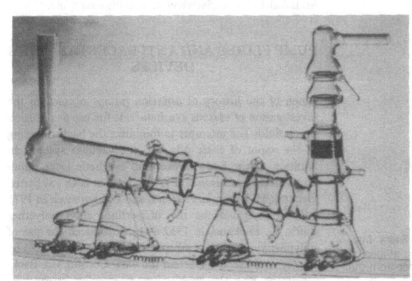

FIG. 7. Hickman's three-stage, self-fractionating diffusion pump. Water-cooled glass model with heating coils inside boilers.

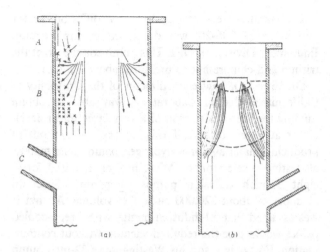

FIG. 8. Embree's streamlined nozzle and vapor conduit model **(b)** compared to previous design **(a)**.

vertical metal fractionating pump, the MCF-40,000. After the decision that the MTA system required mercury pumps, DPI was asked in 1951 to convert the experimental 48-inch oil-diffusion pump for operation

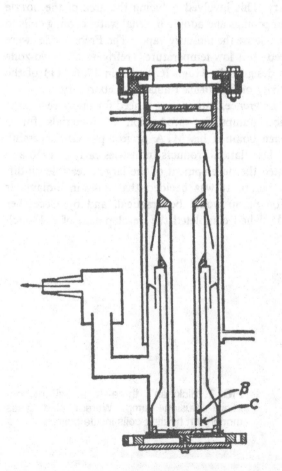

FIG. 9. Malter's all-metal, vertical, concentric cylinder, fractionating, oil diffusion pump with three pumping stages. *B* and *C* are cylindrical walls to separate the vapor from the returning liquid as moves in toward the center.

with mercury. A subsidiary of the Standard Oil Company of California, the California Research and Development Company, was given overall responsibility for the "crash program" of converting the MTA project from oil-diffusion pumps to mercury diffusion pumps. The CRDC set up a pump development facility at Livermore and began designing a giant (6 feet wide), rectangular casing, mercury diffusion pump to permit easier manifolding to MTA-type accelerator chambers. DPI completed the testing of its 48-inch mercury pump, Type MHG-30,000, using 700 pounds of mercury, by February of 1952. However, the MTA project was abandoned by the AEC as other methods of producing the hydrogen bomb from hydrogen and lithium isotopes became available.

The first real thermonuclear device was exploded by the U.S. on 1 November 1952 at Eniwetok atoll, and at the same time the Livermore facility of the University of California was taken over by the AEC and became the Livermore Laboratory. The development of the 32- and 48-inch pumps was not in vain, since pumps of this size later came into use on large environmental chambers for space simulation.

THEORETICAL STUDIES

W. Gaede in 1915 (12) presented a mathematical theory of the operation of diffusion pumps, and since then many papers have been published on the theory and design of diffusion and ejector pumps. Among the authors of these papers are W. Molthan (13), L. Wertenstein (14), T. Ho (3), M. Matricon (15), R. Jaeckel (16), D. Avery and R. Witty (17), B. Dayton (18), S. Oyama (19), R. Witty (20), L. Riddiford and R. Coe (21), V. Skobelkin and N. Yushenkova (22), R. Jaeckel, H. Noeller, and H. Kutscher (23), L. Zobac (24,25), H. Noeller (26), H. Kutscher (27), B. Power and D. Crawley (28), N. Florescu (29), W. Reichelt (30), G. Toth (31,32,33,34), P. Duval (35), M. Wutz (36), M. Hablanian and J. Maliakal (37), K. Sadykov and S. Figurov (38).

PUMP FLUIDS AND ANTI-BACKSTREAMING DEVICES

Much of the history of diffusion pumps centers on the investigation of various synthetic oils for use as diffusion pump fluids and attempts to minimize the backstreaming of the vapor of these oils into the vacuum system. In addition to the hydrocarbon and synthetic ester pump fluids already mentioned, P. Alexander in 1948 (39) tried glycerol, M. Baker, L. Holland, and L. Laurensen in 1971 (40) investigated the use of perfluoroalkyl polyether fluids, K. Hickman in 1962 (41) introduced the use of polyphenyl ethers, and silicone fluids were introduced by the General Electric Co. and the Dow-Corning Co. Backstreaming was thoroughly investigated by researchers

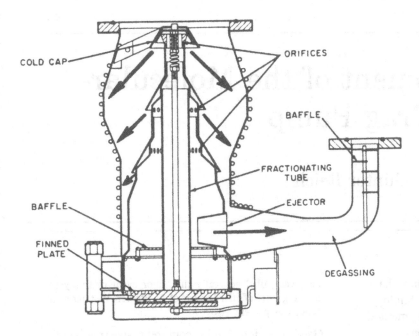

FIG. 10. One version of the anti-backstreaming, cooled cap over the top nozzle of an oil diffusion pump. Cap is cooled by conduction through mounting struts to walls. (Varian VHS 4-stage pump).

from the Balzers Co. in Liechtenstein and by B. Power and D. Crawley (28) of the Edwards Co. in England. These researches resulted in the cooled cowl over the top cap of the nozzle assembly, one version of which is shown in Fig. 10 (42) and the use of Teflon coatings on the cap. However, most systems where oil vapor backstreaming would be a serious problem are now pumped by sputter-ion pumps or cryopumps and by special turbomolecular pumps designed to avoid vapor contamination.

REFERENCES

1. Williams HB. *Phys Rev* 1916; 7:583.
2. Gaede W. *Zeit f Tech Physik* 1923; 4:337–369.
3. Ho TL. *Physics* 1932; 2:386.
4. Copley MJ, Simpson OC, Tenney HM, Phipps TE. *Rev Sci Instr* 1935; 6:265–267.
5. Burch CR. *Nature* 1928; 122:729.
6. Burch CR. The first oil condensation pump. *Chemistry and Industry* (London) 6: February 5, 87.
7. Gaede W. *Z f Technische Physik* 1932; 13:210.
8. Hickman KCD, Sanford CR. A study in condensation pumps. *Rev Sci Instr* 1930; 1:140–163.
9. Dushman S. *Scientific Foundations of Vacuum Technique.* New York: John Wiley & Sons, 1949.
10. Bearden JA. *Rev Sci Instr* 1935; 6:276.
11. Smith HR. The technology of large mercury-pumped systems. *1955 Trans. National Vacuum Symposium, AVS* 1956; 22–30.
12. Gaede W. *Ann der Physik* 1915; 4:357–392.
13. Molthan W. *Z techn Physik* 1926; 7:377, 452.
14. Wertenstein L. *Proc Cambr Phil Soc* 1927; 23:578.
15. Matricon M. *J de Phys Radium* 1932; 3:127.
16. Jaeckel R. *Z f Naturforschung* 1947; 2a:666.
17. Avery D, Witty R. *Proc Phys Soc* (London) 1947; 59:1016.
18. Dayton B. *Rev Sci Instr* 1948; 19:793.
19. Oyama S. *J Phys Soc* (Japan) 1950; 5:192.
20. Witty R. *Br J Appl Phys* 1950; 1:232.
21. Riddiford L, Coe R. *J Sci Instr* 1954; 31:33.
22. Skobelkin VI, Yushenkova NI. *Zh Eksp Teor Fiz* 1954; 24:1879.
23. Jaeckel R, Noeller H, Kutscher H. *Vakuum-Technik* 1954; 3:1.
24. Zobac L. *Slaboproudy Obz* 1954; 15:127.
25. Zobac L. *Slaboproudy Obz* 1955; 16:541.
26. Noeller H. *Z f angew Physik* 1955; 7:218.
27. Kutscher H. *Z f angew Physik* 1955; 7:229, 234.
28. Power BD, Crawley DJ. *Vacuum* 1954; 4:415.
29. Florescu N. *Vacuum* 1960; 10:250.
30. Reichelt W. *Vakuum-Technik* 1964; 13:148.
31. Toth G. *Vakuum-Technik* 1967; 16: 41,193, 215.
32. Toth G. *Vakuum-Technik* 1968; 17:251.
33. Toth G. *Vakuum-Technik* 1970; 19:138, 183.
34. Toth G. *Vacuum* 1982; 32:29.
35. Duval P. *Le Vide* 1969; 24:83.
36. Wutz M. *Molekular Kinetische Deutung de Wirkungsweise von Diffusion Pumpen* Braunschweig: Fr. Viewg. & Sohn, 1969.
37. Hablanian MH, Maliakal JC. *J Vac Sci Technol* 1973; 10:58.
38. Sadykov K, Figurov S. *Vacuum* 1990; 41:2061.
39. Alexander P. *J Sci Instr* 1948; 25:313.
40. Baker M, Holland L, Laurensen L. *Vacuum* 1971; 21:479.
41. Hickman KCD. *Trans. 8th National Vacuum Symposium, AVS.* New York: Pergamon Press, 1962; 307.
42. Power BD. *High Vacuum Pumping Equipment* Chapter 2. Chapman and Hill, London, 1966.

SUGGESTED READINGS

1. Becker W. *Vakuum-Technik* 1958; 7:149.
2. Crawford WW. *Phys Rev* 1917; 10:557–563.
3. Gifford W, McMahon H. *Prog Refrig Sci Technol* 1960; 1:105, Oxford: Pergamon Press.
4. Hall L. *Rev Sci Instr* 1958; 29:367.
5. Hickman KCD. Vacuum pumps and pump oils. *J Franklin Inst* 1936; 221:215–235.
6. Hickman KCD. Trends in design of fractionating pumps. *J Appl Physics* 1940; 11:303–313.
7. Langmuir I. *J Franklin Inst* 1916; 182:719.
8. Langmuir I. *GE Review* 1916; 19:1060.
9. Malter L, Marcuvitz N. A modification of Hickman's distillation pump. *Rev Sci Instr* 1938; 9:92–95.
10. Normand CE, *et al. Vacuum Problems and Techniques* TID-5210, Office of Technical Services, Dept. of Commerce, 1950.
11. Ray K, Sen Gupta N. *Indian J Phys* 1945; 19:Pt. IV, 138.
12. Sullivan HM. Vacuum pumping equipment and systems. *Rev Sci Instr* 1948; 19:1–15.

Early Development of the Molecular-Drag Pump

Günter Reich

During the last 30 years, molecular drag pumps have acquired an increasing importance in vacuum technology. This paper documents the origin of this pumping method and the development of the various modifications.

WOLFGANG GAEDE 1878–1945 (1)

Gaede reported for the first time the principle of drag pumps at the meeting of the Physical Society in Karlsruhe (2) 80 years ago on September 26, 1911. He and the company E. Leybolds Nachfolger, with whom he had been associated since 1906, had traveled the major portion of the long and difficult road to a practical "molecular air pump," the name he had chosen.

His sister mentioned in her biography that her brother had already given this principle some thought in 1905. Several letters between Gaede and Leybold, written between 1907 and 1911, indicate that there had been many disappointments and delays. Finally the first 14 pumps were ready in the fall of 1912. The announcement of the 3 P.M. meeting of the Physical Society in Münster, 16 September 1912, was:

> K. Goes (Cologne), Demonstration of several experiments with the molecular air pump. (3)

(Goes was Gaede's assistant.) A letter written a few days later to Gaede by A. Schmidt, the owner of Leybold at the time, indicates the relief that everything finally worked:

> I went to the meeting with Dr. Goes and two mechanics. The lecture of G. was a big success. He spoke very well and the experiments were performed flawlessly. We were congratulated by everybody. The pump was a sensation

and was called generally the major event of this year's meeting. Mr. Pfeiffer [owner of A. Pfeiffer, Wetzlar at the time] was present also.

Gaede mentioned in his comprehensive treatise of 1913 "The Molecular Pump" (4) how he got the idea to develop such a pump:

> It is initially surprising how much the gas friction affects the speed of the rotary mercury pump in practical application. [This pump built by him in 1905 had a rather narrow intake pipe for technical reasons. (5)] This gave me the idea that the gas friction is of much greater significance for vacuum work than assumed so far. This does not apply only in the negative sense. While the friction in the connecting tubes reduces the pumping speed, I suspect that it should also be possible to utilize the gas friction constructively to generate pumping action.

Gaede had invented the principle of the drag pump.

Gaede applied the physics of highly rarefied gases and developed the theoretical foundation of drag pumps that is still valid today (6). He demonstrated the principle of the operation with a simple arrangement similar to Fig. 1. When the cylinder A, located in the housing C, rotates around the axis B in the direction of the arrow, a pressure difference (i.e., a pumping action) is produced in the pumping groove D between G and F. E is part of the comb, mentioned below. Gaede realized that the path length from F to G was insufficient to attain an adequate pressure difference. Therefore he built a "helix pump." The rotor was a smooth cylinder, while a female thread was cut into the stator. "The threading starts in the middle as a lefthanded thread in one direction and a righthanded one in the opposite direction. Hence the pressure is lowest in the center" (Fig. 2). He could show "that the pressure reduction can become appreciable in vacuum."

This dual flow design was taken over later by Holweck and Becker. The advantage of this arrangement is that the

Editor's note: A slightly different version of this paper was first published in German in *Vakuum in der Praxis* 1992; 4:206. Translated from the German by G. Lewin.

114

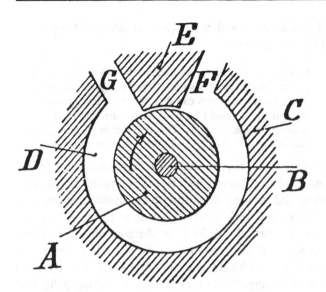

FIG. 1. Basic design of a molecular drag pump. The figure is taken from Gaede's patent (7) "Rotary Vacuum Pump," applied for in 1909.

oil vapor from the bearings cannot penetrate into the high-vacuum region. Gaede was aware of, and emphasized, this feature and the fact that the compression ratio for heavy gases is especially high.

Gaede tried "to enhance the efficiency by increasing the number of threads," but this was impossible because strong oscillations occurred at higher rotational speeds. About 1907 he tried another modification. Based on his theoretical considerations he expected that "the pumping speed could be improved without changing the external dimensions by making the surface area of the rotor larger than the one of the stator." This multistage molecular pump (Fig. 3) had a rotor with grooves of different dimensions. The intake stage had a larger cross-section than the compression stages. The rotor was 50 mm long and had a diameter of 100 mm. The cross-section of the grooves of the intake stage was 6×25 mm and that of the compression stages was 1.5×14 mm. The weight of the pump without the motor was 14.5 kg. A comb C (Fig. 3B) was inserted into the grooves from the top. The connections between the stages are in the housing G. Gas tightness and lubrication of the rotor axle is provided by

oil on the outside K entering through a spiral notch L. The dimensions of this notch were such that the external pressure did not push the oil into the pump housing at the operating speed. This simple solution was satisfactory in continuous operation, but had the disadvantage that the oil penetrated into the housing after a sudden stop. Removal of the oil required the disassembly of the whole pump. The peripheral speed of the rotor was 35 m/s at the specified 8000 rpm, very low by today's standards (larger than 200 m/s). The compression ratio for air was 2×10^5 and the pumping speed was 1.5 l/s below 10^{-4} Torr, a high speed in comparison to the rotary mercury pumps of the time. The claimed ultimate pressure, as measured with a McLeod gauge, was below 1×10^{-6} Torr. Ionization gauges did not yet exist at the time. The pump down time and the quality of the vacuum of x-ray and cathode-ray tubes were cited as proof of the high performance of this pump. Figure 4 shows the molecular pump with a rotary vane pump as forepump demonstrating the arrangement for exhausting an x-ray tube. S. Dushman confirmed Gaede's data in a paper (10) published in 1915.

The particular advantage was "that the molecular pump does not pump gases only but also vapors," contrary to the other pumps known at the time. This advantage was of great significance for many applications, in particular for the evacuation of electron tubes. Great progress was made in this field in a short time, especially in the USA (11,12). Recognizing Gaede's achievements, the Franklin Institute of the State of Pennsylvania bestowed on him in 1913 the Elliot Cresson Gold Medal for his molecular air pump. Leybold sold 289 molecular pumps from 1912 to 1923.

The design was very intricate by the standards of the time and must have caused complications (probably not in manufacturing only). It is not surprising that this pump could not compete in the long run with the mercury diffusion pump, which had no moving parts. This pump had been first described by Gaede in the Leybold catalog of 1913 (pp. 204–205). As far as we know, Gaede did not work on drag pumps any more thereafter, the only exception being his gas ballast patent of 1935 (13), where he describes a molecular pump with

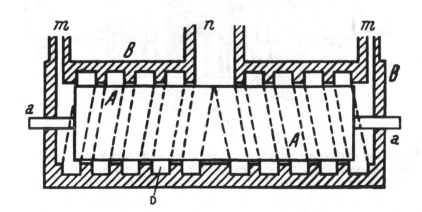

FIG. 2. First design of Gaede's molecular pump, which had two helical grooves. B housing, D grooves, n gas inlet, m gas outlets. This figure is taken from a review article (8) published at a later date.

provision to bleed in a protective gas (Fig. 5) to prevent oil backstreaming from the forepump and bearings.

2. FERNAND HOLWECK (1890–1941) (14)

Holweck must have been intimately aware of Gaede's molecular pump. He writes in his single, comprehensive publication, which does not contain references (15):

> Gaede's molecular pump of 1914 caused tremendous progress in vacuum technology. Unfortunately it is a very delicate piece of equipment, and difficult to take apart for cleaning etc. The lubricating and sealing oil in the bearings can penetrate into the pump housing, either by slow migration (in about 3 months) or by a sudden stop without admission of air.

After mentioning briefly "the mercury containing diffu-

sion and condensation pumps of Gaede and Langmuir," Holweck continues:

> It would be of interest to attempt the development of a pump, which operates without oil, is rugged and easily taken apart and which can directly reach low pressures. In addition it should be possible to turn the pump off without admission of air. This is possible by modifying Gaede's pump while retaining its operating principle.

Some of the requirements are a consequence of the intended application which was the pumping of demountable, high-power transmitting tubes during operation.

Later, Holweck describes a helical pump. This pump does not basically differ from the above-mentioned first version of Gaede's pump (see Fig. 2). He attempted to reach a peripheral speed of 35 m/sec, the speed of Gaede's pump. Hence it can be stated, as Gaede did, that

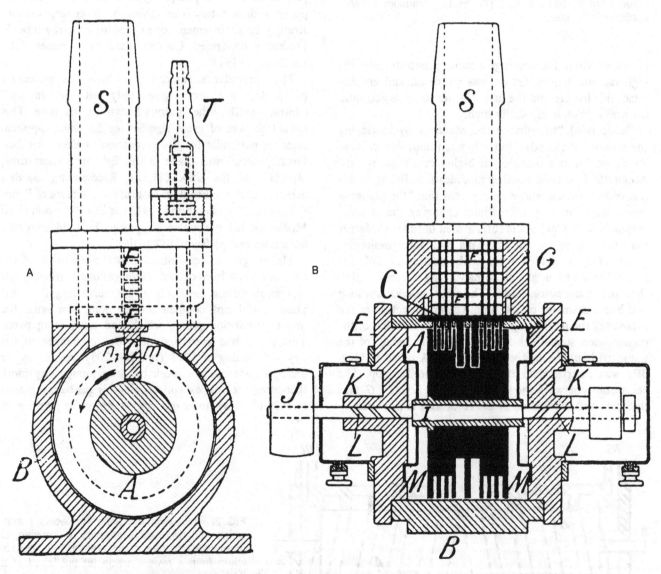

FIG. 3. Cross (A) and longitudinal (B) sections of Gaede's second molecular pump with twice four stages. S and T are the high vacuum and fore vacuum ports. These pictures were taken from a book published in France (9) in 1924. They were drawn to size by Holweck and show more details than those published by Gaede.

"Holweck has returned to the initial molecular pump design of Gaede shown in Fig. 10 [of ref. 8]." Holweck quotes Gaede's contribution (4) in another publication (16), but does not mention its priority.

However, Holweck's pump design constitutes a major technical advance, as shown in Fig. 6. The rotor (150 mm dia., 230 mm long) is a thin-walled tube made of duraluminium. The female thread, cut into the housing, is made of bronze. The gap between the two is 0.03 mm. Holweck noted "The ball bearings are lubricated slightly with a drop of vaseline oil, which lasts several months." An asynchronous motor drives the pump. The rotor is located inside the vacuum and separated from the stator by a thin-walled, but sufficiently stiff, tube. The peripheral speed is 35 m/sec at 4500 rpm. Three types of pumps were built, differing in the depth of the stator grooves. This differing depth decreases from the intake port to forevacuum. The pumps with deeper grooves had a larger speed (up to 4.51 l/s) and a smaller compression ratio (1.6×10^6) for air than those with shallow ones (1.3 l/s and 6×10^6). Holweck's pump was much larger and heavier than Gaede's molecular pump, and probably also more rugged, less subjected to oil problems, and easier to manufacture.

The pump was developed in a remarkably short time. Holweck had been released from military service in July 1919. The pump, described in his patent application (17) of 1 June 1921, already contains essential features of the one built later. Results obtained with a "new molecular pump" were already reported in a short communication (18) in early 1922. His pump has been described in great detail in the above-mentioned paper (15) submitted on 14 November 1922; also it contains the only measurements he ever published.

As to the number of pumps produced and their applications, scant information was given by Holweck. The only exception are the "pumped transmitting tubes" of the twenties (19). To obtain the desired output power, the tungsten filaments had to be operated at such a high temperature that they had to be replaced every 200–300 hours. Of course, this could only be done in continuously pumped demountable tubes. Obviously, there existed a need for such tubes, as the French navy operated 80 of them (20). In 1923 the first 10 kW demountable triode, built by Holweck, was operated at a military transmitter on top of the Eiffel Tower.

Choumoff wrote an article recently at the occasion of Holweck's 100th birthday (21), but did not give any additional information. It is known that a Swiss company began in 1940 to manufacture this pump. The *Neue Züricher Zeitung* published an article on 4 June 1941, "The Molecular Pump" (communication from the laboratory of Trüb, Täubner & Co. AG Zürich) by G. Induni, that states:

> The above mentioned company has now begun to manufacture such pumps in cooperation with Prof. Holweck.... The molecular pump is the ideal pump for cathode ray oscillographs, vacuum spectrographs, electron microscopes, electron diffraction instruments and demountable x-ray tubes.

(Such instruments were manufactured by Trüb-Täubner.) These pumps were very rugged. Rotor and stator were made of special, aged cast iron. The rotor space was not closed off and served as forevacuum container. One type had a pumping speed of 8 l/s and weighed 77 kg (motor included). The peripheral speed was 31 m/s at the rated 3000 rpm.

Holweck also described a spiral pump in a patent addition to ref. 17 in 1921. The rotor is a disc instead of a cylinder, as seen in Fig. 7. A spiral groove is cut into one side of the housing. The inlet port is in the center and the

FIG. 4. Molecular pump with fore pump (rotary vane pump of Gaede). From a Leybold catalogue of 1912/1913.

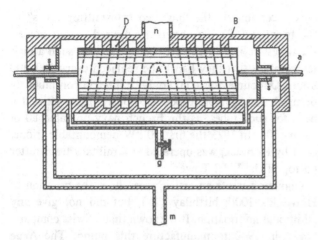

FIG. 5. Pump with helical groove. *g* protective gas inlet, *s* chokes. From Gaede's gas ballast patent (13).

forevacuum outlet at the periphery. The drive is similar to the one of Fig. 6.

Holweck had applied for an additional patent of a molecular pump in 1925 (23). The whole motor is inside the vacuum and arrangements are proposed to increase the pumping speed. It is not known whether such a pump has ever been built.

A Dutch patent, worth mentioning, was applied for on 13 May 1922 by the Mullard Radio Valve Company Ltd., London (23a). The name of the inventor is missing, but the priorities and figures of the Holweck patents of ref. 17 were used. These patents refer to a "vacuum pump with rotor and grooves with special design of the grooves and the drive." Such an application requires the agreement of the inventor and the inventor must have been Holweck. From a remark of Mr. S. R. Mullard in the discussion of C. R. Burch's paper (23b) "Continuously Evacuated Valves" (6 February 1935) we know that there has been a cooperation between Holweck and Mullard.

MANE SIEGBAHN (1886–1978)

Siegbahn designed a spiral pump. A disc instead of a cylinder rotates in a housing that has spiral grooves on

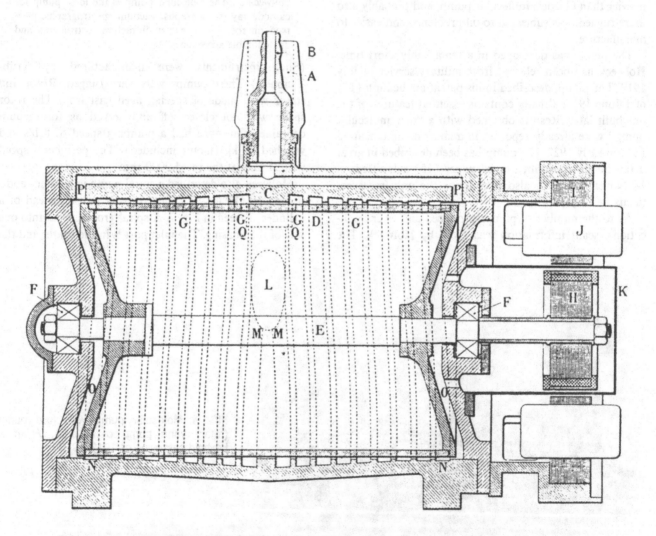

FIG. 6. Holweck's molecular pump. *D* rotor, *C* housing with helical groove *G*, *A* inlet port, *B* forevacuum port, *P-P* gas passages, *F* ball bearings, *H* and *J* rotor and stator of a synchronous motor, *K* vacuum separating tube. From ref. 9, p. 42.

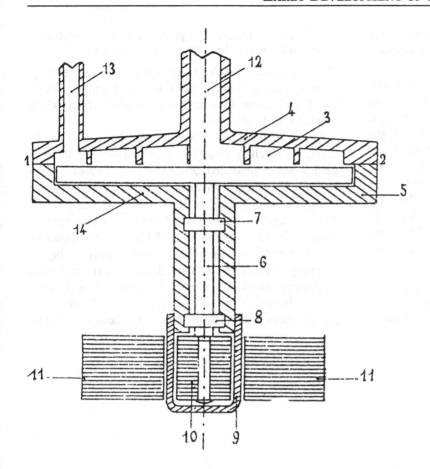

FIG. 7. One-sided spiral pump of Holweck. 14 rotating disc, 4 housing with spiral groove 3, 12 inlet and 13 exit port, 10, 11 drive. From Holweck's patent ref. 17; see also ref. 22.

FIG. 8. Siegbahn's spiral pump. The gas inlet is on top, the forevacuum connection is below the drive shaft. From a 1929 catalog of Leybold, p. 84.

both internal surfaces, making several turns from the periphery to the center—similar to Holweck's arrangement of Fig. 7. It is not known whether Siegbahn knew of Holweck's patent. A smaller pump with a 220 mm diameter disc was built first. The cross-section of the groove was 10×10 mm at the entrance port on the periphery and decreased to 1×10 mm near the axis. The ultimate high-vacuum pressure was 1×10^{-5} Torr at the preferred speed of 6000 rpm and a forevacuum pressure of 0.1 Torr. The stated pumping speed was 2 l/s. The peripheral speed was 70 m/s at the rim and 25 m/s near the axis. The first pump was built in 1926 and is referred to in a paper (24) by Kellström submitted on 25 December 1926. He reports precision measurements with the "*New Siegbahn Tubusspectrometer*":

> The previously used Holweck molecular pump has been replaced by a molecular pump designed by Siegbahn and built in the machine shop of the Physical Institute of the

University of Upsala. This pump has been in operation for 3 months and the performance is excellent.

This pump has been described first in a patent by Siegbahn applied initially in Germany on 4 January 1929. The British patent only has been published (24a); it was granted on 31 July 1930.

About 50 pumps of this type were built (25) in the university machine shop from 1926 to 1940; Fig. 8 shows such a pump produced before 1931 by Leybold under license. The pump was further developed and reached finally a pumping speed of 30 l/s with a disc of 300 mm diameter and a peripheral speed 138 m/s at 8800 rpm, as reported by M. Siegbahn in 1943 (26). In this paper he also described an improved version with a higher pumping speed of 48 l/s (same diameter and rpm). This very complicated pump could be described as a crossbreed between his original pump and Gaede's molecular pump. Noteworthy measurements have been reported by

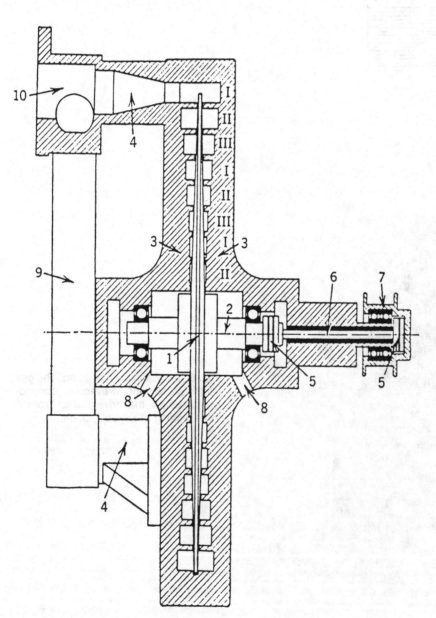

FIG. 9. Large spiral pump. 1 rotating disc, 3 housing with spiral grooves, 8 forevacuum port. From S. Dushman, *Scientific Foundations of Vacuum Technique* (New York, 1949) p. 158.

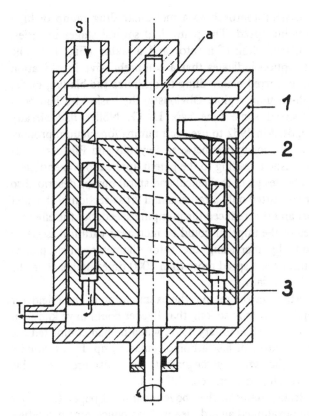

FIG. 10. Molecular pump, from a 1955 patent application of Becker (28). A stationary coil (2) of rectangular cross section is located in the housing (1). Cylindrical rotors (3) surround the coil on the inside and outside. *S* is the high vacuum and *T* the forevacuum port.

S. Eklund (25a). A larger pump with a disc diameter of 540 mm was developed (25) around 1939 for a cyclotron

under construction in Stockholm (Fig. 9). This pump had 3 spiral grooves equivalent to 3 pumps in parallel to maximize the pumping speed. The peripheral speeds were 100 and 30 m/s respectively at 3700 rpm. The size of the grooves decreased from 22×22 mm to 22×1 mm. The pumping speed was 73 l/s.

WILLI BECKER (1919–1986)

Becker writes in his comprehensive paper of 1966 (27):

> I took over the management of the vacuum technique laboratory of the company A Pfeiffer Hochvakuumtechnik GmbH Wetzlar in 1945, and was particularly interested in all processes, which could yield a pumping effect. It was surprising that molecular pumps were described in the literature and particularly in textbooks, but were hardly offered in the market place. The physical properties of the molecular pump were so interesting that we decided to build a molecular pump of Holweck's design, . . . it flew apart after short operation. . . .Since seizure of the pump occurred over and over again, we put the pump quietly into a corner.

However Becker filed a patent application on 22 April 1955 (28) for various molecular pumps, constructed to attain a maximal pumping speed (at equal rotational speed). He had probably realized that the performances of the existing drag pumps were insufficient for modern vacuum technology. Figure 10 shows one model whose manufacture has not been reported.

Leybold applied for several patents (29) on 29 April 1955, naming Prof. C. Pfleiderer of Braunschweig Institute of Technology as the inventor. It indicates the finally

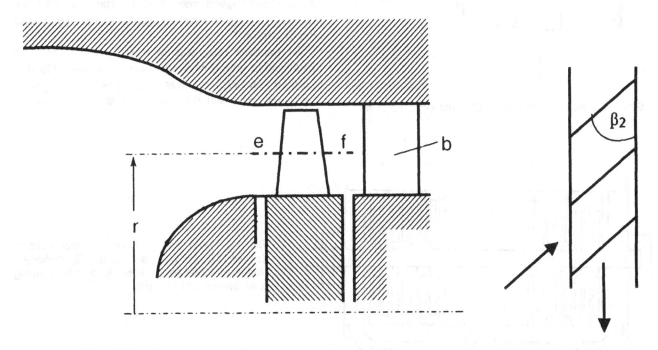

FIG. 11. Cross-section of rotor (*e–f*) and stator (*b*) blades. From a 1955 patent (DE-P 1403049) by Pfleiderer (28).

successful approach to a molecular drag pump of high pumping speed. The pump had several stages in series, each consisting of a rotor and a stator (Fig. 11). The description indicates that a comprehensive investigation had occurred before filing to optimize the blading in the high-vacuum region. Obviously, the basic idea had been conceived some time ago. H. G. Nöller had already proposed in 1952 to use such pumps for vacuum production. A cooperation between Leybold and Pfleiderer was established, leading to the patent applications mentioned.

The design of the helical and spiral pumps described so far permitted only a rather small cross-section of the inlet port and this reduced the pumping speed, but the effective area of the impeller could be increased within reasonable limits by increasing the rotor diameter. Such pumps could have a speed more than 1000 times higher than the previous ones.

Whatever the reason, the examination of the Pfleiderer application took so long that Becker could have known it only after 12 February 1970, when he had completed the development of his turbomolecular pump. He eventually used the same arrangements as Pfleiderer, but his approach had been very different.

Becker states in the above-mentioned paper (27) that he re-examined an old idea in 1955 concerning a rotating baffle for oil diffusion pumps in the shape of a paddle wheel (Fig. 12). Its purpose was "that the air molecules can move from top to bottom without obstruction, while the upward moving oil molecules are caught by the surfaces." Becker continues: "It was noted during the investigations that this rotating baffle produced an appreciable pressure difference in the molecular flow range. This feature was of great interest to us. Because the paddle wheel vaguely resembled a turbine impeller, we placed, in analogy to a turbine, guide channels before and after the rotor."

The pressure ratio was increased thereby, and his turbovacuum pump was the next step. Like Gaede's pump 50 years earlier, it was a dual flow type (Fig. 13). It is noteworthy that the pumping system starts with a

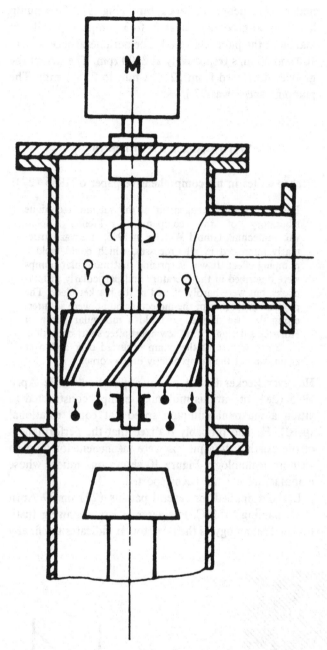

FIG. 12. Rotary baffle of Becker (27) with drive motor *M*.

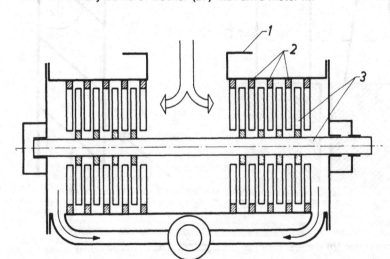

FIG. 13. Two-directional turbomolecular pump of Becker (31). (1) gas inlet flange, (2) stationary blades attached to the housing, and (3) rotating ones fastened to the axis.

FIG. 14. The first molecular pump of Becker (1958) (31). Courtesy of the A. Pfeiffer GmbH.

stator blade. This is obviously disadvantageous and was later changed.

Becker proved by his measurements that the equations derived by Gaede for his molecular pump were also valid for the Becker pump, hence he concluded that his pump was likewise a molecular pump. The patent application (30) was filed on 2 February 1956 and granted on 27 February 1958. After the Pfleiderer patents had been issued (1970), the two companies reached an agreement in 1971.

Becker published information about this pump for the first time at the first International Vacuum Congress in Namur (31) on 12 June 1958. The pump had twice 19 stages (12 high-pumping and 7 high-compression); the diameter of the rotor was 170 mm, the mean peripheral velocity 125 m/s at 16,000 rpm, the pumping speed for air 140 l/s and the maximum pressure ratio for hydrogen 250. In ref. 6, W. Becker described another very large turbomolecular pump, rotor diameter 600 mm, twice 16 stages, peripheral speed 190 m/s at 6000 rpm, pumping speed 4250 l/s for air (Fig. 15).

Undoubtedly the features required for the design of powerful molecular pumps are: several stages of different dimensions in series (as used already by Gaede), and turbine impellers (as first proposed by Pfleiderer). However, it is remarkable that Becker found the correct solution with a completely different approach and independently of Pfleiderer. Figure 14 shows his first molecular pump. It should be noted that at the same conference M. Hablanian reported results obtained with "an axial flow compressor as a high vacuum pump" (32) without having knowledge of the Pfleiderer patents.

FIG. 15. Very large turbomolecular pump, length 1.8m, height 1.2m, (1961). Courtesy of the A. Pfeiffer GmbH.

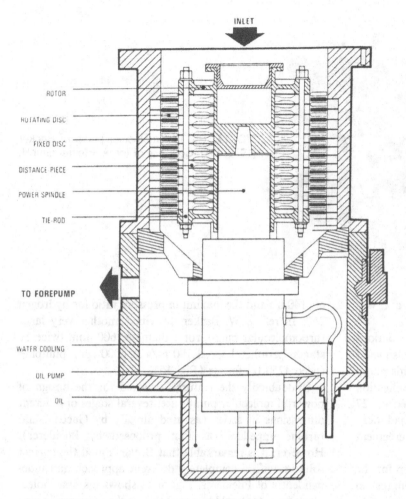

INLET

ROTOR
ROTATING DISC
FIXED DISC
DISTANCE PIECE
POWER SPINDLE
TIE-ROD

TO FOREPUMP

WATER COOLING
OIL PUMP
OIL

FIG. 16. Uni-directional turbomolecular pump from a prospectus of the SNECMA Company, F-Suresnes (1969). The motor is located inside the housing.

The development was more or less completed in 1969 when uni-direction turbomolecular pumps (33) became available (Fig. 16). Their advantages compared to Becker's pump are simpler installation and ease of substitution for oil diffusion pumps.

CONCLUDING REMARKS

The author's intention has been to show how traveling along such a winding road can lead to a new product and that it may often take many years until an optimal solution is found. What counts is the interest of the potential user—namely the growth of telecommunications in the beginning (1913), especially in USA (telephone amplifier tubes); later (1921) military radio communications (pumped transmitting tubes) in France; and finally, in the 1950s, the demand for oil-free, high-speed pumps. It is worth mentioning that initially Gaede's only intention had been the realization of his new pumping method.

The author is obliged to Dr. Adam for the thorough review of the manuscript, Mrs. Paulini for the translation of ref. 15, and others for collecting and making available the voluminous literature and figures and for the loan of documents. I am very much indebted to G. Lewin for translation of this paper.

REFERENCES*

DE-P: German patent
FR-P: French patent
GB-P: British patent
NL-P: Dutch patent
US-P: United States patent

1. Reich G. *Wolfgang Gaede, his life and contributions to vacuum science and technology. History of Vacuum Science and Technology,* 1993; 2:43.
2. Gaede W. Die äußere Reibung der Gase und ein neues Prinzip für Luftpumpen: Die Molekularluftpumpe. *Verh Phys Ges* 1912; 14:775; and *Phys Z* 1912; 13:864; and Gas friction and a new principle for air pumps; the molecular pump. *The Electrician* (London) 1912; 70:48.
3. Goes K. Vorführungen einiger Versuche mit der Gaedeschen Molekularpumpe. *Phys Z* 1912; 13:1105; and 1913; 14:170.
4. Gaede W. Die Molekularluftpumpe. *Ann Phys* 1913; 41:337.
5. Gaede W. Demonstration einer rotierenden Quecksilberluftpumpe. *Verh Phys Ges* 1905; 7:287.
6. Becker W. Zur Theorie der Turbomolekularpumpe. *Vakuum Techn* 1961; 10:199.
7. Gaede W. Rotierende Vakuumpumpe. DE-P 239213 (3.1.1909) and Method and Apparatus for Producing High Vacuum. US-P 1069408 (Dec. 22, 1909).
8. Gaede W. In: Handwörterbuch der Naturwissenschaften: *Luftpumpen* 1931; 6:587-601, S. 592, Fig. 10.
9. Dunoyer L. *La technique du vide* (Paris) 1924; p. 38.

*All papers, patents, and other documents quoted are sampled and stored at the "Gaede Archives" within the "Gaede Foundation." (Enquiries should be addressed to the author.)

10. Dushman S. Theory and use of the molecular gauge. *Phys Rev* 1915; 5:212.

11. Coolidge WD. A powerful Roentgen ray tube with a pure electron discharge. *Phys Rev* 1913; 2:415.

12. Llewellyn FB. Birth of the electron tube amplifier. *Radio Television News* 1957; 57:43.

13. Gaede W. Ein-oder mehrstufige Vakuumpumpe zur Erzeugung tiefer Drücke zum Absaugen von Dämpfen und Gas-Dampf-Gemischen. DE-P 702480 (22.12.1935) Fig. 9 and Method of and Apparatus for Drawing gaseous Fluids from Receptacles. US-P 2191345 (Dec. 21, 1936) Fig. 8.

14. A free French scientist (Obituary of Dr. F. Holweck). *Nature* 1942; 149:163.

15. Holweck F. Pompe moléculaire hélicoidale. *L'Onde Electrique* 1923; 21:497.

16. Holweck F. Pompe moléculaire hélicoidale. *C R Acad Sci* 1923; 177:43.

17. Holweck F. Pompe hélicoidale pour vides élevés. FR-P 536278 (1.6.1921); and Additiv No. 25660 (15.11.1921); and Vacuum Pump, US-P 1492856 (May 11, 1922).

18. Holweck F. Une nouvelle pompe moléculaire. *Jour de Phys* 1922; 3:65.

19. Holweck F. *C R Acad Sci* 1923; 177:164; 1924; 178:1803; 1931; 193:151.

20. Elwell CF. The Holweck demountable type valve. *JIEE* (London) 1927; 65:784; and The Holweck valve. *The Electrician* (London) 1927; 98:647.

21. Choumoff PS. Fernand Holweck et la technique du vide. *Le Vide* 1990; 254:385. Translation: AVS 40th Anniversary Volume, 1993.

22. Duval P, Raynaud A, Saulgeot C, Braunschweig F. *Vakuum Tech* 1988; 37:142.

23. Holweck F. Pompe moléculaire. FR-P 609813 (2.5.1925).

23a.The Mullard Radio Valve Company LTD., Londen: Luchtpomp met rotor en groeven met, bijzondere constructie der groeven en der aandrijfinrichting. NL-P11643 (1.6.1921 and 15.11.1921).

23b.Burch CR, Sykes C. *JIEE* (London) 1935; 77:129, 144–145.

24. Kellström G. *Z Phys* 1972; 41:516.

24a.Siegbahn M. Improvements in or relating to rotary vacuum pumps. GB-P 332879 (4.1.1929).

25. Friesen SV. Large molecular pumps of the disk type. *Rev Sci Instr* 1940; 11:362.

25a.Eklund S. Some measurements of ultimate vacuum and pump speed of molecular pumps. *Ark Math Astron Fys* 1940; 27A:195; 1941; 29A.

26. Siegbahn M. A new design for a high vacuum pump. *Ark Math Astron Fys* 1943; 30B:261.

27. Becker W. Die Turbomolekularpumpe. *Vakuum Technik* 1966; 15:211, 254.

28. Becker W. Molekularpumpe, DE-P 1010235 (22.4.1955).

29. Pfleiderer C. Kreiselpumpe mit axial beaufschlagtem Laufrad zur Erzeugung von Hochvakuum. DE-P 1403049 (29.4.1955). Kreiselpumpe zur Evakuierung von gasgefüllten Behältern, 1063748 und Kreiselverdichter für hohe Luftleere, 1032876.

30. Becker W. Molekularpumpe. DE-P 1015573 (2.2.1956).

31. Becker W. Eine neue Molekularpumpe. *Vakuum Technik* 1958; 7:149; *Adv Vac Science Tech* 1960; 1:173.

32. Hablanian MH. The axial flow compressor as a high vacuum pump. *Adv Vac Science Tech* 1960; 1:168.

33. Mirgel KH. Turbomolecular pump of vertical design of 350 liter/sec pumping speed. *J Vac Sci Techn* 1972; 9:408.

SUGGESTED READINGS

Further Publications of Interest

Andrade EN da C. Molecular air-pumps. *Nature* 1927; 124:657.

Apgar E, Lewin G, Mullaney D. Selective pumping of light and heavy gases with a molecular pump. *Rev Sci Instr* 1962; 33:985.

Beams JW. Bakeable molecular pumps. 7th National Vacuum Symposium, AVS 1960; 1.

Charles CH, Kruger H, Shapiro AH. Vacuum pumping with a bladed axial-flow turbomachine. 7th National Vacuum Symposium, AVS Transact. 1960; 6.

Gondet MH. Étude et réalisation s'une nouvelle pompe rotative à vide moléculaire. *Le Vide* 1948; 18:513.

Jacobs RB. The design of molecular pumps. *J Appl Phys* 1951; 22:217.

Risch R. Diè Berechnung von Molekularpumpen. *Schweizer Archiv* 1948; 279.

Further Patents of Interest

Beams JW. High vacuum pump systems (disc type, magnetically suspended and driven). US-P 3066849 (18.8.1960).

Dorsten ACV, Verhoeff A. Hochvakuumolekularpumpe (rotating cylinder with helix inside and outside). DE-P 910204 (8.4.1951, Dutch priority 12.4.1950).

Grothe ER. Improvements in or relating to rotary vacuum pumps (one disc with two stages-one stage on either side). GB-P 130187 (25.7.1918).

Kichling K. Rotierends Reibungsluftpumps (disc type with more than one stage). DE-P 291268 (11.11.1913).

Leybold Co. Molekularluftpumpe (rotor driven by a fluid turbine). DE-P 625444 (24.2.1934).

Seemann H. Turbohochvakuumpumpe (opposite helices on stator and rotor). DE-P 605902 (8.1.1932).

Reports in Historical Textbooks

Dunoyer L. *Vacuum practice* (English edition) London: G. Bell, London 1926; 30.

Dushman S. High vacuum. *GE Review* (Schenectady, N.Y.) 1922; 51. *Hochvakuumtechnik* (German edition) Berlin: Springer 1926; 59. *Vacuum Technique*. New York: 1949; 151.

Gaede W. in Handb.d.phys.u.techn. *Mechanik* 1927; 6:98–99.

Gaede W. in Handb.d.Experimentalphys. 1930; 4:427, 431. Leipzig

Goetz A. *Physik und Technik des Hochvakuums*. Vieweg Braunschweig, 1922; 32; and 1926; 60.

Jaeckel R. *Kleinste Drucke, ihre Messung und Erzeugung* Berlin: Springer 1950; 119.

Kaye GWC. *High Vacuum*. London: Longmans, Green 1927; 83.

Comments on the History of High Vacuum Turbopumps

M. H. Hablanian

INTRODUCTION

The history of the development of turbine-type, high vacuum pumps can be clearly divided into two sections: before and after 1960. Before 1960, turbomolecular pumps (as manufactured, for example by Trüb, Täuber & Co.) had a very low pumping speed and were used only by a few connoisseurs. An example of an advanced turbomolecular pump manufactured by the Beaudouin Co. in Paris is shown in Fig. 1. This pump was similar to the well-known Siegbahn type. It had a 30 cm single disk rotor and produced a pumping speed of 60 l/s at 8000 rpm, which was rather high for those days. It could maintain a pressure difference of 2 Torr and produced 10^{-7} Torr vacuum when connected in series with a typical rotary vane pump. This was the state of the art until the Pfeiffer Company (now a part of Balzers) introduced a newer molecular pump with a design that was adaptable to larger sizes and higher pumping speeds.

In 1956–57, molecular pumps changed due to two independent but parallel events: the development of the W. Becker pump, and the activities of the National Research Corp., which in 1971 became a part of Varian Associates.

The following sequence of events is mainly taken from a brief history of turbopumps by Edo Rava (1).

In 1958 Becker introduced the turbomolecular pump with a closed pumping cell structure and, in the same year, Hablanian used an axial flow automotive supercharger as a high-vacuum pump, obtaining good results.

In 1960, the Pfeiffer Company, following the work of Becker, commercialized a pump with a closed cell structure, a mid-entry, dual-section body, and a horizontal axis. In the same year, Kruger and Shapiro (Massachusetts Institute of Technology) provided the theoretical basis for the pumping mechanism of axial flow impellers in the molecular flow region.

In 1961, the French company SNECMA (*Societé National d'Étude et de Construction de Moteurs d'Aviation*), with the team of Zelbstein and Rousseau, started a study for the design of an improved vertical-axis, single-sided pump with an open, thin-blade structure. In 1965 the pumps were ready for the market and for serial production. A few years later, SNECMA offered the design to a number of vacuum companies in Europe and the USA.

In 1971, SNECMA granted the license for the improved pump to Leybold for Germany and AIRCO-Temescal for the rest of the world.

In 1972, the pumps were further improved, thanks to the joint effort of Temescal and Elettrorava. The latter was then a company in Turin, Italy, specializing in high-frequency motors and balancing equipment; as such, it participated in the early development and initial production of the new turbopumps.

In 1973, the Alcatel Company (Vacuum Division, Annecy, France) announced a hybrid turbopump (with gas bearings), which could function, after an initial evacuation, without a backing pump. It seems, however, that the gas bearings presented a problem and the anticipated results were not obtained.

This brief sequence of events outlines the main path of development to the point where turbomolecular pumps became generally accepted for a variety of applications in the high-vacuum industry.

In Europe, the companies which continued the manufacture of turbopumps were Balzers-Pfeiffer, Leybold, Alcatel, and Elettrorava. Balzers had an agreement with the Welch Company in the USA. Later, Welch continued the work on their own and established a consulting relationship with Prof. Shapiro at MIT, who, together with his graduate students, provided the theoretical design.

Temescal discontinued their turbopump activities after a few years.

During the next decade, Alcatel introduced varieties of molecular pumps, which became established as versions of Holweck-type pumps, and were also made by other manufacturers. Compound pumps were introduced that combined turbomolecular and turbodrag sections on the same axis, and pumps began to appear with grease-lubricated and also entirely magnetic bearings.

Turbopumps were also made in other countries, notably in Japan, where at one time there were more than a half-dozen manufacturers (Seiko, Osaka, Shimadzu, Rigaku, Hitachi, Ebara, Mitsubishi). Turbopumps were also made in Russia and China. Rosanov (2) lists pumps ranging in pumping speed from 100 to 18,000 l/s, while in the West the largest commercial size appears to be 9000 l/s.

A few years ago, a variety of compound and hybrid pumps have appeared, as well as ceramic ball bearings with sparingly applied special grease lubrication.

A few articles on the history of modern turbomolecular pumps have been published in journals related to high-vacuum technology, notably by J. Henning (3,4) and P. Duval (5). The events related to the axial-flow compressor tests have also been briefly described by the author (6). Henning's review papers trace the development of turbine-type pumps and turbodrag pumps over a 30-year period, summarizing the details of their performance, the principles of operation, theoretical expressions for pumping speed, limiting compression ratios for various gases, and mechanical aspects of bearings, lubrication, vibration, venting requirements, effects of magnetic fields, and pumping of toxic and corrosive gases. In addition, Henning discusses future trends, the compound pumps, the magnetic suspension of the rotor, the increases of peripheral velocities, and drive systems for automatic operation.

Duval and coauthors (5) discuss primarily the development of molecular pumps. They trace the history of various early contributors in the manner of a family tree (see Fig. 2, which is taken from their article). The comments offered below are intended to unravel the clouded part on the left side of Fig. 2 and indicate that modern turbomolecular pumps originated from two somewhat unrelated sources: Gaede molecular pumps and turbomachinery devices well known in the mechanical engineering field. Both began to bear fruit at approximately the same time (1956).

PFEIFFER-BECKER PUMP

In 1958, at the first International Vacuum Congress in Namur, W. Becker (7) introduced the new design of a *Turbo-molekularpumpe* (Fig. 3). Given the tendency in German to combine three or four words to express a

FIG. 1. An example of a 60 l/s molecular vacuum pump in 1957 (Beaudouin Company, Paris).

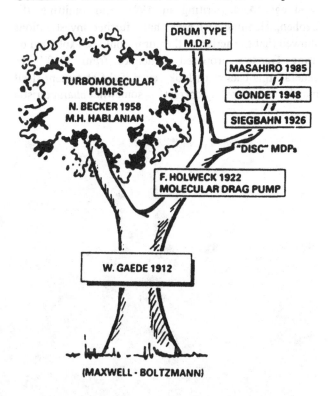

FIG. 2. The development "tree" of molecular pumps. Courtesy of P. Duval (5).

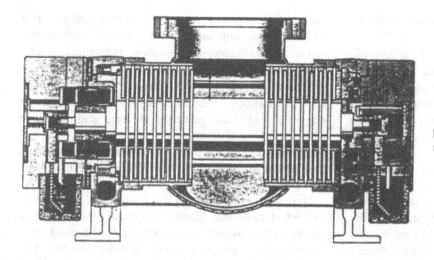

FIG. 3. W. Becker's (7) (Pfeiffer Co.) turbo-molecular pump.

compound meaning, the location of the hyphen in the name of the pump may be significant in regard to the visualization of the design and the principle of the pumping action. The history of Becker's design is interesting because, faced with new vistas, he appears to have been influenced by the design of previously existing molecular pumps and looked back to Gaede and Holweck for inspiration. Writing in 1975 (15 years after the publication of Shapiro's paper [13]), Becker still preserves the hyphen in the German text, while describing improvements of his horizontal, double-sided pump incorporating the thin blades and blade angles according to Shapiro's teachings. And, writing in 1991, and omitting the hyphen, Henning (9) says that "further investigations showed that the action of the pump could be explained by using Gaede's proposed principles from 1913." In general, this backward look (even if the statement itself may be at least partly correct) is somewhat characteristic

for high-vacuum technologists, who often do not relate their pumping devices to existing and well-known compressor machinery.

The last observation should by no means detract from the contribution Becker made to the revival of interest in molecular pumps and from crediting the Pfeiffer company for their long-range view in developing the initial technical and commercial acceptance of the pump. Becker joined Pfeiffer in 1945 as supervisor of the vacuum technical laboratory and became interested in diffusion and molecular pumps. To commemorate his historical achievement, he is shown with his pumps in the photograph of Fig. 4.

Henning (9) relates the development of the initial idea for the new molecular pump to an attempt to reduce the backstreaming from a diffusion pump by installing a rotating baffle at the inlet that is driven by an electric motor, as shown in Fig. 5. During these experiments, it

FIG. 4. Photograph of W. Becker in his laboratory.

was noticed that the rotating baffle produced a pressure difference, and thus was acting as a pump. The pumping action can be explained by noticing that the rotation of a half-chevron configuration (if the velocities are high enough) will produce a baffle that has a higher conductance in the forward direction and therefore will act as a pump. At this point, Becker had essentially discovered an axial flow compressor. However, he related the pump to existing "classical" molecular pumps and thought of his pump as a multi-channeled, parallel-entry Holweck pump. The rotor and stator blades were made thick enough to form closed cells and the role of the stator was to help drag the molecules into the trailing corner of the cell, whence they flow into the next rotor, somewhat analogous to the action of the Gaede pump. (A different view of the role of the stator has been described [6,10,11]). Thus the opportunity to view the pumping action by considering the velocity vectors of moving blades and molecular trajectories seems to have been missed. As a result, Becker's pump was large and heavy for the pumping speed it produced.

Henning (9) presents a graph showing the pumping speed per unit weight of pumps during the period of 1958 to 1990. This graph is shown in modified form in Fig. 6, using only the most advanced designs (in regard to speed/weight ratio) and adding a new pump in 1991. It is interesting to note the nearly 15-year delay at the beginning and the step-wise progress thereafter. Becker's basic pump is still available today, but it appears to be used only by connoisseurs, as the molecular pumps prior to 1958.

The appearance of SNECMA papers (12) and the SNECMA prototype pump played a most important role in promoting the thin-bladed design for modern turbomolecular pumps, and a great majority of turbopumps now follow this basic design.

AXIAL FLOW COMPRESSOR, NATIONAL RESEARCH CORPORATION

Some authors (12,14) associate the development of modern turbomolecular pumps with the appearance of the NRC paper (15) on the use of axial flow compressors in the high-vacuum regime. It appears that the paper had a seminal effect on the progress of turbopump design, although at the time the work was performed this was neither intended nor envisaged. For example, a 1989 letter to the author from Prof. Savada states: "I began a series of studies on turbomolecular pumps about twenty years ago when I happened to find your pioneering work on turbomolecular pumps published in Advances of Vacuum Science and Technology, from which I have learned a lot." Savada has done extensive theoretical work on turbopump function and design (16), helping in the process Japanese manufacturers, and, even allowing for the usual politeness, the connection seems worth noting. The earlier connection via Shapiro and SNECMA was more influential, and it may be worth recounting the associated events in more detail and relating them to the sequence, 30 years later, that led to the pump represented by the last point in Fig. 6.

The National Research Corporation (established in 1941) was a technology driven, adventuresome company using high-vacuum environment in process work and doing associated industry and government-sponsored research. It established an equipment division in 1953 (Newton, Mass.) and began to manufacture vacuum pumps. In the 1955–56 period, primarily driven by the demands of vacuum metallurgy, the company began a review of promising large pumps for use in the 10^{-4}–10^{-2} Torr range. Large, high-throughput diffusion pumps were not yet available, but Roots-type blowers were already in use. The motivation was to find pumps larger and possibly less expensive than the available Roots blowers. An engineering report (NRC No. 57-8-1, 1957, "New Approaches to Vacuum Pumping" by M. Hnilicka and A. G. Walsh), states, among other recommendations: "From all considered means of pumping, the

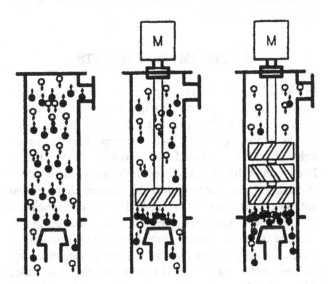

FIG. 5. Rotary baffle for a diffusion pump. Courtesy of Henning (9).

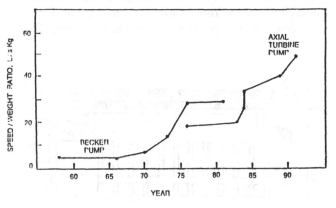

FIG. 6. Progress in the ratio of pumping speed to weight. Courtesy of Henning (9).

development of a high-speed impeller of centrifugal or axial type is most promising from economical and technical point of view for a capacity of 10,000 to 20,000 CFM."

At the end of 1956, H. S. Steinherz, R&D manager (later, president of the American Vacuum Society), had purchased an automobile engine supercharger built by Latham Mfg. Co. (Florida) to be tested as a vacuum pump (Fig. 7). The tests were done at the beginning of 1957 and the Namur paper was the result of the realization that axial flow compressors produce much higher pressure ratios in high vacuum than at atmospheric pressure, which was already known to be the case for the Roots blowers.

The Research Division of NRC was located on MIT land and its engineers and scientists often communicated with MIT professors. Prof. A. H. Shapiro had heard about our experiments; he obtained sponsorship from NASA and the ONR, and together with his graduate students started a study of the pumping mechanism in the molecular flow regime (13). His study (1960) apparently influenced SNECMA to develop the prototype axial flow vacuum pump (12).

From the beginning it was realized that the high-vacuum turbopump could be extended to atmospheric pressure. It was apparent that it is a mistake to use axial flow–type impellers for all 10 or 20 stages of the pump. As the volume flow rate is reduced toward the discharge of the pump, other turbomachinery-type impellers should be used that are lower in pumping speed but capable of producing higher compression ratios. The manuscript of the Namur article (15) even included a section discussing some concepts for a turbine-type vacuum pump that could discharge directly to atmosphere. This section was removed after a reviewer had judged it to be "in the realm of the fantastic" (about 20 years later, M. Maurice of Alcatel achieved this fantasy using a different method).

These possibilities in 1957 were discussed with Prof. W. A. Wilson of MIT's Mechanical Engineering Department. He immediately suggested that the axial-type inlet stages should be followed first by centrifugal impellers and then by a regenerative-centrifugal pump (10) near atmosphere. He warned, however, that near 30 Torr a breakdown of the pumping action would occur. This

often-noted breakdown at low flow rates is also shown in the form of a "surge line" in the performance curve (15) of the Latham compressor near atmospheric pressures. In retrospect, it appears that the surge limits occur when a pump designed for a nearly incompressible flow situation is used in condition of high-pressure ratios.

After these discussions, Prof. M. A. Santalo was asked to analyze the performance of the peripheral regenerative pump under high-vacuum conditions. He at first reported a pressure ratio of 30 per impeller disk, but then came to the conclusion that it would not exceed the ratio of 3, as usually obtained near atmospheric pressure. Subsequently, NRC decided not to continue this effort. After the NRC Equipment Division became a part of Varian Associates (1971), under pressure of interest in oil-free pumping, these issues were revived, but again it was decided not to go into the manufacture of turbopumps because there were already too many companies in the business. The basic ideas for hybrid turbopumps were sketched in my notebook (17) (1974), simply for the record, and remained dormant until about 10 years later when Varian Vacuum Products, with Elettrorava's help, started a new turbomolecular pump effort.

In the meantime, SNECMA had started pilot production of turbopumps. According to S. Choumoff (18), they manufactured on a very small scale a large 9,000 l/s pump with the support of the French Atomic Energy Authority. This pump (12) had twenty stages, a blade tip velocity of 350 m/s, and it was driven by an air turbine. At this early stage of development, because of high peripheral speed and a large number of stages, the pump had a pumping speed for helium 1.7 times higher that for air, while most commercial pumps now have lower speed for helium than for air. SNECMA also made a smaller pump (350 l/s) which was later produced by Leybold under SNECMA license. Similar pumps were also made by Welch and Temescal. Later, Pfeiffer also began to manufacture such so-called vertical pumps and the rest, as it is said, is history.

LATER IMPROVEMENTS

In the 1970s the turbopump design was more or less stabilized. Typically, 9 to 20 stages were used, depending on size and other features, and lubrication was provided to the rotor-supporting ball bearings from a small, light-oil reservoir located at the bottom of the pump. Even though this oil was placed in the forevacuum side of the pump, backstreaming into the vacuum chamber was not always prevented (19). In the 1980s, due to the pressure of the demand for cleaner high- and ultrahigh-vacuum systems, grease lubrication was introduced, which not only greatly reduced the appearance of hydrocarbon contamination at the inlet but had the added advantage of permitting operation in any orientation.

Further improvement was achieved by using ceramic

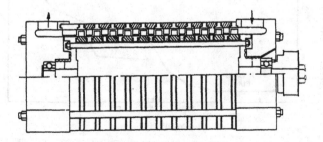

FIG. 7. The axial flow automotive supercharger. Courtesy of Latham Mfg. Company.

ball bearings instead of the usual steel. The major advantage of ceramic ball bearings is their lower weight. When rotors are carefully balanced, the main radial forces in the bearings are due to the weight of the orbiting balls. Lower forces reduce operating temperatures and increase bearing life. Modern pumps, when carefully operated, can last without regreasing two or three years before maintenance.

In response to demands for further cleanliness, magnetic suspension bearings were introduced that provide complete suspension of the rotor. Unlubricated bearings are then used only for supporting the stationary rotor. It is best to use such pumps for applications where the pump is operated continuously for long periods of time. At present only a small percentage of pumps produced have magnetic bearings. There are difficulties in frequent stop-and-go applications and in the higher cost for both the pump and the electronic controllers. Some pumps have predominantly permanent magnets. Opinions are divided in regard to the best practical magnetic bearing design and it may take a few more years before clarity emerges.

Another avenue for cleanliness is to increase the tolerable discharge pressure to permit the use of oil-free backing pumps. Thus, compound turbopumps appeared with various degrees of elegance in their design. Some simply put two turbomolecular pumps in series, sometimes in the same body (20), some added a turbodrag section to the conventional turbopump stages. The most common additional stage was essentially one-half of a Holweck-type pump (again, a backward look?) either with a single or multiple parallel spiral channels. This permits discharge pressures between 10 and 20 Torr, making it possible to use three-stage diaphragm pumps or combinations of diaphragm and piston pumps without reducing the high vacuum levels achieved. Again, such designs increase the cost of the pump and nearly double its size (Fig. 8).

In 1990 another variety of pump appeared, which may be called hybrid pumps to distinguish them from previously mentioned compound pumps. In this case, the pump uses different impellers on the same axis without necessarily increasing the number of stages. An example of such an optimized pump is shown in Fig. 9 (Varian V-250) in which the blade design in the axial stages is changed more frequently than usual and the last three stages are of modified Gaede type (10,21). This pump produces more than ten times the usual compression ratio for all gases without sacrificing other performance features and without an increase in size, or power, and with only a very small increase in manufacturing cost.

CONCLUSION

An argument could be made that the preoccupation with molecular flow may have retarded the development of

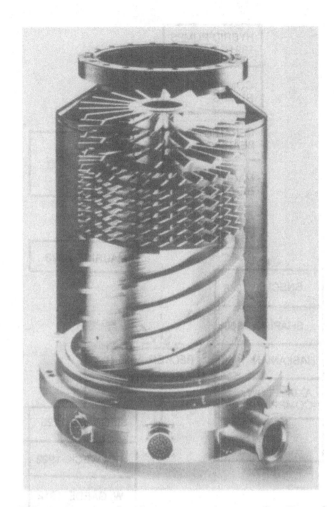

FIG. 8. An example of a compound pump. Courtesy of Edwards Company.

optimized turbine-type high-vacuum pumps. Shapiro did not try to design the best possible pump, but only

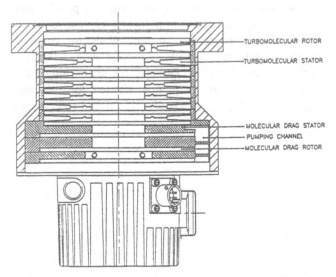

FIG. 9. Schematic view of a hybrid pump. Courtesy of G. Levi (21).

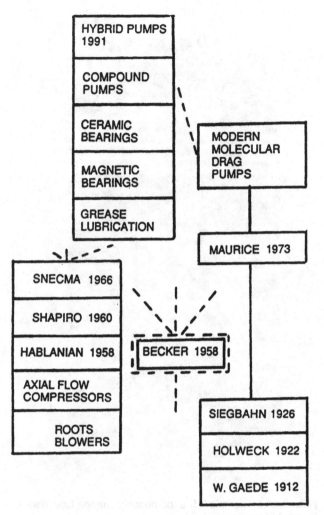

FIG. 10. Turbopump development connections.

high-vacuum pumps and conventional compressors of similar type—for example, diffusion pumps and steam ejectors. In retrospect, even Gaede's molecular pump can be viewed as a "degenerate" case of a regenerative peripheral-channel pump which has submicroscopic-sized buckets or vanes.

The development of high-vacuum turbopumps has followed the general great progress made in high-vacuum technology during the last 35 or 40 years (from Glyptal and De Khotinsky cement to Conflat flanges and extreme high vacuum). Progress has had the usual ups and downs and possibly some blind alleys. The avenue of development could be represented in the manner of Fig. 2, but using a somewhat more current and less romantic vision, as shown in Fig. 10.

Today, turbopumps have become one of the major practical means for obtaining high vacuum. Undoubtedly new developments will follow, and perhaps a complete history of the first 100 years will be written in 2012.

REFERENCES

1. Rava E. Electtrorava Co report (Turin, Italy. 1975) [in French].
2. Rozanov LN. *Vacuum technique.* Moscow: Visshaya Shkola, 1990.
3. Henning J. *Vakuum Technik* 1988; 37:134.
4. Henning J. *J Vac Sci Technol* 1988; A6:1196.
5. Duval P, Raynaud A, Saulgeot C. *J Vac Sci Technol* 1988; A6:1187.
6. Hablanian MH. *Le Vide* 1990; Suppl. 252:19.
7. Becker W. *Advances in vacuum science and technology.* New York: Pergamon, 1960; 173.
8. Becker W, Nesseldreher W. Research report BMFT-FBT 75-33, Ministry of Research and Development, 1975, (in German).
9. Henning J. *Vakuum in der Praxis*, 1991; 3:28.
10. Hablanian MH. *J Vac Sci Technol* 1993; A11:1614.
11. Hablanian MH. *High-vacuum technology* New York: Marcel Dekker, 1990.
12. Rubet L. *Le Vide* 1966; 123:227; and French patent 13046689 (1961).
13. Kruger CH, Shapiro AH. Trans. 7th AVS National Symposium, 1960; 6.
14. Wutz M, Adam M, Walcher W. Theory und Praxis der Vakuumtechnik Braunschweig: F. Vieweg & Sohn, 1982.
15. Hablanian MH. *Advances in vacuum science and technology* New York: Pergamon, 1960; 16.
16. Savada T, *et al. Sci Papers, Inst Phys Chem Res Jap.* 1968; 62:49. *Bull JSME* 1971; 14:48; 1973; 16:312 1973; 16:2; 1973; 16:993.
17. Hablanian MH. Norton-NRC Research Journal. 1974; 21:1.
18. Choumoff S. private communication (1990).
19. Langdon WM, Ivanuski VR. NASA Report No. NASr 65(08), IIT Research Institute Project C6039 (1966).
20. Enosawa H, Urano C, Kawashima T, Yamamoto M. *J Vac Sci Technol* 1990; A8:2768.
21. Levi G. *J Vac Sci Technol* 1992; A10:2619.

presented a model for the pumping action of an axial flow bladed impeller under conditions of molecular flow.

The optimization process for the design of a general-purpose turbopump for high-vacuum use is not complete. It is apparent that further improvements are possible and should be expected in the near future. The conventional turbomolecular pump may be more related to axial flow compressors than to the old molecular pumps (Gaede, Holweck, and Siegbahn). Perhaps they are nothing more than axial flow turbine pumps adapted for pumping highly rarefied gases. Generally, it is useful to compare

The Quest for Ultrahigh Vacuum
(1910–1950)

P. A. Redhead

INTRODUCTION

Ultrahigh vacuum (UHV) technology is now widely used in industry and the laboratory whenever it is essential to keep surfaces clean, gases pure, plasmas undefiled, or charged particles unscattered. This article considers the pre-history of UHV; i.e., the early developments in vacuum technology and physical electronics that led to the ability to produce and measure UHV conditions in the laboratory in the early 1950s. The period covered by this prehistory is from about 1910 until 1950, when the Bayard-Alpert gauge was invented. The development of other UHV components, the engineering required to make UHV equipment commercially available and to introduce UHV into non-laboratory use, was to come in later years. But that will be the subject of another article.

We shall define UHV, for the purposes of this survey, as the region of pressure below 10^{-8} Torr (rather than the usual definition of 10^{-9} Torr) because this was the barrier at which ionization gauges had stuck since they were first introduced and which was thought for some three decades to be the lower limit of pressure that could be reached with existing vacuum pumps. It is also the definition of UHV used by Alpert in the first major review of UHV technology published in 1958 (1). During the period covered by this pre-history, the most commonly used unit of pressure was the millimeter of mercury (mm Hg) which we shall use throughout by its modern name, the Torr.

The need for improved vacuum conditions first arose in the electric lamp industry in the early 1910s. The lifetime of the evacuated lamp bulbs used at that time was severely limited by rapid blackening of the bulb resulting from the transfer of tungsten from the filament to the bulb through the water cycle; i.e., water interacts with the hot tungsten filament to form tungsten oxide and atomic hydrogen, the tungsten oxide readily distills to the bulb, where the atomic hydrogen reacts with it to produce

tungsten and water once again (2). Langmuir's work at the General Electric Laboratories established the processes controlling the blackening of the bulbs and developed greatly improved vacuum technology, leading to the invention of the gas-filled lamp, which solved the blackening problem.

The next impetus to improve vacuum conditions came in about 1920 from the infant vacuum tube industry, with the need to improve performance and life of electron tubes—which up until then had been more like gas tubes than vacuum tubes. Thus the early work on improving vacuum conditions was done in industrial laboratories in reaction to the pressing needs of the high-tech industries of the time.

The need for improved vacuum conditions in research laboratories first arose in the study of thermionic emission (Langmuir again) in the 1910s and then, in the 1920s and 1930s, in the study of surface properties of solids (e.g., work function, photo-electric effects, adsorption, etc.).

In this note we shall examine in some detail the ironic situation that UHV conditions were being produced by some outstanding experimenters for most of the period from 1920 to 1950 but it was not possible to demonstrate by direct measurement that pressures less than 10^{-8} Torr had been obtained. Although the reason for this limit, the x-ray effect in hot-cathode ionization gauges, was suspected by some experimenters in the 1930s, it was not until the development of the Bayard-Alpert gauge in 1950 (3) that the limitation was clearly confirmed and an elegant solution provided. The development of the B-A gauge was the key to unlocking the rapid development of UHV technology following the period of pre-history which is the subject of this note.

OVERTURE: BEFORE 1920

First let us consider briefly the state of vacuum technology in 1910. The commonly used high vacuum pump was the mercury displacement pump such as the Sprengel or Toepler pump, which required the manual raising and lowering of a mercury reservoir by the strong arm of a laboratory technician, or in some cases by water pressure. Edison had used a Sprengel pump in 1879 to improve the vacuum in his experimental lamps and claimed excitedly "we succeeded in making a pump by which we obtained a vacuum of one-millionth of an atmosphere" (4). By 1910, pressures of about 10^{-5} Torr were routinely obtained with these types of pumps. In 1905 Gaede (5) developed a rotary mercury pump which replaced the repetitive raising and lowering of a container of mercury. Gaede also developed in 1907 a rotary oil pump which served as the backing pump for a mercury displacement pump.

The cryopump had been in existence since 1892 when Dewar first used liquid air to evacuate Dewar flasks. By 1904 Dewar was using charcoal immersed in liquid air as a very effective pump.

The use of getters to improve the vacuum in sealed-off incandescent lamps dates back to 1894, when Malignani used a suspension of red phosphorus painted on the inside of the exhaust tube of incadescent lamps. Considerable development of getters had occurred since that time.

The ability of an electrical discharge to clean up the gas in a sealed-off vessel was first reported in 1858 by Plücker, and by 1910 there was a considerable body of work on this topic and electrical discharges were routinely used in lamp manufacture to improve vacuum conditions. Although the cold-cathode discharge in a magnetic field was first demonstrated by Phillips in 1898 (6), its use as a pump or gauge did not occur until the 1930s.

In 1910 only two types of gauge were available for high vacuum measurements. The first was the McLeod gauge, invented in 1874, which could measure to about 10^{-6} Torr if constructed and used with great care. The second was the Knudsen gauge (7) using the radiometer effect, which was developed in 1910 and could measure to about 10^{-6} Torr.

The possibility of measuring pressure in the high vacuum range, by using electrons from a hot cathode to ionize gas and measuring the resulting ion current, was first demonstrated in Germany by von Baeyer (8) in 1909; however, this method of pressure measurement did not reach the USA until 1916 (9).

Thus, in summary, the experimenter in 1910 had vacuum pumps (mercury displacement pumps, cryopumps, and getters) capable of achieving pressures of less than 10^{-6} Torr when used in combination. Pressures down to 10^{-6} to 10^{-7} Torr were measurable with very great care. The need to bake vacuum apparatus very thoroughly to remove water was well understood. Having set the scene, we now return to the pioneering work of Langmuir and the resultant advances in the period 1910–1920.

The history of the quest for very low pressures really starts with the seminal investigations of Langmuir and his colleagues at the General Electric Research Laboratories in 1912 and 1913. The purpose of these investigations was to determine whether improving the vacuum in tungsten filament lamps would improve their life and reduce blackening of the bulb. The tungsten filament lamp had recently been invented by Just and Hanaman as a more efficient replacement for the carbon filament lamp. When Langmuir arrived at the G.E. laboratories in 1909, the drawn tungsten filament had just been developed by Coolidge, replacing the fragile filament of Just and Hanaman. There was considerable evidence that very rapid blackening of the bulb occurred even at the lowest pressure attainable. Langmuir wrote (2) "Various attempts to improve the life of lamps by obtaining a better vacuum than usual, had not been very successful. This failure however could not be taken as proof that a better vacuum would not improve the life. Since the pressures were too low to measure, we had no way of definitely knowing whether one method of exhaust was better than another. . ." A special vacuum apparatus was built to obtain the lowest possible pressures and to reduce the amount of water in the lamps. The system consisted of "a Töpler pump, a sensitive McLeod gauge and a vacuum oven in which lamps could be heated and exhausted without being subjected to atmospheric pressure from without" (2). The lamps were sealed off at a pressure indicated on the McLeod gauge of 10^{-5} Torr and ageing of the lamps reduced the pressure further. "This seemed to demonstrate that even the complete removal of water vapor from the lamp bulb would not lead to a very radical improvement in the life of the lamp . . . among all the causes of blackening that have been suggested, the only one that remains is evaporation from the filament" (2).

Langmuir used similar vacuum techniques in his famous study published in 1913 of the effect of space charge and residual gases on thermionic emission (10). At this time the majority opinion was that electron emission from an incandescent filament was due to the presence of gases. As recently as 1912, Pring and Parker had shown that improving the purity of carbon filaments and improving the vacuum reduced the electron emission. They concluded "the large currents hitherto obtained with heated carbon cannot be ascribed to the emission of electrons from carbon itself, but that they are probably due to some reaction at high temperatures between the carbon or contained impurities, and the surrounding gases, which involves the emission of electrons" (11). Langmuir investigated the thermionic emission from tungsten filaments (the Edison effect), taking extreme care to improve the vacuum conditions; the experimental tube contained two tungsten filaments which could serve as electron emitter and anode. The tube was evacuated with a Gaede rotary mercury pump with a liquid-air trap while being baked at 360 °C for one hour; the filaments were then heated to about 2800 K and the tube sealed off. The filaments were operated at 2400 K for 34 hours and then the tube was immersed in liquid air and the filaments heated to a higher temperature for a short time. Langmuir (12) suggested in 1913 a rotating disk gauge consisting of the rapidly rotating disk spaced closely to a disk suspended on quartz fiber, the deflection of a mirror attached to the stationary disk indicating pressure. Dushman (13) developed this gauge and showed that it was capable of measuring to about 4×10^{-7} Torr, he was also able to show that the pressure in Langmuir's experimental tubes was well below 4×10^{-7} Torr. Langmuir concluded that "with proper precaution the emission of electrons from an incandescent solid in a very high vacuum (pressures below 10^{-6} mm) is an important specific property of the substance and is not due to secondary causes" (12). This was the first time that the need for very low pressures for the measurement of surface properties was appreciated. Langmuir had established in these measurements of thermionic emission the general techniques (thorough baking of the envelope, outgassing of electrodes, gas clean-up by electrical discharge after seal-off, and the use of cryo-trapping or cryo-pumping) which were to be extended in later years to form the basis of modern UHV techniques for surface measurements.

In 1912 Gaede announced a new form of mechanical high vacuum pump, the molecular-drag pump (14) in which "the gas is dragged along from the vessel to be exhausted into the fore vacuum by means of a cylinder rotating with high velocity inside a hermetically sealed casing." This was the first example of a mechanical high vacuum pump depending on momentum transfer, a type of pump which would ultimately lead to the modern turbo-pump. The early molecular pumps of Gaede's design could achieve a pressure of 3×10^{-7} Torr with a forepressure of 5×10^{-2} Torr at a rotational speed of 12,000 rpm.

Measurement of pressures in the high vacuum range was difficult and tedious at this time, the McLeod gauge, the Knudsen gauge, and Dushman's rotating disk gauge being the only practical gauges. In 1916 Buckley (9) described the hot cathode ionization gauge (apparently without being aware of von Baeyer's earlier work) which rapidly replaced all other gauges in high vacuum measurements. It is ironic to note, in the light of later problems in reducing the x-ray effect, that Buckley wrote "The exact form of the electrodes is not important" (9). The first mass spectrometer was announced by Dempster (15) in 1918, although it would not be until 1940 that mass spectrometers would be used to solve vacuum problems, and not until 1948 that the first UHV-compatible instrument would be developed.

Nineteen-fifteen and 1916 were remarkable for advances in vacuum technique; not only was the hot cathode ionization gauge introduced, but also Gaede invented his "diffusion" pump (16) and Langmuir (17) produced the "condensation" pump. Langmuir's pump had a speed of 3–4 l/s, whereas Gaede's design produced only 0.08 l/s. Pumps based on Langmuir's design quickly came into wide use. The great advantage of the diffusion pump over the mercury displacement pump was that there was no theoretical limit to the lowest pressure attainable. This fact was appreciated by both Gaede and Langmuir and it was not until later that measurements indicated (incorrectly as we shall see later) that the pumping speed of diffusion pumps went to zero at 10^{-7} to 10^{-8} Torr. We shall pursue this interesting misunderstanding in a later section.

Thus by the end of the decade there had been enormous advances in the apparatus both for producing and measuring vacuum and in the practical techniques for routinely obtaining pressures in the high vacuum range. Vacuum systems were still almost always made from glass, and the skilled glassblower was an essential member of any vacuum laboratory. By 1920 the advanced research laboratories were achieving pressures as low as 10^{-8} Torr and quite possibly lower. The researchers at the time had no way of knowing and we can only guess today.

THE PROBLEM IDENTIFIED: 1920–1940

As in the previous decade, the requirements for improved vacuum conditions in the laboratory during 1920–1940 were principally for measurements of surface properties of solids; e.g., thermionic and photoelectric emission, work function, contact potential, and adsorption. The industrial motivating force to improve vacuum conditions in the period 1920–1940 shifted from the lamp industry to the vacuum tube industry. In both cases the need was to improve vacuum in sealed-off devices in glass envelopes. The tube industry in the later part of this period also needed to maintain good vacuum in large, high-power tubes (transmitting and x-ray tubes). In Europe the trend was to use demountable, continuously pumped systems for high-power application; in North American sealed-off high power tubes were more often used. Figure 1 shows a diagram of a vacuum system typical of the early 1920s (18).

First we briefly review the developments in high vacuum pumps and pumping methods. Because of the need to improve and understand the clean-up of gases by a hot cathode discharge in a sealed-off device, as in vacuum lamps and vacuum tubes, considerable effort was put into this research in the early twenties. Perhaps the more detailed investigations were those of Campbell and his colleagues at the General Electric Co. in England (19). There was also much work on the use of getters,

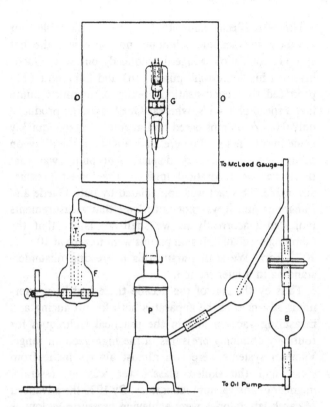

FIG. 1. A vacuum system typical of the early 1920s (27). *O* bake-out oven, *G* hot cathode ionization gauge, *T* liquid air trap, *F* dewar flask, *P* mercury diffusion pump.

both evaporable and non-evaporable, to improve the vacuum in sealed-off devices. Most of those investigations were carried out at pressures in the range 10^{-3} to 10^{-1} Torr. Gas clean-up by electrical discharges at lower pressures (10^{-4} to 10^{-7} Torr) was first studied by von Mayern in 1933 in a hot cathode diode immersed in a magnetic field (20); this was the first true ionization pump achieving pressures of about 10^{-7} Torr.

Development of the molecular drag pump was rapid in this period. A modified form of Gaede's molecular pump was designed by Holweck (21), and in 1927 Siegbahn (22) developed a molecular pump having a single, smooth disc rotating in close proximity to spiral grooves cut in the metal housing. These pumps were widely used in Europe, the Holweck pump being often used for the evacuation of continuously pumped transmitting tubes.

The Gaede pump had a speed of 114 l/s at 10^{-3} Torr, the Holweck pump had a speed of 2.3 l/s, the small Siegbahn pump (22 cms diameter) had a speed of 3 l/s, and the large Siegbahn pump (54 cms diameter) had a speed of 73 l/s. The ultimate pressure achieved with these pumps was limited by the oils and greases used, with the large Siegbahn pump reported to achieve 6×10^{-7} Torr.

In 1928 Burch, of Metropolitan-Vickers in England, showed that mercury could be replaced in diffusion pumps by some organic liquids (23). He was able to achieve pressure of about 10^{-7} Torr without using the usual liquid nitrogen cold trap. The oil-diffusion pump

was further improved by the work of Hickman at Eastman Kodak. The oil-diffusion pump was not used very much for UHV in laboratories, but was later widely used for the achievement of UHV conditions in large systems, such as space chambers.

We now turn to the developments of this period in the measurement of pressure in the high vacuum range. The design of the hot cathode ionization gauge was much improved by the work of Dushman and Found (18), reported in 1921. The ion current in this gauge was found to be linear with pressure to the lowest pressure obtained, about 4×10^{-6} Torr. Dushman states, "Since the linear relation holds down to this pressure, it seems quite justifiable to assume that it would be valid down to the lowest attainable pressures" (18). There was no suspicion at this time that there might be a limiting process in ionization gauges at low pressures. In 1931 Jaycox and Weinhart (24) described an improved design of hot cathode ionization gauge and made measurements of ion current as a function of grid potential (V_g) at an indicated pressure of 10^{-8} Torr. These data have been replotted on a log-log scale in Fig. 2 and compared with similar data measured by Dushman and Found (18) in argon at a pressure of 1.4×10^{-3} Torr. The curve taken at high pressure shows a maximum around 100 volts, similar to a curve of ionization cross-section, whereas the curves at low pressure (10^{-8} Torr or lower) show the typical behavior of a photo-electron current resulting from soft x rays, with the photon current increasing as V_g^n with $n \simeq 2$. Thus as early as 1931 there was clear evidence that the low pressure behavior of hot cathode ionization gauges was limited by a pressure-independent current to the collector. Analysis

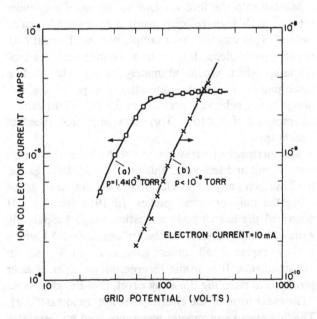

FIG. 2. Ion current in a hot cathode ionization gauge as a function of grid potential with an electron current of 10 mA: **(a)** pressure of 1.4×10^{-3} Torr in the gauge designed by Dushman and Found (21); **(b)** pressure less than 10^{-8} Torr in the gauge designed by Jaycox and Weinhart (24).

of the data of Fig. 2 by the method developed by Alpert (to be described later) indicates that the pressure in Jaycox and Weinhart's experiment was less than 10^{-10} Torr. It was to take another 15 years for the x-ray effect to be appreciated and steps taken to re-design the hot cathode ionization gauge to minimize the effect and permit the measurement of lower pressures. For the time being, the lower limit of attainable pressure was assumed by most experimenters to be about 10^{-8} Torr and to be caused by the pumps.

Penning (25) was experimenting on magnetically confined cold cathode discharges in the mid-thirties and observed that the discharge acted as a pump at pressures below 10^{-2} Torr; this was to lead to the development of Penning pumps suitable for UHV in later years. In 1937 Penning (26) first described the cold cathode gauge that bears his name, which was developed in later years as an UHV gauge.

These were the major developments in pumps and gauges in 1920–1940 that were compatible with UHV operation. Now let us turn to the developments in techniques for achieving very low pressures. These were occurring in two general areas: first, in laboratories studying the properties of solid surfaces, and second, in the tube industry in connection with the development of high-power tubes. We will first examine the vacuum techniques used by experimenters in what was then called physical electronics and included surface science and low-energy gas collision processes.

By 1920 the need for extremely good vacuum conditions had been clearly established by the work of Langmuir described earlier. In 1920 and 1921 Dushman published a series of articles in the *General Electric Review* (27) describing the techniques developed by Langmuir and himself (which were later published as a book, *The Production and Measurement of High Vacuum*). The techniques generally used in 1920 are well described in Dushman's review. The vacuum systems were predominantly made of glass and were rigorously baked as hot as the glass would permit (up to 500 °C) for many hours, Langmuir and O. W. Richardson both baked their systems in a vacuum furnace so that they could be heated above the softening point of the glass without collapsing. Mercury diffusion pumps were used with multiple liquid-air traps (as many as three), sometimes containing activated charcoal. One or two of the traps were baked with the experimental tube. After baking, the metal components of the experimental tubes were heated as hot as possible by electron bombardment or r.f. heating. Getters were frequently used, and these would be evaporated after the experimental tube was sealed-off. The sealed-off experimental tubes were often operated while immersed in liquid air or nitrogen; in some cases a side arm containing charcoal was immersed in liquid air.

At the end of all this elaborate processing, what pressures were achieved in the experimental tubes of the early 1920s? The direct evidence is very slender. Arnold in his work on oxide cathodes (28), claimed "by means of the Knudsen absolute manometer and later by the use of the Buckley ionization manometer, we have measured vacua of the order of 10^{-9} to 10^{-10} millimeters of mercury during the operation of Wehnelt oxide cathodes." The most sensitive design of Knudsen gauge available at the time (29) was claimed to give a measurable indication at 5×10^{-9} Torr with a 150 °C temperature difference between the heater and moving vane. Thus it is possible that Arnold could have estimated operating pressures somewhat less than 5×10^{-9} Torr. Richardson, in the second edition of his treatise on thermionic emission published in 1921 (30), states that the lowest attainable pressures were of the order of 10^{-9} to 10^{-10} Torr.

In 1924 Schirmann (31) published an intriguing paper in which she discussed, in general terms, methods for achieving and measuring *Extremvakuum* of the order of 10^{-12} to 10^{-15} Torr. Pressure measurements were to be made with an *Adsorptionsmanometer* which depended for its operation on the assumption that electron emission from a hot cathode was proportional to pressure. In a 1926 paper, Schirmann (32) claims to have measured pressures below 10^{-11} Torr with the *Adsorptionsmanometer*. In 1930 Molthan (33) noted that the lowest pressure measured by an ion gauge in his system was 5×10^{-8} Torr, which was in contradiction to the claims in the literature (Schirmann [31] and the Lehrbuch der Physik [34]) that pressures from 10^{-12} to 10^{-15} Torr were an attainable target. Molthan noted that no detailed confirmation of these claims had been published. Even if these early efforts were abortive, it indicates that the need for UHV was appreciated and new methods were being sought to produce and measure it at this period.

Thus in the 1920s UHV vacuum conditions were being obtained in several laboratories, but pressure measurements were difficult and unreliable since the available gauges were at their limits. Kaye (35) noted in his textbook *High Vacuum* in 1927, "It is difficult to say what the highest attainable vacuum is, as all our measuring devices ultimately cease to operate. It would appear however, that we shall be safe in assuming that the highest measured vacuum is of the order of 10^{-8} mm. Values less than this may be regarded as somewhat problematic at the present time."

By examining some of the experiments on surface properties in the 1920s and 1930s we can make rough estimates of the best vacuum conditions achieved. As examples of some of the most advanced vacuum techniques we will examine the work of P. A. Anderson at the State College of Washington and W. B. Nottingham at MIT. Anderson published a series of measurements of contact potential between various metals and barium as the reference metal (36a–d) and describes (36a) his vacuum techniques thus, "After outgassing, sealing off, gettering and prolonged immersion of the tube under liquid air or liquid hydrogen the vacuum was such that the galvanometer deflection reading a few seconds after flashing the tungsten at 2800 °K remained unchanged to within 1 mm when the tungsten was allowed to stand cold for as long as 20 minutes; i.e., the work function remained constant to 0.001 volt during this period". If the residual gases in Anderson's experiment were hydrogen and carbon monoxide, then we can estimate that the partial pressures of the gases were below 10^{-11} Torr. Anderson notes that "In the early stage of the work an ionization monometer was used for measurement of residual gas pressures in the tube but it became clear that (the shift in contact potential) was more reliable than measurements with the ionization gauge . . . . Such ionization gauge measurements as were made showed the pressure to be 10^{-8} mm or less." In his 1940 paper (36d) Anderson emphasizes the advantage of using the change in work function of a tungsten surface as an indication of residual gas pressure.

Nottingham was using very careful vacuum techniques in 1936 to measure thermionic emission from tungsten and thoriated tungsten (37) and noted "Although (the ionization gauge reading) corresponds to a pressure of about 10^{-8} mm of mercury, the fact that the condition of the filament would remain very accurately constant for weeks at a time without high temperature flashing proved that the partial pressure of any gas which could contaminate the filament had been reduced to a very small fraction of that indicated by the ionization gauge measurement." In another paper (38), Nottingham states "The time required to bring about this deactivation (of a tungsten filament) served as better indication as to the condition of the tube than any measurements which could be made with an ionization gauge."

Thus by the 1930s it was widely recognized that the hot cathode ionization gauge was not capable of measuring below about 10^{-8} Torr. The reason for this limitation, what we now know as the x-ray limit, does not appear to have been appreciated by experimenters on surface properties, though it was well understood by those in the vacuum tube industry by the early thirties, as we shall see later. First we must briefly examine the research in photoelectric emission resulting from soft x-rays.

In 1921, Richardson and Bazzoni (39) published the first experiments on what was called, at its later reincarnation in the 1960s, *appearance potential spectroscopy*. Electrons bombarded a target, causing the emission of soft x-rays that released photoelectrons from a collector surface, the photocurrent being measured as a function of the energy of the electrons. Breaks observed in those curves were associated with the binding energies of electrons in the target material. Richardson was well aware of the need for good vacuum condition, and was able to use his experimental tube as an ionization gauge; the

measured pressure during the experiments "was of the order 10^{-8} to 10^{-9} mm." This technique was used by Thomas, Compton, Skinner (40) and others in the 1920s and early 1930s to determine binding energies in solids. One can but marvel at the quality of the measurements made with string galvanometers, point-by-point measurements, and graphical differentiation. The method fell into desuetude in the early 1930s when soft x-ray diffraction gratings became available. Thus there was considerable knowledge about the magnitude of photocurrents produced by soft x-rays from electrons with energies in the range 5–500 eV, and various methods had been established to separate the photoelectrons from positive ions produced in the residual gas. Nevertheless, this knowledge did not appear to translate into an understanding of the limitations to the hot cathode ionization gauge.

The situation was different in the vacuum tube industry. It has become standard practice to perform a "gas test" on vacuum tubes during processing. This test involved a measurement of current to the grid when held negative with respect to the cathode (the reverse grid current) under standard operating conditions of anode voltage and current. This measure of gas pressure was called the "backlash ratio" or "vacuum factor" and was defined as the ratio of reverse grid current to anode current. The photoelectric effect in the high-voltage transmitting tubes had been observed in the late twenties, and in 1933 Bell, of the General Electric Co. in England, published his measurements (41) of photoelectric emission from metals caused by soft x rays from the bombardment of copper with electrons of energies up to 20 kV. Bell noted that "Some phenomena observed in large thermionic valves indicated a need for information about the photoelectric emission from tungsten and molybdenum in particular, under the action of x rays from copper at rather higher exciting voltages than those previously studied."

In 1935 C. R. Burch read a paper before the wireless section of the Institute of Electrical Engineers in London describing the development of continuously pumped transmitting tubes (42). In the discussion following the paper, B. S. Gossling (a colleague of Bell) commented, "With regard to the upper limit to the "backlash ratio" which the authors mention, we have proved that the existence of this upper limit is not necessarily to be associated with any limit of the pumps or with the existence of irreducible leaks, but may be due to a quite different effect. The residual "backlash" does not represent an ionic current in the valve, caused by the picking up of gas ions by the grid; it represents a photo-electron emission of electrons from the grid, excited by the x-rays generated at the surface of the anode by the impinging space current."

Burch replied: "Mr. Gossling's statement regarding the upper limit to the backlash ratio is both interesting and important, and solves a problem which has bothered us for some considerable time."

In 1938 Bell and his colleagues at the G.E.C. described their work on transmitting tubes (43) and demonstrated that the reverse grid current contained two components: (1) the positive ion component, and (2) a photoelectric current caused by soft x rays. Figure 3 shows Bell's measurements of the backlash ratio at low pressures as a function of anode voltage. For comparison, the solid lines show Bell's previous measurements (41) of photoelectric emission from tungsten and molybdenum made in a tube designed to eliminate any positive-ion current.

Thus the "x-ray effect" was presumably well known in the vacuum tube industry by the middle thirties, but the extension of this concept to the understanding of the limitation of hot cathode ionization gauges was not publicly made for over a decade.

THE PROBLEM SOLVED: 1945–1954

From 1940 to 1945 the war absorbed all available scientific effort and no significant progress was made toward solving the problem of the lower limit to measurable pressures. The war years produced substantial improvements in vacuum engineering, particularly in the commercial availability of vacuum equipment. Figure 4 is a diagram of a glass vacuum system used to produce very low pressures in the late 1940s. One of the most important advances of this period was the development of the mass spectrometer leak detector based on the design of Nier (44).

It was Nottingham, well aware of the lower limit of ionization gauges, who suggested in the spring of 1947 at the Physical Electronics Conference at MIT (commonly known until his death as the Nottingham Conference)

that the lower limit to ionization gauge measurements was caused by a pressure-independent current of photoelectrons from the ion collector caused by soft x rays resulting from electron bombardment of the grid (anode). This photocurrent was indistinguishable in the measuring current from the current of positive ions arriving at the ion collector. Experimenters at several laboratories then tried to devise methods to test Nottingham's hypothesis or, assuming the hypothesis to be correct, to devise methods to minimize the x-ray effect.

There appeared to be four possible ways to reduce the x-ray effect:

1. Locate the ion-collector so that it was not in direct line of sight with the source of x rays (the grid).

2. Reduce the area of the collector exposed to x rays.

3. Provide a negative electric field at the surface of the ion collector that would return photoelectrons to the collector.

4. Greatly increase the ionizing efficiency by increasing the electron path length, either by a magnetic field or by an appropriately shaped electric field. This would result in an increased positive-ion current per unit electron current, while the x-ray photocurrent per unit electron current would not increase.

As might be expected, Nottingham, in 1948 (45) was the first to test a new design of ionization gauge shown

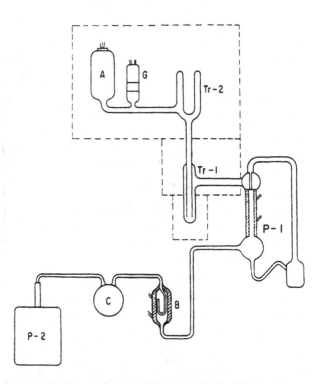

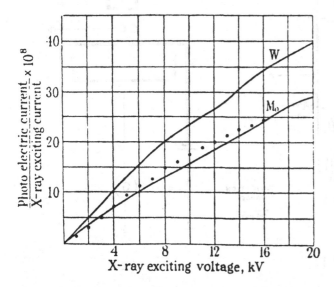

FIG. 3. Photoelectron current from *W* and M_0 as a function of the potential on the Cu anode producing x rays. The dots are measurements of the "backlash ratio" in a triode with a copper anode and a molybdenum grid. The data were normalized at 16 kV (43).

FIG. 4. Typical glass vacuum system for very low pressures in the late 1940s. *A* experimental tube, *G* ionization gauge, *Tr*−1 and *Tr*−2 liquid nitrogen traps, *B* and *C* traps to prevent oil reaching the diffusion pump, *P*−1 mercury diffusion pump, *P*−2 backing pump.

schematically in Fig. 5(a). This design uses method 1 above to reduce the x-ray effect, and the ion collector is hidden behind the shadow electrode so that soft x rays from the grid cannot directly strike the ion collector. This design appears to have been unsuccessful, probably as the result of reflection of soft x rays.

The next advance in understanding the x-ray limit came in 1949 when Metson completed his measurements of the "vacuum factor" of oxide-cathode tubes (46). The British Post Office Research Station had started a program to develop long-life amplier tubes for submerged repeaters. Metson was studying the effects of vacuum condition on life and was determining pressure, in the same manner as vacuum-tube engineers in the thirties, by measuring the ratio of reverse grid current to anode current, the vacuum factor. Metson's curves of vacuum factor versus anode voltage were similar in shape to those obtained by Jaycox and Weinhart in 1931. Metson, commenting on the anomalous form of the variation of vacuum factor with anode voltage, wrote that "The linear component is probably due to the loss of electrons by the negative collector under irradiation by soft x-rays—this theory has a bearing on the interpretation of ionization gauge measurements at pressures below 1×10^{-6} mm of mercury" (46).

The most convincing proof that the x-ray effect was the cause of the low pressure limit of the hot cathode ion gauge appeared in 1950 when the Bayard-Alpert gauge was described in a presentation by Alpert at the Physical Electronics Conference in March of that year (47); an all-metal UHV valve and a capacitance manometer suitable for UHV were also described. A paper on the Bayard-Alpert gauge was published shortly thereafter (3). The solution of the problem was method 2 above—the reduction of the area of the ion collector exposed to the source of soft x rays. Figure 6(c) shows the Bayard-Alpert gauge (BAG), which was an inversion of the "conventional" hot cathode ion gauge design, i.e., the ion collector was on the axis of the grid cylinder and the filament(s) outside the grid. The collector was a fine wire, reducing the radiation from the grid that struck the collector by a factor of about 100 compared to the conventional ion gauge. The proof of the effectiveness of the design is shown in Figs. 6(a) and 6(b), which compare the collector current versus grid potential curves for the BAG with a conventional gauge. The x-ray effect (straight line on the log-log plot) completely dominated the gas ionization curve in a conventional gauge at a pressure of less than 10^{-8} Torr. The same dominance did not occur in the BAG until a pressure of about 5×10^{-11} Torr was reached.

Lander, of the Bell Telephone Laboratories, submitted a paper only one week after the Bayard and Alpert paper, describing another ion gauge design based on the same principle of reducing the collector area (48). This gauge was shown in Fig. 5(b), where P_2 is the ion collector used at low pressures. Launder wrote "It was found that the wire support of P_2 alone is practically as sensitive as wire

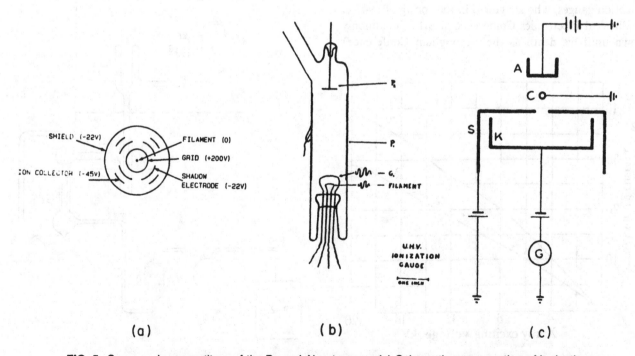

(a)   (b)   (c)

FIG. 5. Some early competitors of the Bayard-Alpert gauge. **(a)** Schematic cross-section of ionization gauge designed by Nottingham in 1948 to reduce the x-ray effect (44), **(b)** Ultrahigh vacuum gauge developed by Lander in 1950 (47). P_1 ion collector for UHV, P_2 ion collector for higher pressures, **(c)** Suppressor gauge developed by Metson in 1950 (48). C filament, A anode, S shield electrode, K ion collector.

plus plate. Thus by removing the plate, a gain of more than two decades (in the low pressure limit compared to a conventional gauge) is expected" (48).

The motivation for the work on ultrahigh vacuum (as it was christened by Alpert) at Westinghouse laboratories was because research on entrapment of resonance radiation required very pure gases (less than 1 part in 10^9 of impurities). The need at Bell Telephone laboratories arose from several experiments on surface science (adsorption on metals, electron ejection from metals by ions, etc.) requiring monolayer formation times of several hours. Although the motivations for pursuing UHV were quite different, the results were similar. The motivation for Metson's work has already been mentioned. His approach to the design of a UHV ion gauge was method 3 above. Figure 5(c) showed schematically the gauge design that Metson (49) developed: electrons from the cathode (C) strike the anode (A), and the ion collector (K) is shielded from the x rays by the suppressor/shield (S), which is held negative with regard to the ion collector to return any photoelectrons to the collector. Metson wrote "Providing suppression is complete and no electron emission occurs on the inner surface of the suppressor shield, there appears to be no obvious limit to measurement" (49).

That the time was ripe for the development of a UHV ion gauge is evidenced by the fact that the Bayard and Alpert, Lander, and Metson papers were all *submitted* to their respective journals within a period of five weeks in March/April 1950.

The Bayard-Alpert gauge was the most elegant solution to the problem of measuring very low pressures, the problem having existed since about 1915. Because of its simplicity, high sensitivity, and ease of outgassing, the Bayard/Alpert design became the most widely used UHV gauge and made possible the rapid development in UHV equipment and techniques that occurred in the following two decades.

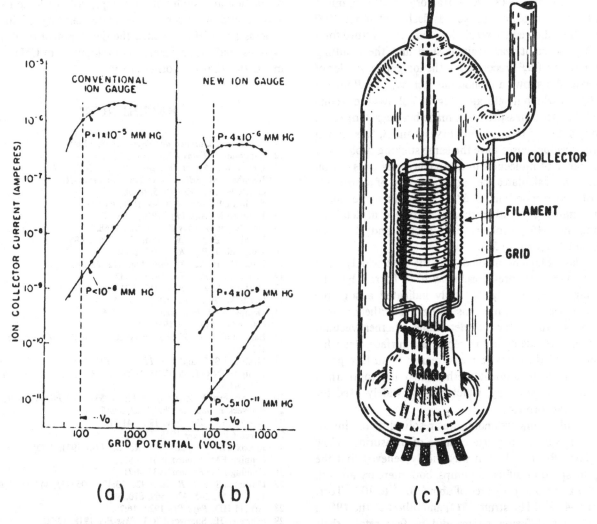

(a)   (b)   (c)

FIG. 6. The elegant solution. **(a)** Ion collector current versus grid potential for a conventional (large collector) ion gauge. **(b)** Collector current versus grid potential for the Bayard-Alpert gauge. **(c)** Drawing of an early Bayard-Alpert gauge (3).

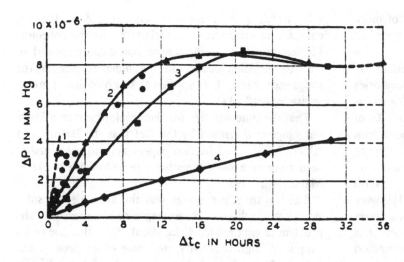

FIG. 7. Pressure burst (ΔP) versus cold time (Δt_c) for a tungsten filament flashed in residual gases. Curve (1) one day after bake-out. Curve (2) about two weeks later. Curve (3) after one month. Curve (4) after exposure to atmosphere and baking again (51).

Before the invention of the Bayard-Alpert gauge, a method was developed to estimate the pressure of gases which would adsorb on a tungsten surface (e.g., H_2, H_2O, CO, N_2, O_2, etc.). The technique, called the flash-filament method, appears to have been first developed by Apker (50) in 1948. A refractory metal filament mounted near a vacuum gauge was flashed at about 2500 K, and then allowed to remain at room temperature for a time Δt_c, when it was again flashed and the resulting pressure burst (ΔP) recorded by the ion gauge. A plot of ΔP versus Δt_c shows a monotonic increase of ΔP until a time Δt_m at which it saturates. Figure 7 shows the results of flash filament measurements made by Hagstrum (51), showing a Δt_m of 14 hours for curves 2 and 3, "Assuming that the residual gases have the same sticking probability and number of molecules in a monolayer as does N_2 on tungsten one calculates the partial pressure of adsorbable impurity corresponding to $\Delta t_m = 14$ hours to be above 2×10^{11} mm Hg." This method was studied in detail by Nottingham (45), Molnar (52), Becker and Hartman (53), and Hagstrum (51).

The flash filament method was the first widely used method to measure pressures below 10^{-8} Torr. It had the advantage of simplicity, but only indicated adsorbable gases and was very time consuming. Nevertheless, it is the most useful method to determine whether vacuum conditions are adequate to keep a test surface clean for the time needed to make a measurement of surface properties. The flash filament method also led to thermal desorption spectroscopy, which is now widely used for studies on adsorption.

One final misunderstanding was resolved by the invention of an ionization gauge capable of measuring below 10^{-8} Torr. Prior to 1950 it was widely believed that the pumping speed of diffusion pumps, both mercury and oil, dropped to zero at pressures of about 10^{-6} to 10^{-7} Torr. Alpert (54) and Hagstrum (51), and others in the 1950s, showed that diffusion pumps did in fact retain their pumping speed to pressure of 10^{-10} Torr or below.

* * *

The invention of an ionization gauge with a low x-ray limit was the culmination of a long, and sometimes myopic, quest for the cause of the lower limit to pressure in vacuum systems. The Bayard-Alpert gauge made possible the rapid developments of other UHV components (ion and sublimation pumps, cold cathode gauges, valves, and hardware), the understanding of limiting processes in UHV systems, the development of all-metal systems, and the commercial development of UHV equipment. But that is another story.

REFERENCES

1. Alpert D. *Handbuch der Physik* 1958; 12:609.
2. Langmuir I. *Trans AIEE* 1913; 32:176.
3. Bayard RT, Alpert D. *Rev Sci Instr* 1950; 21:571.
4. Josephson M. *Edison* New York: McGraw-Hill, 1959.
5. Gaede W. *Phys Zeit* 1905; 6:758.
6. Phillips CES. *Proc Roy Soc* 1898; 64:172.
7. Knudsen M. *Ann Phys* 1910; 32:809.
8. von Baeyer O. *Phys Zeit* 1909; 10:168.
9. Buckley OE. *Proc Natl Acad Sci* 1916; 2:683.
10. Langmuir I. *Phys Rev* 1913; 2:450.
11. Pring JN, Parker A. *Phil Mag* 1912; 23:192.
12. Langmuir I. *Phys Rev* 1913; 1:337.
13. Dushman S. *Phys Rev* 1915; 5:212.
14. Gaede W. *Phys Zeit* 1912; 13:1105.
15. Dempster AJ. *Phys Rev* 1918; 11:316.
16. Gaede W. *Ann Phys* 1915; 46:357.
17. Langmuir I. *G. E. Review* 1916; 19:1060; *J Frank Inst* 1916; 182:719.
18. Dushman S, Found CG. *Phys Rev* 1921; 17:7.
19. Campbell NR, Ryde JWH. 1920; 40: 585; Campbell NR. *Phil Mag* 1921; 41:685.
20. von Meyern W. *Zeit Phys* 1933; 84: 531; *Zeit Phys* 1934; 91:727.
21. Holweck F. *Jour de Phys* 1922; 3:645.
22. Kellstrom G. *Zeit Phys* 1927; 41:516.
23. Burch CR. *Nature* 1928; 122:729.
24. Jaycox EK, Weinhart HW. *Rev Sci Instr* 1931; 2:401.
25. Penning FM. *Physica* 1936; 3:873.
26. Penning FM. *Physica* 1937; 4:71.
27. Dushman S. *G. E. Review* 1920; 23:493, 605, 731, 847; *G. E. Review* 1921; 24:58, 245, 436, 669, 810, 890.
28. Arnold HD. *Phys Rev* 1920; 16:70.
29. Schrader JE, Sherwood RG. *Phys Rev* 1918; 12:70.
30. Richardson OW. *The emission of electricity from hot bodies*, 2nd ed. London: Longman Green, 1921.
31. Schirmann MA. *Phys Zeit* 1924; 25:631.

32. Schirmann AM. *Phys Zeit* 1926; 27:748.
33. Molthan W. *Zeit techn Phys* 1930; 11:522.
34. Senftelen H. In: Lehrbuch der Physik. Braunschweig: Vieweg, 1929; 1:1238.
35. Kaye GWC. *High vacua* New York: Longmans Green, 1927.
36. Anderson PA. (a) *Phys Rev* 1935; 47:958. (b) *Phys Rev* 1936; 49:320. (c) *Phys Rev* 1938; 54:753. (d) *Phys Rev* 1940; 57:122.
37. Nottingham WB. *Phys Rev* 1936; 49:78.
38. Nottingham WB. *J Appl Phys* 1937; 8:762.
39. Richardson OW, Bazzoni CB. *Phil Mag* 1921; 42:1015.
40. Skinner HWB. *Proc Roy Soc* 1932; A135:84.
41. Bell J. *Proc Roy Soc* 1933; A141:641.
42. Burch CR, Sykes C. *JIEE* 1935; 72:129.
43. Bell J, Davies JW, Gossling BS. *JIEE* 1938; 83:176.
44. Nier AO. *Rev Sci Instr* 1940; 11:212.
45. Nottingham WB. In: Report on Physical Electronics Conference, M.I.T., 1948. E. B. Callick, British Admiralty Report Sci. Ad. 4/48.
46. Metson GH. *Br J Appl Phys* 1950; 1:73.
47. Alpert D, Bayard RT. In: Report on 10th Annual Physical Electronics Conference, 1950, p. 88.
48. Lander JJ. *Rev Sci Istr* 1950; 21:672.
49. Metson GH. *Br J Appl Phys* 1951; 2:46.
50. Apker L. *Ind Eng Chem* 1948; 40:846.
51. Hagstrum HD. *Rev Sci Instr* 1953; 24:1122.
52. Molnar JP. In: Report on Physical Electronics Conference, M.I.T., 1949, p. 74.
53. Becker JA, Hartman C D. *J Phys Chem* 1953; 57:157.
54. Alpert D. *J Appl Phys* 1953; 24:860.

Ultrahigh Vacuum Technology at Westinghouse Research Laboratories
(1947–1957)

Daniel Alpert

INTRODUCTION

In his well-documented historical essay in this volume,* Paul Redhead has shown that industrial laboratories made major contributions to the science and technology of high vacuum in the decades preceding World War II. Much of the motivation for such R&D was driven by the development and widespread commercial dissemination of new products such as incandescent lamps, vacuum tubes, and x-ray sources, soon to be followed by fluorescent lighting, gaseous switches, vacuum metallurgy, photocells, and so forth. Irving Langmuir, who personified the importance and value of the industrial research laboratory (General Electric) after the turn of the century, also provided a dominant influence on high vacuum technology from 1910 to 1940, both in basic science and in the invention of practical devices. Another key industrial contributor, and a prominent historian in the field, was Saul Dushman, who was doing research on ionization gauges before I was born. Over the course of almost five decades, Dushman worked with Langmuir, Found, Coolidge, Apker, Lafferty, and others at the General Electric Research Laboratory. In the early 1920s, he wrote a series of related articles on vacuum technology in the *General Electric Review*, and in 1949 he published the classic volume, "Scientific Foundations of Vacuum Technique." (1) It is a great tribute to his personality and scientific contributions that when his GE colleagues carried out a major revision of this classic in 1960, they still listed Saul Dushman as the principal author.

For me, and others at Westinghouse, the research and development leading up to a technology of ultrahigh vacuum got under way shortly after World War II,

*"The Quest for Ultrahigh Vacuum: 1910–1950," in Section II.

initially motivated by scientific requirements rather than the manufacture or improvement of commercial products. I had had an extensive encounter in using existing high vacuum technology as a graduate student with Prof. W. W. Hansen at Stanford University, completing my Ph.D. thesis in 1941, just before the attack on Pearl Harbor. At that time, I accepted a Westinghouse Research Fellowship at the corporate research laboratories in East Pittsburgh, to carry out basic research in a program headed by Edward U. Condon, a prominent theoretical physicist recently recruited from Princeton University. Immediately upon the entry of the U.S. as an active protagonist in World War II, we all turned our attention away from physics and became totally absorbed in weapons-systems design. My group, headed by Dr. Sid Krasik, was engaged in the design of key components for airborne radar systems (then being designed at the MIT Radiation Laboratory). After the successful design, testing, and operational installation of our radar components, we were assigned to the Manhattan Project at the Berkeley campus of the University of Caliornia where we worked until the war was over.

In late 1945, a number of us who had been recruited as research fellows returned to Westinghouse Research Laboratories to resume our scientific careers. The company management had been tremendously impressed with the accomplishments of the nation's physicists during the war and was prepared to make a major corporate investment in basic science. We physicists carried with us the pride—some would call it unwarranted arrogance!—associated with distinctive wartime achievement, and we turned our attention to peacetime research. In my late twenties at the time, I was one of the senior members of this contingent. During the same period, many young turks with similar experience were resuming or initiating basic research in other major industrial research laboratories, and there was an upsurge of federal support for academic research. It was a time when many talented young graduate students were going into science.

At Westinghouse we were assured of substantial company resources and given the freedom to initiate research projects in fields of our own choosing. Our small group selected research aimed at enlarging our understanding of electrical discharges in gases, and proceeded to build experimental equipment to examine a variety of fundamental processes taking place in such discharges. By the early 1950s, this initiative had expanded to a departmental program with a substantial effort devoted to theoretical as well as experimental advances in physical electronics, gaseous electronics, and surface phenomena. We all shared a need for the best available high vacuum systems, and for the improvement of existing techniques for achieving atomically clean surfaces and equipment for handling very pure gases.

It is not easy today to visualize the state-of-the-art of vacuum technology in the mid-forties. To carry out an experiment, the researcher typically designed and built a measuring apparatus and attached it to a one-of-a-kind vacuum system assembled from home-made or off-the-shelf components, whose performance was often erratic and seldom reproducible. Although Dushman's classic 1949 book (1) provided an expansive and useful compendium of scientific and technical information, the equipment of the time was poorly understood and the specifications often erroneous—especially at pressures below 10^{-7} Torr. The typical research scientist rarely had the time to sort out the many ambiguities that were encountered, and conventional vacuum laboratory practice included major components of trouble-shooting, unresolved symptoms, and "black magic."

During this period, the annual MIT Conference on Physical Electronics was a popular meeting place for the exchange of ideas and information relating to physical and gaseous electronics, with participating scientists from industrial, government, and university laboratories. Coordinated by MIT Professor Wayne B. Nottingham, the conference agenda was loosely structured and included extensive informal discussions and exchanges of unpublished results as well as formally prepared papers. Many of Nottingham's personal contributions were in the form of informal comments or extemporaneous presentations on topics under consideration. In addition to scientific and technical exchanges, the MIT conferences offered unscheduled opportunities for comparing notes about the management of research—as well as occasions for some hilarious excursions which continued far into the night (2). I especially recall valuable conversations with colleagues in other industrial laboratories in the US and elsewhere, including Saul Dushman, Leroy Apker, and Jim Lafferty at GE, Julius Molnar, J. B. Johnson, Joe Becker, and Homer Hagstrum at Bell Laboratories, Lou

Malter at RCA, and many others. Dushman was a wonderful storyteller, and I learned about life with Langmuir, Coolidge (who at the time was in his 80s and still going strong) and the great founding director of the GE Research Laboratory, Willis Whitney.

It was during the 1947 conference that Nottingham presented an explanation for the widely observed lower limit to the achievement of pressures as measured by the conventional ionization gauges. Like other researchers who had tried to reach pressure readings below 10^{-8} Torr, I was wary of the reliability of ion gauges in this range. While I considered the Nottingham x-ray hypothesis quite plausible, it was not at all clear that lower pressures could be achieved with existing vacuum equipment. At the 1948 conference, Roy Apker described his use of the "flash filament method" to demonstrate the low-pressure limitations of conventional ionization gauges, and to provide a simple, if time-consuming, technique for estimating the partial pressure of adsorbable gases (3).

A SYSTEMS APPROACH TO HIGH VACUUM

Starting in 1946, I was a participant in collaborative research projects on microwave discharges in noble gases (4) and on the imprisonment of resonance radiation in mercury vapor (5), experiments which demanded gas or vapor purity of very high order. To assure gas purity, I embarked on a series of efforts to redesign and improve available vacuum and gas-handling equipment for these and follow-on experiments. I was soon to appreciate the need and the opportunity for substantial advances in the state-of-the-art of vacuum technology.

During the next few years, the Westinghouse Research Laboratories provided an ideal place for making such advances. We were in the process of expanding our programs in gaseous and physical electronics, recruiting a number of highly qualified physicists to carry out self-initiated, basic research projects in these and related fields. We formed a cohesive group, often exchanging ideas and comparing notes on research methodologies. As a result, my initial project evolved into a broadened effort, aimed at achieving a high-performance vacuum system design that would be assembled from "standard" modules, with sufficient flexibility to serve the diverse needs of our Westinghouse researchers. I could justify the added investment in vacuum R&D in terms of its potential value to a sizeable group of investigators.

By the late 40's, we had already made significant progress in developing a "standard" modular bench-top vacuum system, which is briefly described below. We had also developed operational procedures that were based on the extensive experience of Westinghouse researchers, and incorporated the work of Apker, Nottingham, Molnar, Hagstrum, and others. In such systems, investigators in our group were routinely achieving very low pressures of adsorbable gases—below 10^{-8} Torr as indicated by the flash filament method—before the development of the Bayard-Alpert gauge.

THE BAYARD-ALPERT GAUGE

In late summer or fall of 1949, Robert Bayard joined our Westinghouse research group and asked me to suggest a project that would also provide a research topic for his Master's degree thesis at the University of Pittsburgh, where I held an appointment as an adjunct professor of physics. I suggested he carry out an experiment to verify the Nottingham x-ray hypothesis or to provide an alternative explanation for the low-pressure limit on ion gauge pressure measurement. After identifying the pertinent literature in the field, I proposed the following steps:

assuming the validity of the x-ray hypothesis, design and build an experimental apparatus to verify this hypothesis or explore other possible explanations;

as a first attempt, design a version of the ionization gauge that would operate in the usual manner, but whose ion collector would intercept only a small fraction of the x-rays produced at the grid;

on one of our "standard" bench-top high vacuum systems, install the new gauge and achieve a background pressure below the saturation limits of a conventional ionization gauge, as indicated by Apker's flash-filament method;

compare the low-pressure characteristics of a low-intercept gauge with those for a standard gauge in the manner described. . . .

The ensuing research project extended over a short time interval, at most a few months. As I recall, Bayard's very first idea for designing a collector with a low interception cross-section was to invert the standard ionization gauge design. He placed the cathode outside the positive grid, while suspending the negatively charged ion collector, a fine tungsten wire, at the axis of the electron-collecting grid (Fig. 1). In this configuration, now very familiar, if the ionizing efficiency were comparable to that of a standard gauge, the cross-section for incident x-ray photoelectrons based on solid-angle geometry should be expected to be at least 100 times lower.

After installing the inverted gauge and following "normal" Westinghouse vacuum preparation, the grid and collector were outgassed by electronic bombardment to visible temperatures, and the potentials then adjusted for normal ionization gauge operation. The gauge readings in the first several runs indicated pressures well below 10^{-8} Torr. Bayard then introduced gas at various pressures, comparing the readings of the inverted gauge with those of the "standard" gauge, and plotting collector current vs. grid voltage for the two gauges. As shown in

Fig. 2, these plots followed a now-familiar pattern. For pressures above 10^{-7}, the measured ion collector current i_c plotted as a function of grid potential v_g resulted in curves typical of the gas ionization process for both standard and inverted gauges. At lower pressures the i_c vs. v_g curve is radically different for the two gauges, the inverted gauge showing a much lower saturation current. These straightforward experiments provided incontrovertible evidence for the x-ray hypothesis. At the same time, they provided an elegant design for an ionization gauge with an x-ray limit two to three orders of magnitude lower than previously measurable.

After he had plotted the initial runs, Bayard rushed

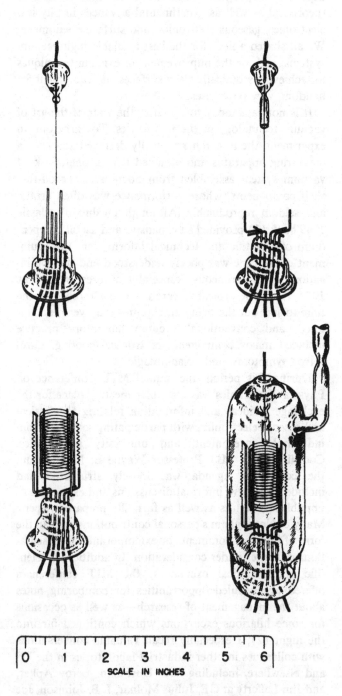

FIG. 1. Construction of Bayard-Alpert ionization gauge.

into my office, graph paper in hand. "Well," he exclaimed, "I'm afraid Nottingham's x-ray hypothesis is correct." So committed had he become to question or disprove the x-ray hypothesis, that this clear proof of its validity left him somewhat ambivalent. "Don't feel bad about it," I replied, "either the results are in error, or you have just demonstrated the best ionization gauge known to man!" Within a few weeks after his first experimental run, Bayard had assembled all the essential data, and we were preparing the paper, first presented at the 1950 Nottingham conference in March and submitted for publication on 13 April 1950 (6). Although the original experimental design was later to require a few minor changes, the simplicity of Bayard's ingenious and elegant design was hard to improve upon. The basic design of the Bayard-Alpert gauge remains relatively unchanged, more than 40 years after its original public disclosure.

Within a matter of a few weeks of our submission of the Bayard-Alpert paper, J. J. Lander (7) of the Bell Telephone Laboratories and G. H. Metson (8) of the Post Office Research Station, London, submitted papers which independently provided corroboration of the x-ray hypothesis and proposed new ion gauge designs to extend the pressure measurement range. Prior to reading their articles (Metson's appeared in 1951) we were unaware of their efforts.

The timing of these contributions poses an interesting historical question: What prevented these advances from taking place much earlier? By hindsight, we have evidence that pressures below 10^{-8} Torr had been previously achieved by several different investigators. For example, by re-plotting their ionization gauge data reported in 1931 (9), I was later to show that Jaycox and Weinhart had achieved pressures well below the x-ray limits of their ion gauge measurements. Anderson and Nottingham had also demonstrated such vacuum conditions in the mid-1930s. In other words, evidence for the achievement of ultrahigh vacuum preceded the invention of a suitable gauge for measuring such pressures by almost 20 years. Furthermore—as I was much later to learn—Bell, Davies, and Gossling (10) had observed and explained the x-ray effect in vacuum tubes in 1938. Yet within a matter of a few months, three different designs of improved ionization gauges were reported.

In retrospect, it seems reasonable to believe that Nottingham's presentation of the x-ray hypothesis at the 1947 MIT conference came at a critical time, and stimulated my curiosity and that of other investigators. At Westinghouse, the availability of what we were later to refer to as our "standard ultrahigh vacuum system" made it feasible to assemble a system and to reliably attain the ultrahigh range within a few days. The conditions for the Bayard-Alpert experiment were uniquely available, and the motivation for doing the experiment was enhanced by our overall goal: the design of a total system.

SYSTEMS FOR ULTRAHIGH VACUUM (UHV)

The Bayard-Alpert gauge provided a major step forward in vacuum technology, and was quickly adopted by scientists at the cutting edge of physical and gaseous electronics. It was described at the 1950 Nottingham conference in early spring, and a number of such researchers had built and tested such gauges before the paper appeared in print in June 1950. However, a much longer time interval would elapse before the gauge was widely adopted in the typical laboratory or in generally available, commercial high vacuum equipment. The reason for the delay is not hard to identify. Since much of the vacuum equipment in general use was not capable of achieving pressures below the range of conventional gauges, many researchers still greeted claims of lower pressures with a long-standing skepticism (11). For such users, it would require a significant step forward in vacuum system design to justify or require the improved gauges.

In 1953, I published a paper entitled "New Developments in the Production and Measurement of Ultra High Vacuum Systems" (12) for which I was to receive requests for more than a thousand reprints in this country and abroad. The paper gave the field a name that was widely adopted, and represented a milestone in achieving widespread acceptance of the idea that proper systems

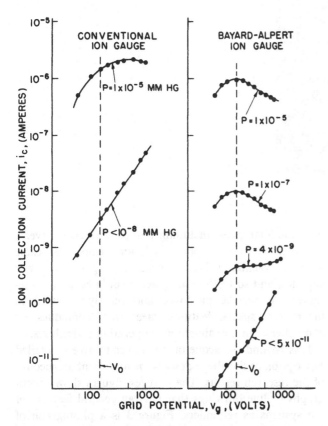

FIG. 2. Ion current versus grid potential for a conventional gauge (RCA type 1949) and a Bayard-Alpert gauge. V_0 is the normal operating voltage (150 V).

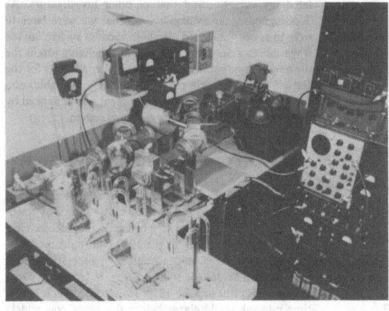

FIG. 3. Ultrahigh vacuum systems in the Westinghouse physics department laboratories in the mid-1950s.

design could permit the achievement of ultrahigh vacuums without the use of low-temperature refrigerants, and with less effort than was required on conventional systems to achieve pressures a thousand times higher. This was followed, in 1954, by a paper by Alpert and Buritz (13), which examined the limiting factors in the achievement and measurement of very low pressures, and described the physical mechanisms that would demand attention if we were to proceed further to extend the UHV range.

These papers described the design of flexible, modular vacuum systems using several new ultrahigh vacuum components, and a set of operational procedures which permitted the reliable achievement of pressures at or below 10^{-10} Torr. The bench-top systems were designed on a modular basis, including pumps, outgassing ovens, room-temperature copper isolation traps, bakeable all-metal vacuum valves, ionization gauges and power supplies, and so forth. The systems could be custom-designed for specific purposes, and readily modified to incorporate special features, measuring techniques, or components, or to validate new operating procedures.

It is beyond the scope of this paper to give a detailed description of the above designs, or to present an account of the research results. The next few figures show photographs or line drawings that portray general features of the systems as reminders. Figure 3 is a photograph of vacuum systems in one of the Westinghouse physics department laboratories in the mid-1950s. Figure 4 shows a schematic diagram and line drawing of the UHV system use by Alpert and Buritz (11) for demonstrating linearity of the Bayard-Alpert gauge over seven decades of pressure. Figure 5 shows the expandable modular bake-out furnaces in place.

In 1958, I wrote a review article (14) summarizing the above contributions as well as research carried out in the intervening period. This paper included a discussion of work on the removal of gases by ion-pumping to adjacent surfaces (15), and contributions of Lange and Alpert to the design of copper gaskets for demountable all-metal vacuum systems (Fig. 6) (16). By 1957 we had built a number of systems that departed from those depicted above, and were also building sections made up of demountable stainless steel. Clearly, another transformation of UHV systems was about to take place—calling for the successful design and implementation of large-scale demountable systems—to be used for preparation and design of semiconductor materials and devices, for high energy accelerators, thermonuclear research reactors, and a wide range of other applications.

In 1957, I left Westinghouse. Soon thereafter I was named the first editor of the *Journal of Vacuum Science and Technology*, and was to witness the flourishing of a major vacuum technology industry, which incorporated many of the advances that had first taken place at Westinghouse. This was not the first time in industrial history that a company did not reap a substantial payoff from its own basic research efforts. Nor was it to be the last. Although I have given considerable thought to such ques-

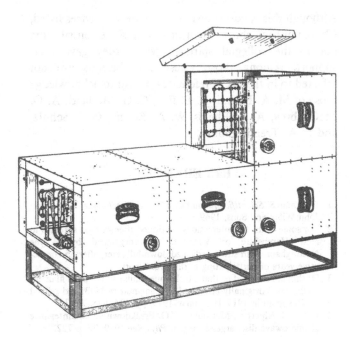

FIG. 5. Standard bake-out furnaces in place on modular UHV system.

tions since my days at Westinghouse, these issues relate to another story.

In closing, I want again to emphasize that the Westinghouse contributions in the development of UHV represented a collaborative effort of many participants.

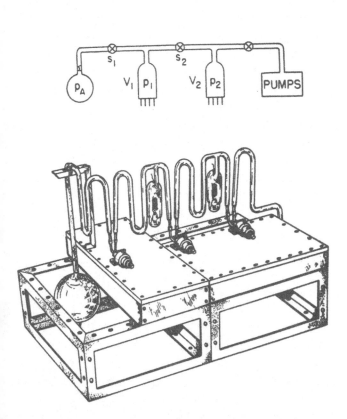

FIG. 4. Line drawing and schematic diagram of UHV system for experiment to check ionization gauge linearity (11).

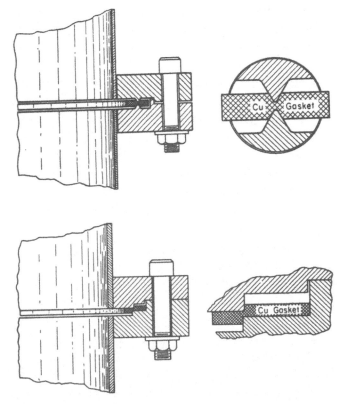

FIG. 6. Typical knife-edge flange using a flat copper gasket and a step-type demountable joint.

Although this is quite apparent from the references listed, I have not been able to do justice to all of the contributors nor to the collegial spirit in which they gave their support. Among those whose contributions are not referred to in the list of references, I wish to acknowledge those of M. A. Biondi, R. E. Fox, C. G. Matland, A. O. McCoubrey, A. V. Phelps, W. A. Rogers, G. J. Schulz, and E. A. Trendelenburg.

REFERENCES

1. Dushman S. *Scientific foundations of vacuum technique.* New York: John Wiley and Sons, 1949.
2. For some of us, the interactions continued afterward on ski trips to New Hampshire or Vermont, also organized by Wayne Nottingham. I am still a dedicated down-hill skier, after 55 consecutive years of participating in this great sport.
3. Dushman reminded us that a technique similar to the flash filament method had originally been used by Langmuir in 1913, and refined by Dushman in 1915. [Dushman 1949, pp. 624, 675, 732.]
4. Krasik S, Alpert D, McCoubrey AO. Breakdown and maintenance of microwave discharges in argon. *Phys Rev* 1949; 76: p 722.
5. Alpert D, McCoubrey AO, Holstein T. Imprisonment of resonance radiation in mercury vapor. *Phys Rev* 1949; 76. p. 1257.
6. Bayard RT, Alpert D. Extension of the low pressure range of the ionization gauge. *Rev Sci Instr* 1950; 21:571.
7. Lander JJ. *Rev Sci Instr* 1950; 21:726.
8. Metson GH. The physical basis of the residual vacuum characteristic of a thermionic valve. *Br J Appl Phys* 1951; 2:46. (This paper first received 14 March, 1950, and in its final form 5 December, 1950.)
9. Jaycox EK, Weinhart HW. *Rev Sci Instr* 1931; 2:401.
10. Bell J, Davies JW, Gossling BS. High power valves: Construction, testing, and operation. *JIEE*, 1938; 83:176.
11. Such skepticism had persisted for years. H. Schwarz, for example, argued that "Everyone who measures pressures under 10^{-4} Torr with an ionization gauge must regard his measurements with more or less skepticism." *Z Physik* 1943; 122:437.
12. Alpert D. New developments in the production and measurement of ultra high vacuum. *J Appl Phys* 1953; 24:860.
13. Alpert D, Buritz RS. Ultra-high vacuum II; Limiting factors on the achievement of very low pressures. *J Appl Phys* 1954; 25 (2):202.
14. Alpert D. Production and measurement of ultrahigh vacuum. In Flugge, ed. *Handbuch der Physik*, Vol. 12. Berlin: Springer-Verlag, 1958; 609–663.
15. Varnerin LJ, Carmichael JH. *J Appl Phys* 1955; 26.
16. Lange WJ, Alpert D. *Rev Sci Instr* 1957; 28:726.

Section III

Reproductions of Historical Papers
(1900–1960)

This chapter contains reproductions of a dozen classic papers published in the period 1900–1960 describing important inventions, or the solution to significant problems, related to vacuum technology. Several of these papers were cited in Sections 1 and 2. Most of these papers were chosen because of their importance and are well known to workers in the field, a few were chosen both for their importance and because they are not widely known.

About slow cathode rays

O. von Baeyer

Institute of Physics, Berlin University

(Received October 19, 1908)

Measurements of the specific charge of the electron, which are performed on slow cathode rays, agree poorly with those made on high-velocity beams. This applies especially to J. J. Thomson's[1] method for the determination of the path length of electrons moving in a uniform electric field in combination with a perpendicular magnetic field. As described below, I investigated the behavior of slow cathode rays, to find the reason for these discrepancies. In particular, the following topics were treated:

(1) The production of slow cathode rays of high intensity and uniform velocity by the use of a hot oxide cathode.
(2) Ionization by such cathode rays.
(3) Determination of their velocity (in volts).
(4) Secondary emission and/or reflection.

Lenard[2] has thoroughly investigated slow cathode rays, which were produced by irradiation with light. The cathode rays emitted from electrodes which were irradiated with ultraviolet light, had appreciable velocities, about 70% of them had initial velocities greater than zero. Hence this method does not lend itself to the production of beams of rather uniform, low velocity.

I. INTRODUCTION

The use of a hot oxide cathode as described by Wehnelt[3] appears to be a better approach. However, the electrons emitted by such electrodes have also some initial velocity as shown already by Elster and Geitel.[4] But a plot of the saturation curves of such electrodes indicates that the fraction of the electrons of noticeable initial velocity is negligible.

Following Lenard's reasoning, the measurement of the current flowing from the hot cathode to the surroundings in the absence of a charge, yields the number of particles leaving with initial velocity. All saturation curves of this type have a steep decline toward zero. A more exact investigation of this point is not possible without further studies. Consider an electrically heated wire located on the axis of a cylinder. If one measures with a galvanometer the current flowing from the wire to the cylinder, the cylinder will attain a positive or negative potential depending on which end of the heated wire is connected with the galvanometer. The potential difference corresponds to the potential drop along the wire. (This is the

so-called *Edison Effect* of incandescent lamps.) In addition, the appreciable heater current produces a magnetic field at the filament surface. This field influences the very slow cathode rays. To avoid these complications, I employed a motorized commutator with mercury contacts during all measurements. (Also in those cases, where it is not mentioned when describing the experimental set up.) This commutator switched on alternately the heater current or the galvanometer (5–10 times every second). This way, only the quantity of electrons is measured which are formed while the wire is still hot, but no heater current is flowing. The arrangement for plotting a saturation curve is shown in Fig. 1.

C is the cylinder located in an evacuated glass tube, *D* is the heater wire coated with calcium oxide. The battery *B* is connected with the filament via the commutator. The other contact connects the filament with the galvanometer. The galvanometer is connected with *C* via the potentiometer *W*, which adjusts the voltage supplied by the Battery B_1. The proper operation of the commutator can be ascertained by the absence of any indication on the galvanometer when the heating current is reversed. This must not change the galvanometer reading. The galvanometer was made by Siemens and Halske. (It had a sensitivity of 5×10^{-10} Amp per division.) A shunt was used for the measurement of larger currents. A saturation curve, curve 1, is plotted in Fig. 2. The current at $V=0$ was specially checked, it was always less than 1/2% of

Editor's note: This paper was originally published in Verh d D Phys Ges 1908; 10:96 and reprinted in Phys Zeit 1909; 10:168. The first two sections of this paper have been translated from the German by G. Lewin. It describes the measurement of pressure using the ionization of gases by electrons from a hot cathode 8 years before the paper by Buckley [Proc Natl Acad, Sci U.S., 1916; 2:683], who is usually credited with the invention of the hot-cathode ionization gauge.

the value at 22 V where the saturation current was reached approximately. The setup also permitted the determination of the number of electrons which escape when D has a positive charge. An effect can be found from 0 down to about 2 V, as was to be expected in view of Elster's and Geitel's results on glowing wires without calcium oxide coatings. Naturally, these experiments were performed in a vacuum, as high as possible, to avoid ionization effects. A Kaufmann pump was used for evacuation. In addition a U-shaped part of the tube leading to the discharge tube was cooled with liquid air to remove mercury and grease vapors.

Nevertheless it appeared to be desirable to determine the amount of ionization occurring in such experiments.

II. IONIZATION OF SLOW CATHODE RAYS

Lenard[2] has shown that electrons produced by light irradiation, have a maximum velocity of ionization of about 10 V. It was not obvious that the same relationships existed here, because the rays have a much higher intensity. The experiment was performed in an analogous manner to Lenard's. The glowing wire was surrounded by a coaxial cylinder (Fig. 3) of 1 cm diameter. It was a wire screen made of brass (wire size 0.1 mm, mesh size 0.3 mm); a second coaxial brass cylinder had a diameter of 2 cm. Cathode rays of various velocities were produced between wire and screen by a variable potential difference V_1. These rays traversed the screen, but could not reach the second cylinder because it had a negative potential ($V_2 = 20$ V) relative to the screen and higher than the filament potential. Positive charge must be transported

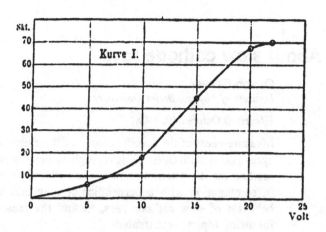

FIG. 2. Curve of electron current as a function of anode voltage. One division (skt.) is 5×10^{-10} Amps.

from the screen to the cylinder, if the gas is ionized between the cylinder and the screen. It must be possible to measure this current with a galvanometer connected between cylinder and screen, if the ionization is large enough in proportion to the passing electrons. No current could be found flowing between screen and cylinder for V_1 between 0 and 10 V. A positive current was flowing from cylinder to screen for V_1 larger than 10 V, and this current rose rapidly with increasing V_1. This positive current did not rise when the negative potential V_2 was enlarged. Hence saturation existed already. This behavior is analogous to the one of the much weaker cathode rays produced by irradiation with light. It was necessary to measure the amount of electrons entering the space between the screen and the outer cylinder to determine

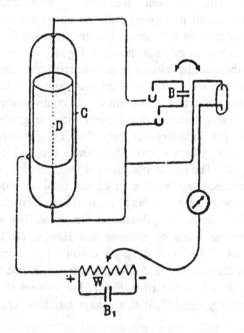

FIG. 1. Experimental arrangement. D is the hot filament, C the anode, B is the battery supplying the heater current.

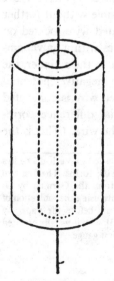

FIG. 3. Electrode arrangement for measurement of ion current.

TABLE I.

Voltage between hot filament and screen (V)	Electron current from screen to out cylinder ($V_2 = 0$) i_- (μA)	Ion current formed between screen and cylinder ($V_2 = 20$) i_+ (nA)	Ions formed per electron i_+/i_-
14	10	8	8×10^{-4}
13	8	3.5	4×10^{-4}
12	6.5	2.0	3×10^{-4}
11	4.5	1.5	3×10^{-4}

the amount of ions produced per electron. Therefore, I also measured the current between the outer cylinder and screen while both had the same potential, i.e., $V_2 = 0$. The ratio of the currents between screen and cylinder for $V_2 = 20$ and 0 V indicates the amount of ions produced by an electron travelling from screen to the cylinder and back. The measurements are tabulated in Table I.

This measurement was performed at a pressure of about 0.01 mm. However, an exact pressure measurement is impossible in the presence of the glowing wire. I also could not investigate different gases for the same reason. It is hardly possible to state the composition of the gas, owing to the continual release of gas from the wire and, on the other hand, to the adsorption which always occurs. The measurement as performed here, indicated that no ionization could be found below 10 V. This does not require the assumption that a still weaker ionization is absent below this limit.

According to Stark,[5] ionization in mercury vapor starts at 10 V. It was important to check this point. For the removal of mercury vapor would permit higher velocities in all cases where a minimal ionization is required. However, I could not reduce the ionization substantially by condensation of the mercury vapor which was done by cooling a U-shaped portion of the connecting tube with liquid air.

For example, the discharge tube was evacuated to a pressure of 0.005 mm with the mercury pump and closed off with a stop-cock. For rays which left the screen with a velocity of 14 V, the current of positive ions was (at $V_2 = 20$)

$$i_+ = 50 \text{ divisions,}$$

and the negative electron current was (at $V_2 = 0$)

$$i_- = 86\,000 \text{ divisions.}$$

When the U tube was cooled with liquid air, the corresponding values were

$$i_+ = 85, \quad i_- = 106\,000 \text{ divisions.}$$

As can be seen, the number of positive ions produced per electron is approximately the same in both cases.

If the vacuum was improved by the application of charcoal cooled with liquid air, a situation was easily reached where the ionization could not be measured any more by these methods, when the velocity was not too high.

[1] J. J. Thomson, Phys. Mag. **48**, 547 (1899).
[2] Lenard, Ann. Phys. **2**, 355 (1990). Ann. Phys. **8**, 149 (1902). Ann. Phys. **12**, 449 (1903). Ann. Phys. **12**, 714 (1903). Ann. Phys. **15**, 485 (1904).
[3] Wehnelt, Ann. Phys. **14**, 425 (1905).
[4] Elster and Geitel, Wien. Ber. **97**, 1175 (1888).
[5] Stark, Phys. Zeit. **5**, 51 (1904).

XXII. *The Emission of Electrons from Tungsten at High Temperatures: an Experimental Proof that the Electric Current in Metals is carried by Electrons.* By O. W. RICHARDSON, *M.A., D.Sc., F.R.S.* [*]

THAT the carriers of the negative thermionic current from incandescent solids are negative electrons was first established by J. J. Thomson [†]. In 1901 [‡] the writer developed the view that this emission of negative electrons occurred by virtue of the kinetic energy of thermal agitation of some of the electrons in the solid exceeding the work which was necessary to overcome the forces which tend to retain them in the body and which prevent them from escaping at lower temperatures. This conception has proved a very fruitful one, and its consequences have been verified in a number of ways. It has provided a quantitative explanation of the variation, with the temperature of the body, of the number of electrons emitted. It led to the prediction of a cooling effect when electrons are emitted by a conductor, and a corresponding heating effect when they are absorbed. Both these effects [§] have since been detected experimentally, and found to be of the expected magnitude, within the limits of experimental error. The magnitude and distribution of energy of the emitted electrons has been found by experiment to be that given by Maxwell's law [||], in accordance with the requirements of the theory. Finally, the same general train of ideas has led to useful applications in the direction of the

[*] Communicated by the Author.
[†] Phil. Mag. vol. xlviii. p. 547 (1899).
[‡] Camb. Phil. Proc. vol. xi. p. 286 (1901) ; Phil. Trans. A. vol. cci. p. 497 (1903).
[§] Richardson and Cooke, Phil. Mag. vol. xx. p. 173 (1910), vol. xxi. p. 404 (1911). Cooke and Richardson, Phil. Mag. vol. xxv. p. 624 (1913).
[||] Richardson and Brown, Phil. Mag. vol. xvi. p. 353(1908); Richardson, Phil. Mag. vol. xvi. p. 890 (1908), vol. xviii. p. 681 (1909).

Phil. Mag. S. 6. Vol. 26. No. 152. *Aug.* 1913.   2 A

Originally published in *Phil Mag* 1913; 26:345.

Editor's note: This paper is included both for its clear description of the best vacuum techniques of the time and also for its importance in finally laying to rest the erroneous notion that electron emission from hot bodies (thermionic emission) resulted from some sort of chemical action.

346   Prof. O. W. Richardson *on the Emission of*

theory of metallic conductors*, contact potential†, and photoelectric action‡.

It has long been known that ions are emitted in a number of cases in which solids react chemically with gases. The recent experiments of Haber and Just§ indicate that the alkali metals liberate electrons when they are attacked by certain gases. It seems likely, from various considerations‖, that effects of this nature would account for most of the emission from heated sodium which was measured by the writer¶. In consequence of this conclusion, together with the results of a number of experiments which are at first sight in conflict with the theory referred to at the beginning of this paper**, the view appears to have become rather prevalent that the emission of electrons from hot bodies is invariably a secondary effect, arising in some way from traces of chemical action. That this view is a mistaken one is, I think, conclusively shown by the following experiments which I have made with tungsten filaments.

The tests to be described were made with experimental tungsten lamps carrying a vertical filament of ductile tungsten which passed axially down a concentric cylindrical electrode of copper gauze or foil. The tungsten filaments were welded electrically in a hydrogen atmosphere to stout metal leads. These in turn were silver soldered to platinum wires sealed into the glass container. The lead to the copper electrode was sealed into the glass in the same way. The lamps were exhausted with a Gaede pump for several hours, during which time they were maintained at a temperature of 550–570° C. by means of a vacuum furnace. The exhaustion was then completed by means of liquid air and charcoal, the tungsten filament meanwhile being glowed out by means of an electric current at over 2200° C. Most of the tests were made after the furnace had been opened up and the walls of the lamps allowed to cool off. The walls were always considerably above the temperature of the room on account of the heat radiated by the glowing filament.

The processes described are extremely well adapted for

* Richardson, Phil. Mag. vol. xxiii. p. 594 (1912), vol. xxiv. p. 737 (1912).

† Richardson, Phil. Mag. vol. xxiii. p. 263 (1912).

‡ Richardson, Phil. Mag. vol. xxiv. p. 570 (1912); Richardson and Compton, Phil. Mag. vol. xxiv. p. 575 (1912).

§ *Ann. der Phys.* vol. xxx. p. 411 (1909), vol. xxxvi. p. 308 (1911).

‖ *Cf.* Fredenhagen, *Verh. der Deutsch. Physik. Ges.* 14 Jahrg. p. 384 (1912); Richardson, Phil. Mag. vol. xxiv. p. 737 (1912).

¶ Phil. Trans. A. vol. cci. p. 497 (1903).

** *Cf.* Pring and Parker, Phil. Mag. vol. xxiii. p. 192 (1912).

getting rid of the absorbed gases and volatile impurities which form such a persistent source of difficulties in experiments of this character. Unless some such treatment is resorted to, the metal electrodes and glass walls of these tubes continue to give off relatively large amounts of gas under the influence of the heat radiated from the filaments, and it has always been possible that this evolution of gas might have played an important part in the electronic emission. The mode of treatment used, for which I am largely indebted to the experience and suggestions of Dr. Irving Langmuir, of the General Electric Company's Research Laboratory at Schenectady, N.Y., seems very superior to anything in this direction which has previously been published.

Tests have been carried out covering the alternative hypotheses, as to the possible mode of origin of the electronic emission, which are enumerated below:

(1) The emission is due to the evolution of gas by the filaments.

The lamp and McLeod gauge were cut off from the rest of the apparatus by means of a mercury trap, the volume being then approximately 600 c.c. A filament 4 cm. long, giving a thermionic current of ·064 amp., was found to increase the pressure from zero to $< 1 \times 10^{-6}$ mm. in five minutes. The number of molecules N_1 of gas given off is therefore $2 \cdot 13 \times 10^{13}$. The number of electrons given off is $N_2 = 1 \cdot 2 \times 10^{20}$. The number of electrons emitted for each molecule of gas evolved is thus $N_2/N_1 > 5 \cdot 64 \times 10^6$.

In the above experiment a liquid-air trap was interposed to keep the mercury vapour off the filament. In another experiment with a filament 8 cm. long this was not the case, and with a current of ·050 amp. the pressure rose in thirty minutes to a value which was too small to measure, but which was estimated as less than 10^{-7} mm. The corresponding value of N_2/N_1 is $2 \cdot 6 \times 10^8$. In this case the current was unaffected when the mercury vapour was subsequently cut off by liquid air (a change of 0·4 per cent. would have been detected).

The magnitude of the above numbers effectually disposes of the idea that the emission has anything to do with the evolution of gas.

(2) The emission is caused by chemical action or some other cause depending on impacts between the gas molecules and the filaments.

In a tube with a filament 1·4 cm. in length and having $1 \cdot 65 \times 10^{-2}$ cm.² superficial area, the pressure rose to

348   Prof. O. W. Richardson *on the Emission of*

$< 2 \times 10^{-6}$ mm. in 5 minutes with an emission of ·050 amp. If the gas is assumed to be hydrogen, which makes most impacts, using a liberally high estimate of the temperature of the copper electrode which determines the temperature of the gas, I find that the maximum number N' of molecules impinging per second during the interval would be $< 7·0 \times 10^{13}$. The number of electrons emitted per second would be $N_2 = 3·13 \times 10^{17}$. The ratio N_2/N' is thus $> 4·47 \times 10^3$. If the putative hydrogen atoms simply turned into a cloud of electrons whose total mass was equal to that of the hydrogen, the value of N_2/N' would be only $3·68 \times 10^3$. The data already referred to for the tube with the filament 8 cm. long give an even larger ratio for N_2/N', namely $1·5 \times 10^4$. Moreover, in some of the experiments the changes in gas pressure were much larger than those recorded above, but they were never accompanied by any change in the electronic emission : also the admission of mercury vapour at its pressure (about 0·001 mm.) at room temperature produces no appreciable change in the emission. Thus there is no room for the idea that the emission of electrons has anything to do with the impact of gas molecules under the conditions of these experiments.

(3) The emission is a result of some process involving consumption of the tungsten.

To test this question some of the lamps were sealed off after being exhausted in the manner described. The filaments were then heated so as to give a constant thermionic current which was allowed to flow for long intervals of time. In this way the total quantity of negative electricity emitted by the filament was determined. The wire was placed in one arm of a Wheatstone's bridge so that the resistance could be recorded simultaneously. The cold resistance was also checked up from time to time.

At these high temperatures the resistance of the filaments increases slowly but continuously. This increase is believed to be due to evaporation of the tungsten. It was found to be proportional to the time of heating when the thermionic current was kept constant, in the case of any particular filament. In the case of one filament which gave 0·05 amp. for 12 hours, the increase in the resistance of the hot filament was 9 per cent. The accompanying proportionate increase in the cold resistance was slightly lower, namely 7 per cent. The latter may probably be taken as a fair measure of the amount of tungsten lost by the filament. The increase in resistance of the hot filament, which is less favourable for our

Electrons from Tungsten at High Temperatures. 349

case, will be considered instead in the following experiment for which the other data are lacking.

A filament 3 cm. long gave 0·099 amp. electronic emission continuously for 2·5 hours. The resistance when hot rose from 4773 to 4787 in arbitrary units. The number of atoms of tungsten lost by the filament in this time was $= 5·66 \times 10^{15}$ whilst the number of electrons emitted $= 5·57 \times 10^{21}$. The number of electrons emitted per atom of tungsten lost was $9·84 \times 10^{5}$. The mass of the electrons emitted in this experiment was thus very close to *three times* the mass of the tungsten lost by the filament.

This tube gave 0·1 amp. electronic emission on the average for 6 hours altogether. By that time the mass of the electrons emitted was approximately 2 per cent. of the mass of the tungsten filament. The tube came to an end owing to an accident: the filament gradually became deformed until it touched the copper electrode and broke. The hardness of the tube was then tested with an induction-coil and the equivalent spark-gap was found to be 3·3 cm. The discharge through the tube gave a bright green fluorescence on the glass around the negative wire, but there was no indication of a glow or the faint purple haze which is obtained when traces of gas are present in tubes of this kind. There is thus no appreciable accumulation of gas even when the filaments are allowed to emit a large thermionic current continuously for a long time.

Another tube with a wire 2·7 cm. long giving 0·050 amp. lost $1·19 \times 10^{17}$ atoms of tungsten in 12 hours as measured by the change in the cold resistance. The number of electrons emitted for each atom of tungsten lost was thus $1·13 \times 10^{5}$, and the mass of the emitted electrons about one third of the mass of the tungsten lost. This tube ran altogether for about 23 hours giving various currents, and finally gave out owing to the local loss of material near one end caused by the sputtering or evaporation. Local over-heating is very apt to occur in these experiments as the thermionic leakage causes the heating current in the wire to be bigger at one end than the other. The mass of all the electrons emitted by this filament was equal to 4 per cent. of its total mass. Under a low-power microscope the filament did not appear to be much changed except in the region where it had burnt out, where it was much thinner than elsewhere.

There is no known reason for believing that the loss of tungsten is due to anything more profound than evaporation. But, in any event, the fact that the mass of the emitted

350   *Emission of Electrons from Tungsten.*

electrons can, under favourable circumstances, exceed that of the tungsten lost proves that the loss of tungsten is not the cause of the electronic emission.

(4) The only remaining process of a similar nature to those already considered which has not been discussed is the bare possibility that the emission is due to the interaction of the tungsten with some unknown condensable vapour which does not affect the McLeod gauge. This possibility is cut out by the fact that the thermionic emission is not affected when the liquid air and charcoal is cut off and the vapours allowed to accumulate in the tube, and by the fact that very considerable changes in the amount and nature of the gases present (as by the admission of mercury vapour) have no effect on the emission.

Taken together, these experiments prove that the emission of electrons does not arise from any interaction between the hot filament and surrounding gases or vapours, nor from any process involving consumption of the material of the filament. It thus follows that the emission of electrons from hot tungsten, which there is no reason for not regarding as exhibiting this phenomenon in a typical form, is not a chemical but a physical process. This conclusion does not exclude the possibility that, under other circumstances, electrons may be emitted from metals under the influence of various chemical reagents—a phenomenon which would be expected to exhibit the same law of dependence upon temperature; but it does involve a denial of the thesis that this emission is invariably caused by processes involving changes of material composition.

The experiments also show that the electrons are not created either out of the tungsten or out of the surrounding gas. It follows that they flow into the tungsten from outside points of the circuit. The experiments therefore furnish a direct experimental proof of the electron theory of conduction in metals.

I wish to express my appreciation of the assistance I have received from Mr. K. K. Smith, Instructor in the Laboratory, in the preparation of the tubes and in carrying out some of the measurements. Mr. Smith and I are engaged in a more detailed quantitative study of the emission of electrons from tungsten, the results of which we hope shortly to publish. I also wish to thank Dr. W. R. Whitney and Dr. I. Langmuir, of the General Electric Company, both for supplying the specimens of ductile tungsten used and also for giving me the benefit of their invaluable experience.

Palmer Physical Laboratory,
    Princeton, N.J.

THEORY AND USE OF THE MOLECULAR GAUGE.

BY SAUL DUSHMAN.

SOME time ago Dr. I. Langmuir described the construction of a "molecular" gauge for the measurement of very small gas pressures.[1] At the suggestion of Dr. Langmuir the writer undertook a more detailed study of the theory and use of the instrument, and the following paper contains the results of a number of measurements that were carried out with the aid of this gauge.

THEORETICAL.

If a plane is moving in a given direction with velocity u relatively to another plane situated parallel to it at a distance d, there is exerted on the latter a dragging action whose magnitude may be calculated from considerations based on the kinetic theory of gases.

At comparatively higher pressures where the mean free path of the gas molecules is considerably smaller than the distance between the plates, the rate of transference of momentum across unit area is given by the equation

$$B = \frac{\eta u}{d},\qquad (1)$$

where η denotes the coefficient of viscosity.

According to the kinetic theory of gases this coefficient ought to be independent of the pressure. The confirmation of this deduction over a very large range of pressures has been looked upon as one of the most striking arguments for the validity of the assumptions on which the kinetic theory of gases is based.

It was found, however, by Kundt and Warburg,[2] that at very low pressures, where the mean free path of the molecules becomes of the same order of magnitude as the distance between a moving and stationary surface placed in the gas, there is distinct evidence of a slipping of gas molecules over the planes. The amount of this slip was found to be inversely proportional to the pressure.

[1] PHYSICAL REVIEW, *1*, 337 (1913). See also abstract, PHYS. REV., *2* (1913).

[2] Pogg. Ann., *155*, 340 (1875). Poynting and Thomson, Properties of Matter, p. 220.

Originally published in *Phys Rev* 1915; 5:212.

Denoting the coefficient of slip by δ, it may be defined by the relation

$$v_g - v_s = \frac{\eta}{\zeta} \frac{\partial v}{\partial x} = \delta \cdot \frac{\partial v}{\partial x}, \tag{2}$$

where v_g = velocity of gas molecules at the surface,

v_s = velocity of surface,

ζ = coefficient of external viscosity.

It follows from hydrodynamical considerations that the amount of momentum transferred per unit area is

$$B = \frac{\eta u}{d + 2\delta}. \tag{3}$$

Thus, owing to slip there is an apparent increase in the thickness of the gas layer between the two surfaces. This increase amounts to $\delta = \eta/\zeta$ for each surface.

Experiments on the conduction of heat at low pressures led to similar observations in this case. According to the kinetic theory the heat conductivity should be independent of the pressure. Accurate determinations showed that at very low pressures the conductivity apparently decreases. This led to the conception that at the surface there occurs a very steep temperature gradient (Temperatursprung) so that the amount of heat, Q, conducted between two surfaces maintained at temperatures T_1 and T_2 is given by

$$Q = \frac{T_1 - T_2}{d + 2\gamma}, \tag{4}$$

where d is the distance between the two plates and γ represents the apparent increase in thickness of the layer of gas at each surface.

The definition of γ may be expressed by the following relation, which by its analogy with equation (2) helps to exhibit the complete parallelism of the phenomena observed in both the case of heat conduction and that of viscosity effect. Denoting the drop in temperature at the surface by ΔT, and the temperature gradient there by $\partial T/\partial x$, the definition of γ follows from the relation

$$\Delta T = \gamma \cdot \frac{\partial T}{\partial x}. \tag{5}$$

An interpretation of this temperature drop on the basis of the kinetic theory of gases was first advanced by Maxwell and subsequently developed still further by Smoluchowski.[1]

It is assumed that of the molecules striking a heated surface only a fraction f is absorbed and then emitted with an average kinetic energy

[1] Ann. Phys., *35*, 983, where references to previous literature are given.

corresponding to that of the surface. The remainder $1 - f$ is reflected according to the laws of elastic collision. If T_1 denote the temperature of the molecules striking the surface, T_2, the temperature of the latter, and T_2^1 the temperature of the molecules leaving the surface, then:

$$T_2^1 - T_2 = (1 - f)(T_1 - T_2). \qquad (6)$$

The constant f is known as the coefficient of equalization (Smoluchowski) or accommodation (Knudsen),[1] for it is evident that f is unity when the average temperature of the molecules leaving the heated surface corresponds to the temperature of the latter.

In consequence of this lack of complete equalization of temperatures, there is produced an apparent temperature drop at the surface, which is related to the coefficient of equalization by the following equation:

$$\gamma = \frac{2 - f}{f} \cdot \frac{15}{4\pi} L, \qquad (7)$$

where L is the mean free path of the molecules at the given pressure.

The same method of interpretation was extended to the case of transference of momentum from one surface to another at very low gas pressures. Of the molecules striking a moving surface a portion β is "absorbed" and then "emitted" with velocities that range according to Maxwell's distribution law. The direction of emission is perfectly independent of the direction of incidence. On the other hand, the fraction $1 - \beta$ is "reflected" according to the laws of elastic collision.[2]

A number of investigators have concerned themselves with the mode of determination of these coefficients β and f, which may be designated as the coefficients of accommodation for viscosity and heat conduction respectively.

Knudsen,[3] who has carried out a large number of investigations on the behavior of gases at very low pressures, concludes that while the value of this coefficient is less than unity for heat conduction (and differs with the nature of the gas) it is equal to unity in all those cases where transference of momentum is concerned. He assumes, in other words, that all the molecules are emitted from a moving surface in directions which are absolutely independent of the original directions of incidence and that these molecules then obey Maxwell's law of distribution of velocities.

Timiriazeff[4] makes the assumption that the coefficient of accommodation has the same value, both for viscosity measurements and for the

[1] *Ann. Phys.*, *34*, 593 (1910).
[2] A. Timiriazeff, *Ann. Phys.*, *40*, 978 (1913).
[3] *Ann. Physik*, *28*, 75 (1908); *31*, 205 (1909); *33*, 1435 (1910); *34*, 593, 823 (1911); *35*, 389 (1911); *36*, 871 (1911).
[4] *Ann. Physik*, *37*, 233 (1912).

determination of heat conduction in gases at low pressures and deduces from this assumption the relation

$$\delta = \frac{8}{15}\gamma \qquad (8)$$

where δ has the significance assigned to it in equation (3) above.

From equations (7) and (8) it follows that

$$\delta = \frac{2}{\pi} \cdot \frac{2-f}{f} \cdot L. \qquad (9)$$

More generally, we can write

$$\delta = a \cdot L$$

where a is a constant whose exact value depends upon the particular assumptions made regarding the value of the accommodation coefficient.

At extremely low pressures, where L is large compared to d, equation (3) reduces to

$$B = \frac{\eta u}{2aL} \qquad (10)$$

or, since

$$\frac{\eta}{L} = 0.31 p \sqrt{\frac{8M}{\pi RT}},$$

$$B = \frac{2 \times 0.31}{a} pu \sqrt{\frac{M}{2\pi RT}}. \qquad (11)$$

Substituting for a the value deduced by Timiriazeff, see equation (9), it follows that

$$B = \frac{f}{2-f} \times 0.31\pi up \sqrt{\frac{M}{2\pi RT}}. \qquad (12)$$

A relation of the same form as this may also be deduced by means of considerations similar to those used by Knudsen. This method of derivation has the advantage that it does not involve any extrapolation of equation (3), but starts from fundamentally different premises.

At very low pressures, the mass of gas striking unit area of a surface per unit time is equal to

$$\frac{1}{4}\rho\Omega = p \sqrt{\frac{M}{2\pi RT}}$$

where

ρ = density of gas,

Ω = average (arithmetical) velocity.

Assuming, as Knudsen does, that the coefficient of accommodation is unity, it follows that the rate of transference of momentum per unit area from a surface moving with velocity u is

$$B = up \sqrt{\frac{M}{2\pi RT}}. \tag{13}$$

Equations (12) and (13) agree in the conclusion that at very low pressures B is proportional to $p\sqrt{M/RT}$.

According to Gaede,[1] Knudsen's assumption that the accommodation coefficient is equal to unity in the case of viscosity measurements at very low pressures is justified only at pressures below about 1.33 bars (.001 mm. of mercury). In a very recent paper, Baule has discussed the work of Smoluckowski, Knudsen, Timirazeff and others in detail[2] and by introducing some very plausible assumptions as to the actual mechanism by which a gas molecule exchanges energy with a molecule of the surface against which it strikes, he arrives at the relation

$$B = up \left(\frac{(1 - \alpha'\nu)}{(1 + \alpha'\nu)} \right) \sqrt{\frac{M}{2\pi RT}}, \tag{14}$$

where $\alpha'\nu$ is a function of the masses and diameters of the molecules of the gas and solid, and the distances between the molecules in the plane surface. It is thus evident that no two writers are agreed upon the manner in which the coefficient f is to be calculated.

We are, however, justified in concluding that there exists a relation between B and $p\sqrt{M/RT}$ of the general form

$$B = kup \sqrt{\frac{M}{RT}}, \tag{15}$$

where k is a *constant* whose value depends upon the nature of the gas and that of the surface with which it is in contact.

This is the fundamental relation upon which is based the construction of the "molecular gauge" described in the following section.

DESCRIPTION OF GAUGE.

The construction of the gauge is shown in Fig. 1. It consists of a glass bulb B in which are contained a rotating disc A and, suspended above it, another disc C. The disc A is made of thin aluminum and is attached to a steel or tungsten shaft mounted on jewel bearings and carrying a magnetic needle NS. Where the gauge is to be used for measuring the pressure of corrosive gases like chlorine, the shaft and disc may be made of platinum. The disc B is of very thin mica, about

[1] Ann. Physik, *41*, 289.
[2] B. Baule. Ann. d. Physik, *44*, 145 (1914).

Vol. V]
No. 3.] *THEORY AND USE OF THE MOLECULAR GAUGE.* 217

.0025 cm. thick and 3 cm. in diameter. A small mirror, M, about 0.5 cm. square is attached to the mica disc by a framework of very thin aluminum. This framework carries a hook with square notch which fits into another hook similarly shaped, so that there is no tendency for one hook to turn on the other. The upper hook is attached to a quartz fiber, about 2×10^{-3} cm. diameter, and 15 cm. long.

"The lower disc can be rotated by means of a rotating magnetic field produced outside the bulb. This field is most conveniently obtained by

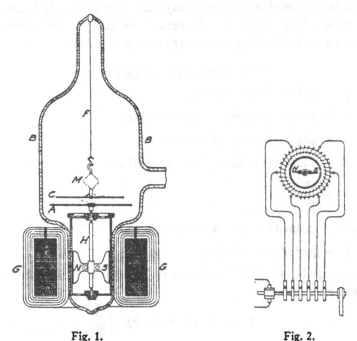

Fig. 1.   Fig. 2.

Rotating Commutator and Connections to Gramme Ring *G–G* of Fig. 1.

a Gramme ring (GG) supplied with current at six points from a commutating device run by a motor (see Fig. 2). In this way the speed of the motor determines absolutely the speed of the disc, since the two revolve in synchronism. The speed of the disc may thus be varied at will from a few revolutions per minute up to 10,000 or more."

In constructing the gauge, the lower part of the bulb is made to fit an aluminum spring holder which supports the spun aluminum cylinder. The latter contains the upper and lower jewel bearings on which the shaft (H) rotates. The bulb is then cut across the widest portion (at BB) and the two discs are introduced, care being taken to see that the framework which carries the mirror is not bent during the subsequent re-sealing of the two parts. The quartz fiber is threaded through the hook and a little glass bead attached on the lower end, while the upper end is fastened to a platinum wire by means of sealing-in glass. The last operation consists in "fishing" for the mica disc by means of the hook on the quartz fiber, and after the distance between the two discs has been adjusted, so that the upper disc hangs centrally over the lower disc in a perfectly horizontal plane at a distance of less than 1 cm., the glass at the top of the bulb is closed up around the platinum wire.

One of the great advantages of the gauges as constructed in the above manner is *the complete absence of any parts that cannot be heated up to a temperature of about 300° C. No cement, shellac or other source of vapor should be used in attaching the mirror or the quartz fiber.*

Calibration of Gage.

Let r = radius of rotating disc,

ω = angular velocity of rotating disc,

α = angle of torque of upper disc,

D = "Direktions-kraft" on upper disc

$= \dfrac{\pi^2}{t^2} K,$

where K = moment of inertia of upper disc,

t = period of oscillation.

From equation (15) it follows that the momentum transferred per unit time to upper disc is

$$B_0 = \int_0^r \omega r^2 \cdot 2\pi r dr \cdot kp \sqrt{\frac{M}{RT}}$$

$$= \frac{k\pi r^4 \omega p}{2} \cdot \sqrt{\frac{M}{RT}} \tag{16}[1]$$

$$= \alpha D = \frac{\alpha \pi^2}{t^2} \cdot K.$$

Consequently

$$\alpha = \left(\frac{kt^2 r^4}{2\pi_k}\right) pw \sqrt{\frac{M}{RT}}. \tag{17}$$

Hence the torque on the upper disc is proportional to the product of the speed of rotation of the aluminum disc and the function $p\sqrt{M/RT}$. Upon this equation depends the use of the instrument as a sensitive vacuum gauge.

[1] This equation is only rigorously true if the diameter of the rotating disc is very large compared with that of the upper disc, so that errors due to "edge effect" are avoided.

By properly designing the dimensions of the discs it is evident that equation (17) could be used for *very accurate determinations of the value of k.* In this manner the conclusions of Knudsen, Smoluckowski and Baule on the correction for slip could readily be tested. As the present investigation was carried out mainly with the view of determining the utility of the instrument as a gauge, no such accuracy was attempted so that definite conclusions could not be drawn regarding the value of *k.* In each case the gauge was calibrated at pressures of about .001 to .01 mm. of mercury against a McLeod gauge. The following data give, however, some idea of the degree of sensitiveness to be expected (and actually ob' 'ned) from a gauge constructed on the above principles.

For this particular gauge, the weight of the mica disc was 0.1 gm., $r = 2$ cm., $t = 12$ seconds.

Consequently

$$K = \tfrac{1}{2} W^2 = \tfrac{1}{2} \times 0.1 \times 4 = 0.2.$$

Assuming a speed of 1,000 r.p.m.,

$$\omega = \frac{2\pi}{60} \times 1,000.$$

In the case of air at a pressure of *1 bar*[1] and 300° Abs., it is found by making the proper substitutions in equation (17) and assuming $k = 1/\sqrt{2\pi}$ that

$$\alpha_{calc.} = 150° \text{ per bar.}$$

By illuminating the mirror and using a similar arrangement to that used for galvanometers, it is possible to detect a deflection of 1 mm. at a distance of 50 cm. or

$$\frac{1}{500} \times \frac{180}{\pi} \times \frac{1}{150} \text{ bar} = 0.8 \times 10^{-3} \text{ bar.}$$

Increasing the speed to 10,000 r.p.m. increases the sensitiveness tenfold and under these conditions it ought, therefore, to be possible to measure a pressure of about 10^{-4} bar.

CORRECTION FACTORS.

In using the instrument there are, however, several points regarding which special care ought to be taken.

1. *Correction Due to Eddy Currents in Metal Parts of Mica Disc.—*

[1] In accordance with most recent practice, we have adopted in this paper as unit of pressure 1 dyne per cm.². This is known as a *bar.* The relation between this unit and the conventional unit ($1\mu = 10^{-3}$ mm.) is very simple. For all purposes the relation $1\mu = 4/3$ bar is accurate enough. The exact relation is that 1 micron of mercury at 45° latitude and sea-level is equal to $1.01327/.76 = 1.33325$ bar.

Owing to the rotation of the magnetic field produced by the Gramme ring, eddy currents are set up in the metal framework used to hold the mirror on the mica disc. Denoting the current through the commutator and Gramme ring by i, the torque actually produced on the upper disc may be expressed as additively composed of two terms, one due to the gas molecules from the rotating disc, and the other due to eddy currents in the metal parts of the upper disc. Consequently, equation (17) assumes the form

$$\alpha = \left(\frac{k l^2 r^4}{2\pi K} p \omega \sqrt{\frac{M}{RT}} \right) + k_1 i^2 \omega, \qquad (18)$$

where k_1 is a constant for the gauge.

The magnitude of the correction term may be diminished by using metal parts whose electrical resistance is very high and by placing the Gramme ring at a greater distance below the upper disc.

On the other hand, there is really no need for any metal parts whatever in connection with the upper disc. The mirror could be supported in a mica or glass holder, and where extreme accuracy is desired such a construction could no doubt be worked out in detail. For ordinary purposes where it is desired to measure pressures that are not less than 0.001 bar, a framework of thin aluminum wires for holding the mirror introduces no measurable errors.

2. *Synchronism.*—From the construction of the apparatus, it is evident that the aluminum disc rotates five times as fast as the magnetic field. In order to maintain the disc and commutator in synchronism, a rotating sector with five slots in it may be attached to the commutator so as to enable the operator to view a mark on the aluminum disc which should obviously appear to remain stationary if the two are in synchronism. An equally good check is to take readings of the deflection at different speeds. If the speed of the commutator is increased very slowly, there is no difficulty in maintaining the disc and commutator in synchronism.

4. *Relative Position of Discs.*—At a pressure of 1 bar and ordinary temperatures, the mean free path for air is about 10 cm. Consequently, in order that equation (17) should be valid at this pressure, the discs ought to be placed at a distance of less than 1 cm. apart. Care should also be taken to see that the upper disc is located centrally over the lower one. Regarding which disc should be the larger, the following considerations are of interest. In the operation of the gauge there is always a tendency for the upper disc to start swinging or at least get away from its symmetrical position with respect to the lower disc. If the latter is large compared to the mica disc, there obviously results a much greater torque on one side of the disc than on the other and the tendency to

swing is increased until finally the disc hits the walls of the bulb. As the damping at low pressures is very feeble, it is very difficult to stop the oscillation when once started, except by imparting to the bulb itself an opposing motion by hand. After a little experience it is easy in this manner to stop any tendency for the disc to vibrate.

Where it is intended to use the instrument as an absolute gauge or for the determination of k, it is obviously necessary to have the rotating disc much larger. On the other hand, for most purposes, that is where the instrument can be calibrated against say a McLeod gauge at pressures above 1 bar, and used to extrapolate the indications of the latter for very low pressures, it is more advantageous to have the upper disc larger since, in this manner, the tendency to swing is diminished considerably. It must be remembered, however, that as the area of the upper disc is increased beyond that of the lower disc the sensitiveness is decreased.

EXPERIMENTAL.

1. *Preliminary Experiments.*—The gauge used in these experiments contained a much heavier mica disc (weight about 0.5 gm.) and a phosphor-bronze suspension similar to those used in galvanometers. The deflection was determined directly by noting the position of a mark on the mica disc with respect to a circular scale outside the bulb. The molecular gauge was connected in series with a liquid air trap to a Gaede mercury pump and ordinary McLeod gauge.

The following data show that the deflections observed are proportional to the rate of rotation of the aluminum disc. Under p is given the pressure in bars; under r, the rate of rotation in r.p.m., and under A the deflection in degrees. The fourth column gives $D = (A/r) \times 1{,}000$, while the last column gives $D_0 = (A/r) \times 1{,}000/p$, that is, the *deflection per bar at 1,000 r.p.m.*

TABLE I.

p	r	A	D	D_0
0.97	850	18	21	21.3
	1,200	28	23	23
	2,000	44	22	22.5
5.74	1,050	80	76	13.5
	1,750	120	70	12

The reason for the larger value of D at low pressures is probably due to the presence of water-vapor and other condensible gases in the gauge, as the bulb had not been previously baked out.

After allowing dry air to enter the system until the pressure was over

222   *SAUL DUSHMAN.*   [SECOND
                                                           [SERIES.

5 mm. of mercury, readings were taken of both the McLeod and molecular gauges, as the pressure was decreased by pumping.

The results of the observations are recorded in Table II.

TABLE II.

p	D	D_0
1,707	990	0.6
960	870	0.9
540	840	1.6
304	820	2.7
171	745	4.0
96	655	6.8
55	490	9
31	360	11.6
24.5	280	11.4
14.0	180	12.8
8	106	13.5
4.7	67	14.3
3.5	46	14.3

It will be observed that up to about 20 bar the deflection was proportional to the pressure. At this pressure the mean free path in air is about 0.5 cm., and this was about the distance between the two discs.

2. *Vapor Pressures of Mercury and Ice.*—For the observations recorded in this section, the sensitiveness of the molecular gauge used was such that $D_0 = 9°$ corresponding to 180 mm. on scale.

The gauge was baked out for one hour at 330° C. and observations then taken on both the McLeod gauge and the molecular gauge under different conditions.

TABLE III.

Press. in McLeod.	D	Press. in Mol. Gage (Calc.).	Remarks.
0.49 bar	23°.7	2.6 bars	Liquid air on trap. Pump not exhausting.
0.8	105.	11.2	Removed liquid air.
0.27	36.4 mm.	0.27	Liquid air on trap and pump exhausting.
1.33	84°	9.3	Removed liquid air.
0.27	36.4	0.27	Liquid air on trap.
0.033	4.5	0.033	Liquid air on trap and pump exhausting.
Removed liquid air.		Stopped exhausting.	
	170 mm.	1.26	At end of 1 minute.
	100°	11.3	At end of 8 minutes.
Put on liquid air again			
	4.5 mm.	0.033	At end of 5 minutes.

It will be noted that the molecular gauge followed changes in pressure which were altogether lost as far as the McLeod was concerned. The pressure of about 10 bar observed on removing the liquid air is evidently due to mercury vapor and non-condensible gases. Allowing about 1 bar for the pressure of the latter (indicated on McLeod) it follows that the pressure due to mercury vapor alone was about 9 bar at room temperature (298° Abs.).

According to Smith and Menzies[1] the vapor tension of mercury between 20° and 30° C. is as follows. (The pressure in bar was obtained by multiplying the pressure in mm. by $4/3 \times 1,000 \times \sqrt{200/28.8}$.)

Temperature.	Press. in Mm.	Press. in Bar.
20°	.0013	4.57
24	.00183	6.53
28	.00254	8.92
30	.00299	10.5

The determination of the vapor pressure of mercury as given above is in fair accord with the data for 28°–30° C.

A determination was also made of the vapor tension of ice at − 78° C. A bath of acetone with solid carbon dioxide was put around the liquid air trap and the pressure in the gauge measured while the pump was exhausting. The average of three determinations was 0.9 bar. Allowing for the difference in the temperature of gauge and liquid air trap, the pressure in the latter must have been $0.9\sqrt{195/298} = 0.78$ bar.

Extrapolating from the data given by Scheel and Heuse[2] for the vapor tension of ice at temperatures down to − 68° C., and allowing for the difference in molecular weight of air and water vapor, the pressure at − 78° C. is calculated to be about 0.2 bar. The pressure due to non-condensible gases was not over .05 bar in the above measurement.

3. *Calibration of the Gauge with Hydrogen.*—The theoretical conclusion that the indications of the gauge at constant pressure ought to vary with the square root of the molecular weight was tested by introducing hydrogen into the gauge instead of air.

The gauge used gave a deflection, with air, of 135 mm. per bar at 1,000 r.p.m. The sensitiveness with hydrogen should therefore have been $135 \times \sqrt{2/28.8} = 35$ mm. per bar at 1,000 r.p.m. The actual experiments gave values ranging from 37 to 42 mm.; the discrepancy being probably due to the presence of small quantities of air in the hydrogen used.

4. *Pressure in Tungsten Lamp.*—An ordinary 60-watt type Mazda lamp bulb was connected to a molecular gauge and after exhausting

[1] Jour. Am. Chem. Soc., *32*, 1447 (1910).
[2] Ann. Phys., *29*, 723 (1909).

them for one and a half hours at 250° C. they were sealed off and pressure observations taken at intervals during the life of the lamp.

The gauge used had a sensitiveness of 1,100 mm. per bar at 1,000 r.p.m. After sealing off and before lighting the filament, the pressure indicated was about 0.8 bar, but as soon as the filament was lighted the pressure decreased and inside of less than an hour it went down to below 10^{-3} bar. The lowest pressure was certainly well below 5×10^{-4} bar, and this determination may be regarded as an upper limit of the probable pressure in a tungsten lamp. It must be noted that the lamp used in these measurements was not given nearly as good a heat treatment as in the usual lamp exhaust.

The fact that the vacuum in a tungsten lamp improves when the filament is lighted has been known for some time, and the causes of this "clean-up" effect have been discussed in a number of papers published during the past three years by I. Langmuir.[1]

As the volume of the gauge was only slightly greater than that of the lamp bulb, we can conclude that the pressure obtained in a well-exhausted tungsten lamp, when the filament is lighted, is certainly well below 10^{-3} bar.

5. *Experiments with the Gaede Molecular Pump.*—A Gaede molecular pump[2] was run in series with an oil pump which in turn was connected to the "rough vacuum" line. A McLeod gauge was inserted between the oil pump and molecular pump in order to read the pressure on the rough side of the latter.

A liquid air trap was arranged between the gauge and the pump so that the diffusion of vapor of stopcock grease or of water could be prevented.

The following table shows the results obtained under different conditions:

No.	Press. on Rough Side of Molecular Pump.	Press. on Fine Side, Read by Molecular Gauge.	Conditions of Experiment.
1	13.3 bar	0.20 bar	After exhausting for 1 hour. No heating of gage; no liquid air.
2	13.3	0.09	Put on liquid air.
3	13.3	0.033	Heated gage to 300° C. for 1 hour, but did not heat glass tubing between gage and pump.
4	1,333	0.033	Let in dry air on rough side. Press. on fine side remained constant. Ratio = 40,000:1.
5	20,000	0.4	Let in more air on rough side. Ratio = 50,00:1.
6	20	≪0.0007	Ratio 30,000:1.

[1] J. Am. Chem. Soc., *35*, 107 (1913), et sub.
[2] W. Gaede, Ann. Physik, *41*, 337 (1913).

It is evident from experiment (3) that heating the gauge alone was not sufficient to reduce the pressure in it owing to the constant diffusion of water-vapor from the tubing between the gauge and liquid air trap. But after this tubing had been also heated to 330° C. (experiment 6) the pressure in the gauge went down below 7×10^{-4} bar.

The sensitiveness of the gauge employed was such that 1 bar gave a deflection of 525 mm. at 1,000 r.p.m. At very low pressures a correction term had to be introduced for the eddy current effect in the framework of the mirror. The equation connecting pressure (p) and deflection (D) may be written in the form

$$D = \frac{525}{1000} \cdot p.r. + ki^2r, \qquad (19)$$

where r = revolutions per minute, and k is a constant. See equation (18).

By noting D for different values of r and i, while p is maintained constant, it is possible to determine the value of k, and hence introduce the proper correction into the calculation for p. A special series of experiments showed that in the case of the above gauge, the equation for calculating p was of the form:

$$p = \frac{D}{0.525r} - 29 \times 10^{-5}i^2.$$

The value of i varied from 3 to 5 amperes. It is evident that the presence of this eddy current effect limited the sensitiveness of the gauge, for even at zero pressure, the deflection at 4 amperes and 10,000 r.p.m. would be 24 mm. At 7×10^{-4} bar, the deflection at 10,000 r.p.m. and 4 amperes would be 27.6 mm.

Under the best vacuum conditions, that is, using liquid air, and heating the gauge and all connecting parts to over 330° C. for about one and a half hours or longer, the deflections actually obtained were only slightly greater (1 or 2 mm. more) than the correction due to the eddy current effect. Allowing for experimental errors and for the difficulty in reading to an accuracy of 2 mm. when the disc was rotated at very high speeds, it is probably correct to conclude that the vacuum obtained was less than 7×10^{-4} bar. Assuming the ratio of 50,000 : 1 as holding down to the very highest vacuum conditions, the vacuum attained in the gauge, with a pressure of 20 bar on the rough side should have been 4×10^{-4} The experimental observations are in satisfactory agreement with this calculation.

It is worth noting in this connection that in his paper describing the construction of the molecular pump, Gaede states that (at 8,200 r.p.m.)

with a rough pump pressure of 1 mm. he obtained a pressure of .02 μ on the fine pump side corresponding to a ratio of 50,000 : 1. Both pressures were read by means of McLeod gauges.

OTHER VACUUM GAUGES.[1]

In this connection it might not be amiss to mention briefly some of the other vacuum gauges that have been suggested for the measurement of pressures below 1 bar.

1. The radiometer has been used by a large number of investigators. Dewar has stated the case for this instrument as follows:[2] "The radiometer may be used as an efficient instrument of research for the detection of small gas pressures. For quantitative measurements the torsion balance or bifilar suspension must be employed."

Some years ago Mr. W. E. Ruder, of this laboratory, developed a method of using the radiometer for the measurement of the gas pressure in incandescent lamps. "It was found that when exhausted to the degree required in an incandescent lamp the radiometer could not be made to revolve, even in the brightest sunlight. In order to get a measure of the vacuum, the radiometer vanes were revolved rapidly by shaking the lamp and the time required to come to a complete stop was therefore a measure of the resistance offered to the vanes by the gas, together with the frictional resistance of the bearing. The latter quantity was found to be so small in most cases that a direct comparison of the rates of decay of speed of the vanes gave a satisfactory measure of the degree of evacuation. In this manner a complete set of curves was obtained which showed the change in vacuum in an incandescent bulb during its whole life and under a variety of conditions of exhaust.

"The chief objections to this method of measuring vacua were the difficulty in calibrating the radiometer and the difference in frictional resistance offered by differnet radiometers. For *comparative* results, however, the method was entirely satisfactory."[3]

2. Scheele and Heuse[4] devised a manometer which has been used successfully for the accurate determination of the vapor pressures of mercury and ice at very low temperatures. This gage consists of two chambers separated by a copper membrane. One of the chambers is

[1] A good description of some of the gages mentioned in this summary is given in K. Jellinek's recently published "Lehrbuch der Physikalischen Chemie," I, 1.

Shortly after this paper was sent to the printer, a description of a modified Knudsen manometer was published by J. W. Woodrow, Phys. Rev. 4, 491 (1914). The sensitiveness of this gage is stated to be about 4×10^{-8} bar.

[2] Proc. Roy. Soc., *A. 79, 529* (1907).

[3] This account was kindly prepared by Mr. Ruder at the request of the writer.

[4] Ber. d. deutsch. phys. Ges., 1909, 1–13.

maintained at constant pressure while the other is connected to the system under investigation. The membrane presses on a glass plate and the variation in the thickness of the film is measured by noting the number of interference bands.

3. The McLeod gauge can be constructed so that it is sensitive to 0.01 bar. Its field of application is however necessarily limited.

4. Pirani[1] has suggested a resistance manometer which depends upon the fact that at low pressures the heat conductivity of gases is a function of the pressure. In consequence of this change in the heat conductivity, the apparent resistance of the wire changes with pressure of the gas surrounding it. The method has been improved by C. F. Hale[2] and the manometer has been found to give reliable results down to 0.00001 mm. of mercury, that is to about 0.01 bar.

5. Very recently W. Rohn has described a vacuum meter based on almost the same principles.[3] In this case the effect on the thermo-electromotive force of varying gas pressure is used as a method of determining very low gas pressures. The instrument is most sensitive between about 100 and 1 bar (0.075 mm. and 0.00075 mm. of mercury) the electromotive force varies approximately linearly with the logarithm of the pressure. At lower pressures, the sensitiveness diminishes quite rapidly.

6. Haber and Kerschbaum have used vibrating quartz fibers to measure the pressure of mercury and iodine.[4] This method was originally suggested by I. Langmuir[5] for measuring the residual gas pressure in sealed-off tungsten lamps and has been in use in this laboratory for about three or four years. As the pressure decreases the duration of the oscillations increases. Haber and Kerschbaum have deduced a relation between the pressure p and the interval t in which the vibration decreases to half its original amplitude, as follows:

$$\Sigma(p\sqrt{M}) + a = \frac{b}{t},$$

where a and b are constants and Σ denotes that the sum of the products of partial pressure and square root of the molecular weight is to be taken for each gas present. The lowest pressures actually measured by the above authors were about 0.015 bar, but the method has been used in this laboratory by Dr. Fonda to measure pressures considerably smaller than this.

[1] Ber. d. deutsch. physikal. Ges., 1906, 686.
[2] Trans. Am. Electrochem. Soc., *20*, 243 (1911).
[3] Z. f. Elektrochem., *20*, 539 (1914).
[4] Z. f. Elektrochem., *20*, 296, 1914.
[5] J. Am. Chem. Soc., *35*, 107, 1913.

7. W. Sutherland[1] and, subsequently, J. L. Hogg[2] derived simple relations between the pressure and the logarithmic decrement of a vibrating mica disc. The gauge based on this principle requires calibration at two known pressures, and is not very sensitive below about 0.01 bar.

8. M. Knudsen as the result of an elaborate study of the laws of "Molekularströmung" devised an absolute form of manometer[3] which depends upon the fact that at very high vacua there exists a very simple relation between the pressure and the torque imparted to a movable surface by the molecules flowing to it from a hotter surface. This relation has the form

$$K = \frac{p}{2}\left(\sqrt{\frac{T_1}{T_2}} - 1\right),$$

where K denotes the force of repulsion between two surfaces maintained at temperatures T_1 and T_2 respectively in a gas at pressure p. From the dimensions of the movable disc and the period of oscillation of the suspension, the value of K may be calculated and the gauge may therefore be used without any previous calibration. Knudsen uses this guage to indicate pressures as low as 2×10^{-5} bar.

9. Still more recently[4] Knudsen has devised a simplified form of vacuum guage, based on the same principles as the above, which he states to be sensitive to 2×10^{-6} bar.

CONCLUDING REMARKS.

The vacuum gauge described above might obviously be used to determine the magnitude of the "accommodation coefficient" for different gases, and thus test out the deductions advanced by different investigators.

Another line of investigation for which the gauge would be useful is the determination of the vapor tension of oils, waxes, etc., such as are used in connection with vacuum work. Owing to pressure of other work the writer has been prevented till now from carrying on such an investigation; but the results would be of great practical utility, as these materials are being constantly used by experimenters in connection with so-called "high vacuum" experiments.

In conclusion the author desires to express his appreciation of the

[1] Phil. Mag., *43*, 83 (1897).

[2] Proc. Am. Acad., *42*, 6 (1906); Phil. Mag., *19* (1906); Proc. Am. Acad., *45*, No. 1, Aug., 1909.

[3] Am. Physik *32*, 809 (1910).

[4] Am. Physik *44*, 525 (1914).

kindly interest shown by Dr. I. Langmuir during the progress of the investigation and for helpful suggestions.

SUMMARY.

A theoretical consideration of the behavior of gases at very low pressures shows that a rotating disc exerts a torque on a disc suspended symmetrically above it, that is proportional to the quantity $\Sigma(p\sqrt{M/RT})$. Here p denotes the partial pressure and M the molecular weight of each constituent present in the gas and R and T have their usual signification.

The paper contains the description of a vacuum gauge based upon this principle, and also the results of a number of measurements carried out with its aid.

It was found that in order to obtain the best possible results with a Gaede molecular pump, it is necessary not only to heat the vessel to be exhausted and connecting tubing to a temperature at which most of the moisture adsorbed in the walls is driven out, but also to insert a liquid air trap to prevent the diffusion backwards of condensible vapors.

RESEARCH LABORATORY,
GENERAL ELECTRIC CO.,
SCHENECTADY, N. Y.

A HIGH VACUUM MERCURY VAPOR PUMP OF EXTREME SPEED.

By Irving Langmuir.

ASPIRATORS or ejectors in which a blast of steam or air is used to produce a partial vacuum have been in use many years. For example the vacuum in the automatic vacuum brake system is obtained by such a device. The Parsons' vacuum augmenter used in the condensers of steam turbines produces a pressure as low as a few centimeters of mercury.

In these devices the high velocity of the jet of steam causes, according to hydrodynamical principles, a lowering of pressure, so that the air to be exhausted is sucked directly into the jet. If, however, the jet were surrounded by a perfect vacuum it can be readily seen from the principles of the kinetic theory that there would be a blast of gas molecules escaping from the jet into this vacuum. It is therefore not possible to obtain a very high vacuum by means of an instrument based on the principle of the ordinary ejector or aspirator.

The action of the aspirator or ejector, however, really consists of two processes.

1. The process by which the air is drawn into the jet.

2. The action of the jet in carrying the admixed air along into the condensing chamber.

The aspirators cease operating at low pressures because of the failure of the first of these processes. If air at low pressures could be made to enter the jet, and if gas escaping from the jet could be prevented from passing back into the vessel to be exhausted, then it should be possible to construct a jet pump which would operate even at the lowest pressures.

Gaede[1] has recently described a pump (called the diffusion pump) in which a blast of mercury vapor carries along the gas to be exhausted into the condenser in much the same way as the steam aspirator does (Process 2). In order to introduce the gas into the blast of mercury vapor (process 1), Gaede has used diffusion through a porous diaphragm, or through what amounts to the same thing, a slit of a width comparable with the mean free path of the mercury atoms in the blast. A portion of the mercury blast escapes through the slit, and the gas to be exhausted

[1] Ann. Phys., 46, 357 (1915).

Vol. VIII.]
No. 1. *A HIGH VACUUM MERCURY VAPOR PUMP.* 49

diffuses in against this blast of mercury vapor. This renders it necessary to make the slit very narrow (about 0.1 mm.) and for this reason the speed of the pump is necessarily relatively low.

Gaede defines the speed of a pump S by the equation

$$S = \frac{V}{t} \ln \frac{p_0}{p},$$

where V is the volume of the vessel being exhausted and t is the time required for the pressure to fall from p_0 to p.

Gaede gives some interesting data on the speed of his various pumps when exhausting air at about .001 mm. pressure.

Rotary mercury pump S = 120 cc. per sec.
Molecular pump S = 1,300 c.c. per sec.
Diffusion pump S = 80 c.c. per sec.

The great advantage of the diffusion pump is that the value of S remains constant down to the lowest pressures, whereas with all other forms of pump the value of S rapidly decreases when the pressure becomes much less than .001 mm.

Some time ago it occurred to the writer that there should be other methods by which the gas to be exhausted may be introduced (Process 2) into the blast of mercury vapor. The serious limitation of speed imposed by the slowness of diffusion through narrow slits may thus be overcome.

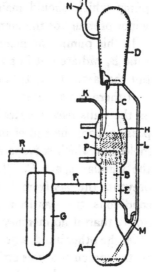

Fig. 1.

Several methods have been found by which this may be accomplished. One of the types of pump which has given satisfactory results is that shown in the sketch (Fig. 1).[1]

[1] A preliminary announcement of the development of this pump was made April, 1916, by Miss Helen Hosmer (General Electric Review, *19*, 316, (1916)). Dr. H. B. Williams showed before the New York meeting of the American Physical Society (Feb. 26, 1916) a mercury vapor pump constructed entirely of glass which, unlike Gaede's pump, did not depend on diffusion through a narrow slit. This pump was subsequently described by Dr. Williams in an abstract in the PHYSICAL REVIEW, *7*, 583, (1916). In regard to this abstract Dr. Williams has requested me to make the following announcement in his behalf:

" In describing recently before the Physical Society a mercury vapor pump, I stated that before carrying out the experiments I had in mind with a view to improving the Gaede type, I learned through Professor Pegram of Columbia that Dr. Irving Langmuir was already working with a pump in which the slit had been discarded and diffusion took place through a wide opening. I therefore constructed my own pump with a wide opening instead of experimenting with various widths as had been my original intention. In writing an abstract I omitted to make this acknowledgment owing to a desire to condense

In this device a blast of mercury vapor passes upward from the heated flask A through the tubes B and C into the condenser D. Surrounding B is an annular space E connecting through F and the trap G with the vessel to be exhausted. The tube C is enlarged into a bulb H just above the upper end of the tube B. This enlargement is surrounded by a water condenser J from which the water is removed at any desired height by means of the tube K which is connected to a water aspirator. The mercury condensing in D and H returns to the flask A by means of the tubes L and M. The tube N connects to the " rough " or " backing " pump which should maintain a pressure considerably lower than the vapor pressure of the mercury in A.

In this pump, the mercury atoms escaping from the upper end of the tube B, radiate out in all directions. A part of them passes up into C, but the larger part strikes the walls of the enlargement H.

If there is no water in the condenser J the mercury which condenses on the walls reëvaporates nearly as fast as it condenses. The molecules passing from the end of the tube B towards the walls H collide with the molecules which reëvaporate and may then be deflected downward into the annular space E. This blast of mercury vapor down through E prevents the gas from F from passing up into H so that under these conditions the gas from F may pass through the pump much more slowly than if no mercury vapor were produced in A.

On the other hand, when cold water circulates through the condenser J, *all* the mercury atoms striking the walls of H are condensed, so that no mercury passes down through E. The gas from F thus passes freely up through E and when it meets the mercury vapor blast at P is blown outward and upward along the walls of the condenser H, and finally forced into the main stream of mercury vapor passing up through C into the condenser D.

The main advantages of this pump are:

1. Simplicity and reliability.
2. Extremely high speed.
3. Absence of lower limits to which the pressure may be reduced.

The speed of a pump like that shown in the sketch has been determined for air and hydrogen.

A vessel of eleven liter capacity was connected to R. Air was admitted to the vessel, and the pressure was found to decrease as follows:

the abstract and because I expected to make such an acknowledgment later in print. The publication of a description of Dr. Langmuir's pump makes it seem desirable that this statement be made at the same time.   H. B. Williams."

Time, Seconds.	Pressure in Bars.		Time Interval, Seconds.	Speed S Cc. per Sec.
	at R.	at D.		
0	1,470.	1,160	> 30	590
30	294	720	> 30	1,150
60	12.8	218	> 20	3,700
80	0.015	18		

The rough pump used had a speed of about 200 c.c. per second at pressures of 400 bars, but this speed fell off to 60 c.c. at 40 bars, and became zero at about 10 bars' pressure. The speed of the mercury vapor pump increased rapidly as the pressure decreased and reached a limit of about 4,000 c.c. per second at pressures below 10 bars. Theoretical considerations would indicate that this speed should remain constant down to the very lowest pressures.

Several experiments with hydrogen showed that the maximum speed with this gas was about 7,000 c.c. per second.

The pump shown in the sketch is made of glass. Many pumps operating on these principles have also been constructed of electrically welded sheet iron.

In a subsequent paper the writer will describe in more detail other modifications of mercury vapor pumps, some of which have marked advantages in simplicity of construction and in reliability of operation over that shown here. One particularly efficient type of pump is made wholly of metal.

The writer is greatly indebted to Dr. S. Dushman for assistance in the development of the new pump.

RESEARCH LABORATORY,
  GENERAL ELECTRIC COMPANY,
    SCHENECTADY, N. Y.

NOVEMBER 10, 1928]   *NATURE*   729

Oils, Greases, and High Vacua.

IN the course of some work (which I hope shortly to publish) on the evaporative distillation of petroleum derivatives, I became aware of the possibility and advantages of using oil in place of mercury as working fluid in condensation pumps. I was distilling lubricating oil in an apparatus similar in principle to that used by Brönsted and Hevesy to separate the isotopes of mercury. The saturation pressure of the oil vapour could be deduced from the observed rate of distillation and the estimated molecular weight of the oil : in a particular case the saturation pressure was about one dyne/cm.^{2} at 118° C., that is, about the same as the saturation pressure of mercury at room temperature. No decomposition could be detected. Clearly, if this oil could be heated until its vapour pressure was, say, 100 dynes/cm.^{2}, without decomposition, it could be used as working fluid in a condensation pump and might be expected to give a performance, without artificial cooling, comparable with the performance of a mercury condensation pump with a cold trap 100° C. below room temperature. I therefore prepared by fractionation a quantity of this oil and evacuated ionisation gauges (large and small thermionic valves), on oil condensation pumps. I have been unable to measure the lower limit of pressure reached by these pumps. 10^{-3} dynes/cm.^{2} has been reached without ovening the glasswork : when the glass was ovened, the ionisation current could not be detected with the instruments available—the pressure probably did not exceed 10^{-4} dynes/cm.^{2}.

Not all oils can be distilled to dryness in the evaporative still. Decomposition usually begins at 320-340° C. I was able to prepare a grease with a vapour pressure of not exceeding 1 dyne/cm.^{2} at 320° C. (as deduced from distilling speed) : this grease was used to lubricate the ground joint between the ionisation gauge and the pump in the above experiments. Mr. J. D. Cockcroft, at the Cavendish Laboratory, found the vapour pressure of this grease to be less than 10^{-3} dynes/cm.^{2} at 70° C. As was to be expected, it was too small to be detected by the evaporation method used. Joints made with this grease may in fact be employed freely, even at temperatures as high as 70° C. (This substance is not a 'grease' in the sense used by the oil technologist, *i.e.* it does not contain a soap, but is simply a petroleum jelly residue.)

It has been customary to regard with grave suspicion the introduction of oil or grease into systems in which high vacua are to be produced. This attitude represents a generalisation which must now be subject to many reservations.   C. R. BURCH.

Research Laboratories,
Metropolitan-Vickers Electrical Co.,
Trafford Park, Manchester.

"Oils, Greases, and High Vacua" was first published in *Nature* 1928; 122:729.

CHEMISTRY AND INDUSTRY February 5, 1949   87

THE FIRST OIL CONDENSATION PUMP

An Example of the Impact of one Technology on another

By Prof. C. R. Burch, F.R.S.

Warren Research Fellow in Physics, University of Bristol

(*Read at the High Vacua Convention at Gleneagles, October 12–13, 1948*)

TWENTY years ago it was my privilege to carry out, in the research department of Metropolitan-Vickers Electrical Co., some simple experiments on vacuum distillation which have had profound and far-reaching repercussions on vacuum technology, and particularly on vacuum electrotechnics. I want to try to recapture for you today the story of the start of that work, for basically it is you who are responsible for it—you and your predecessors of the Society of Chemical Industry, who by your work and your writings aroused in me an interest in chemical technology in general, and in particular in distillation.

It was our Process and Insulation Section which provided the immediate stimulus when they asked me to make a vacuum impregnator, to see whether the use of really high vacuum in impregnating pressboard with transformer oil would give improved dielectric strength. It seemed reasonable to aim at a permanent gas pressure of 0·001 mm., and, indeed, a McLeod reading as low as this was easily obtained on a cold impregnator. No improvement was found in dielectric properties, and someone suggested that the impregnator should be steam-jacketed. To aim at 0·001 mm. with a hot impregnator seemed to me absurd, for certainly the total pressure would be 10 or 100 times higher until all the oil had distilled out of the impregnator into the pump. This thought awoke in me an interest in distillation latent since schooldays, and I resolved to work on vacuum distillation.

It is with considerable diffidence and hesitation that I quote to this conference of distinguished chemical engineers the thoughts that then took shape; at that date, however, they had, I think, a considerable measure of justification.

" No chemist," I told myself, " knows how to make a vacuum, and no physicist is interested in vacuum distillation—except Brønstedt and Hevesy, and they are concerned with isotopes rather than organic chemistry. Let me therefore apply molecular distillation to organic chemistry, and something ought to come out of it."

And so we made our pot-still, consisting of an electrically heated tray some 40 sq. cm. in area, supported inside a nearly horizontal 4 cm. diameter water-cooled glass tube, providing a cold condensing roof 1—2 cm. above the tray. It is a pleasure to acknowledge the co-operation of Mr. Bancroft, who made the tray, and Mr. Watts, who made the glasswork, a task far beyond my powers.

I considered how to pump it. Should we arrange to bake out the glass and use liquid air traps on our mercury condensation pump, or would it do to omit the bake-out and the trap? If we followed the simpler plan, the " vacuum " would consist of 0·001 mm. or thereabouts of mercury vapour. If we regarded the distilling organic molecule as having the same order of cross-section as a mercury molecule, its free path in the residual mercury vapour would be a few cm., so that distillation would proceed about as fast through this mercury vapour as it would if the vacuum was perfect. Accordingly, we put the still on a mercury pump of 4 litres per sec., and though we did use a cold trap, we put it not between the pump and the still, but on the rough side of the mercury pump, for I reasoned (with sound instinct) that there might be enough " light ends " in some of the things we distilled to upset the fore-vacuum by dissolving in the oil of the rotary pumps on compression.

The experiment was bound to be a fruitful one—I reasoned, before we had distilled anything at all. For either the substances we tried would distil at negligible rate, less than a few drops per minute (in which case their vapour pressures at distilling temperature would be less than about 0·001 mm., and so presumably considerably less than 0·001 mm. at room temperature—in which case they would be potential low-vapour greases and waxes) or they would be distillable at only a few drops per minute, in which case they would not be distillable except in a still of this type, or they might distil 10—100 times faster, and in that case perhaps they would replace mercury in condensation pumps. We started by distilling some of the oil used in the rotary pump. Appreciable distillation began very little above room temperature, but it was necessary to go to 120°—150° to bring the main bulk over at a few drops per minute. This, then, was a liquid 100° less volatile, to put it crudely, than mercury, and so we could reasonably hope might give a performance when used as working fluid in a condensation pump without refrigerant, comparable with that given by a mercury pump and a trap cooled 100°.

To try this, we set up a 150-watt triode—by way of ionization gauge—on a pump filled with the middle fraction, obtained in the molecular still, from the rotary pump oil, and after baking out and bombarding in the usual way, found a smaller ion current than is usual in sealed valves of that type.

We had made, in fact, a 150 watt demountable valve. High power valves were of special interest to us, so this was speedily followed by a water-cooled demountable triode of 25 kW, and by a series of triodes and tetrodes of larger powers, used for wireless transmission and later for radar. It became necessary to make a continuous molecular still to make oils and greases for this work—to which the general name " Apiezon products " was given. In this we collaborated with the Shell Co. To investigate the possibility of refining commercial lubricants, a pilot plant of 6 tons per day was put up at the Shellhaven refinery; molecular distillation, however, compares unfavourably with existing methods in cost. I need not touch in detail on this development, which has been described by van Dijck and myself (J.S.C.I., 1934) or on the joint work with B.D.H. on the distillation of vitamin-*A*, which has been described by Carr and also by Burrows (J.S.C.I., 1934). I should like to pay tribute to the independent and substantially simultaneous work on the use of organic substances—the phthalates and later the sebacates—as pump fillings by K. C. D. Hickman, and to his very much more extensive work on molecular distillation, including the devising of analytical molecular distillation techniques, and of the first centrifugal still and the large-scale distillation of vitamins by Distillation Products, Inc.

The cobbler must stick to his last, and it is perhaps but natural that we should have pushed our follow-up further on vacuum electrical rather than chemical applications. It is perhaps worth recording how nearly our work came to an end shortly after we had made our first oil condensation pump. This pump had been dismantled—I think for incorporation into a demountable valve assembly—and I tried to make a second pump work. Only with the greatest difficulty and careful adjustment of the heat input could I get it to achieve even the poor reading 10⁻⁴ mm. on an ionization gauge, whereas the first

"The First Oil Condensation Pump" was first published in *Chem Ind;* 87 (February 5, 1949).

pump had achieved 10^{-8} mm. without difficulty, nor had its heat input been critical.

After some days' fruitless effort, I happened to heat it with the identical bunsen that I had used to heat the first pump. Immediately it worked properly. Like Röntgen on a famous occasion, " I did not think : I investigated." Of all the bunsens in the laboratory, only the one would make that pump work properly.

And then the reason became clear. The first bunsen was the one I normally used, and its air supply was adjusted to give a pointed green cone ; the others had inadequate primary air, and gave soft woolly flames. The pointed flame gave a good thermal circulation in the oil ; the woolly flame superheated the oil in the crevices of the re-entrant weld between the base and the main body of the pump. On so small a point can success or failure of a novel venture hang.

Looking back, I think that we (and also Dr. Hickman's Group) did understand correctly from the first, many of the points arising out of the use of organic fluids as pump fillings. We realized that in general these fillings could not be regarded as single pure liquids, but would contain traces of " light ends " of almost all degrees of volatility, and that the pump must be regarded as two opposed fractionating columns, the relative efficiencies of which governed the equilibrium pressure ratio— and that therefore the fore-vacuum pipe should be allowed to run hot. We realized the difficulty of making any form of absolute pressure measurement at pressures of the order of 10^{-6} mm. I was not certain—I am still uncertain—what correction for radiation pressure should be applied to Knudsen gauge readings, but I think it may be comparable with the lowest

pressure—if memory serves, about 10^{-7} mm.—recorded on a Knudsen gauge (by Schræder and Sherwood). Partly for that reason, and partly because I had no experience of Knudsen gauges, I decided not to attempt absolute measurements, but to use ionization gauges, and to calibrate these by continuous flow methods in which the absolute pressure of air admitted by a leak was ascertained simultaneously with the pump speed. It was relatively easy to obtain readings of 10^{-8} mm. " equivalent air pressure " with an oil condensation pump, on baking out and bombarding one of these gauges, and it seemed that the better one's bake-out and general cleanliness precautions, the more quickly was this attained. The lowest reading I ever attained in this way was 10^{-7} mm. I realized that one could not necessarily conclude that the pump would give this pressure in the absence of the gauge, for I envisaged the possibility that bake-out and bombardment might crack up volatile " light ends " if these were present in the fine vacuum, perhaps due to oxidation products in the pump filling, and so indirectly " condition " the pump filling. What I did not realize—indeed, " went blind to "—was that, apart from any cracking, of volatiles into carbon plus permanent gas during bombardment, an ionization gauge can continue to clear up (absorb) residual gas (including oil vapour) for a time long compared with the few hours' duration of my individual experiments, and that in the absence of this " continuous clean-up " effect, pressures may be higher by a factor of 10 or 20. The reality and magnitude of this effect were recently established conclusively by J. Blears, whose work explains the mystery of the discordant values for oil condensation pump ultimate pressures, recorded by various authors during the past 15 years.

The Flow of Highly Rarefied Gases through Tubes of Arbitrary Length

P. Clausing

Natuurkundig Laboratorium, N. V. Philips' Gloeilampenfabrieken, Eindhoven, The Netherlands
(Received 26 November 1931)

Editor's Note: Translated from the German [Ann. Physik (5) 12, 961 (1932)]. Since this classic paper is referred to so frequently by workers in vacuum technology, it has been decided to publish an English translation. The translation was made available through the courtesy of Veeco Instruments, Inc., who also supported its publication.

Contents

(I) The expressions of Knudsen, v. Smoluchowski, and Dushman for steady-state molecular flow; (II) transformation of the equations to kinetic variables, transmission probability; (III) derivation of the Dushman formula; (IV) the problem of the short, round cylindrical tube; (V) the problem of the long, round cylindrical tube; (VI) the application of the flow equation in high-vacuum engineering; summary.

I. The Expressions of Knudsen, v. Smoluchowski, and Dushman for Steady-State Molecular Flow

Knudsen[1] has given the following expression for the steady flow of a highly rarefied gas through a tube of arbitrary cross section

$$J = \frac{8}{3}\left(\frac{2}{\pi}\right)^{\frac{1}{2}} \cdot \frac{S^2}{sL} \cdot \frac{1}{(d)^{\frac{1}{2}}}(p_1 - p_2), \qquad (1)$$

where J is the amount of gas flow per second, measured by the volume it would occupy at unit pressure, d is the density of the gas at unit pressure, S is the cross-sectional area of tube, s is the circumference of the cross section, and L is the length of the tube. p_1 and p_2 are the pressures in the two volumes connected by the tube, and we assume that the first is always to the left of the second (the indices 1 and 2 in this article refer exclusively to the two volumes).

For the derivation it is assumed that L is very large in comparison both to the transverse dimensions of the cross section of the tube as well as to the mean free path of the molecules, and further, that the molecules leave the wall of the tube diffusely in accordance with the cosine law.[2]

For a long circular cylindrical tube with radius r,

Eq. (1) gives

$$J = \frac{4(2\pi)^{\frac{1}{2}}}{3} \cdot \frac{r^3}{L} \cdot \frac{1}{(d)^{\frac{1}{2}}}(p_1 - p_2). \qquad (2)$$

For an orifice in a very thin wall, Knudsen[6] found

$$J = S/(2\pi)^{\frac{1}{2}} \cdot 1/d^{\frac{1}{2}}(p_1 - p_2), \qquad (3)$$

which gives for a circular opening

$$J = (\pi/2)^{\frac{1}{2}} \cdot r^2/d^{\frac{1}{2}}(p_1 - p_2). \qquad (4)$$

M. v. Smoluchowski[7] has shown that Eq. (1) is not correct and that it should be replaced with

$$J = \frac{1}{2(2\pi)^{\frac{1}{2}}} \cdot \frac{A}{L} \cdot \frac{1}{(d)^{\frac{1}{2}}}(p_1 - p_2),$$

where

$$A = \int_s \int_{-\pi/2}^{+\pi/2} \tfrac{1}{2}\rho^2 \cos\vartheta \, d\vartheta \, ds, \qquad (5)$$

in which ρ represents a chord of the cross section, which forms an angle ϑ with the normal at ds.[8] It is probably accidental that Eq. (2) is correct; for the circular cylinder the error of the Knudsen equation is removed.

As stated, Eqs. (2) and (5) apply only for relatively long tubes. When $L=0$ the result would be that $J=\infty$ in both cases instead of changing to Eq. (3).

Dushman[10] was the first to propose an expression which applies also to short tubes and which, for $L=0$, does not give $J=\infty$, but rather Eq. (3). (For practical reasons Dushman considered only circular cylindrical tubes and round orifices.)

A second expression has been derived by the author.[9,11,12] Although it gives the same results as the Dushman formula for $L/r=0$ and $L/r \approx \infty$ [Knudsen's Eqs. (2) and (4)], the two equations deviate considerably from each other in the intermediate region. In

First published in *Ann Physik* 1932; 12:961.
English translation published in *J Vac Sci Technol* 1971; 8:636.

the present paper the author wishes to show that the Dushman expression can only be regarded as a rough approximation, whereas the author's expression represents a rigorous solution to the problem of short tubes.

Dushman proceeds in the following manner. Equations (2) and (4) can be written

$$J = \frac{p_1 - p_2}{\omega'}; \quad \omega' = \frac{3}{4(2\pi)^{\frac{1}{2}}} \cdot \frac{L}{r^3} \cdot d^{\frac{1}{2}}, \tag{6}$$

$$J = \frac{p_1 - p_2}{\omega''}; \quad \omega'' = \left(\frac{2}{\pi}\right)^{\frac{1}{2}} \cdot \frac{1}{r^2} \cdot d^{\frac{1}{2}}. \tag{7}$$

The ω's are to be regarded as resistances to gas flow. "Hence where we have a tube of diameter $2r$ and length L connecting two vessels at low pressures, the total resistance to flow, of this tubing, is"

$$\omega = \omega' + \omega'', \tag{8}$$

and thus

$$J = (p_1 - p_2)/(\omega' + \omega''). \tag{9}$$

It will be shown in Sec. III that there is a physical basis for the addition of the two resistances; however, it requires no special proof that it is not permissible in the case of a short tube to use a formula which has been derived for $L \gg r$.

II. Transformation of the Equations to Kinetic Variables; Transmission Probability

Using the known kinetic expressions,

$$p = \frac{\pi}{8} \cdot nmu^2; \quad d = \frac{8}{\pi u^2}; \quad J = \frac{\pi}{8} \cdot mu^2 K, \tag{10}$$

and[13]

$$\nu = \tfrac{1}{4}nu, \tag{11}$$

where n is the number of molecules per cubic centimeter; m is the mass of a molecule; u is the mean molecular speed; K is the number of molecules per second entering the second volume; ν is the incidence rate which equals the number of gas molecules per second per cubic centimeter striking the wall, Eqs. (2)–(5) may be transformed into

$$K = (8r/3L) \cdot \pi r^2 \nu_1, \tag{12}$$

$$K = 1 \cdot S \nu_1, \tag{13}$$

$$K = 1 \cdot \pi r^2 \nu_1, \tag{14}$$

$$K = (A/2LS) \cdot S \nu_1. \tag{15}$$

For simplification we have assumed $p_2 = 0$. If $p_2 \neq 0$, it is only necessary to superimpose the two flows $(p_1 = p_1, p_2 = 0)$ and $(p_1 = 0, p_2 = p_2)$ in order to obtain the desired result.

We can now interpret Eqs. (12)–(15) as follows.

The second portion of the right-hand sides of these equations represents the number of molecules per second striking the left opening of the tube. The first portion of these terms gives, therefore, the fraction which indicates how many of the molecules entering from the left reach the second volume. We shall call this fraction the *transmission probability* and designate it by W. W is the probability that a molecule entering from the left reaches the right-hand volume without having returned to the left-hand volume. Equations (12)–(15) can thus be summarized by

$$K = W \cdot S \nu_1. \tag{16}$$

As a result of this interpretation, however, it is evident that this equation has a far more general range of validity than the four Eqs. (12)–(15), which refer only to infinitely long or infinitesimally short tubes. Equation (16) is valid for any connection of two volumes, and one may even define the entrance orifice of a given connecting tube at will within certain limits.

The problem now is to determine W, a quantity without dimensions, which depends only on the shape of the tube. This problem is therefore a purely geometric one.

If diffuse scatter[14] at the wall is assumed for all molecules, the problem will have a specific solution W for any tube. The physical proof is superfluous, unless it is desired to test whether the scatter is actually diffuse.[15] According to Gaede[16] and the author,[3] the second law of thermodynamics requires the validity of the cosine law for the angular distribution of molecules leaving a wall. If there is no reflection and only scatter, this must follow the cosine law and thus be diffuse. The experimental proof of the transmission probability represents, in this case, a test of the second law in the region of very low pressures.

The experimental determination of K from Eq. (16) amounts to a determination of ν_1, or, since we must assume the number of molecules per cubic centimeter to be known, to a measurement of the mean molecular speed u. This follows from the independence of Eq. (11) from the speed distribution. On the other hand, u can be calculated from the Maxwellian speed distribution of the molecules and the experimental determination of K is therefore nothing but a partial check of the speed distribution.

The author emphasized this argument earlier by publishing his experimental work on the transient molecular flow of argon and neon through a long circular tube under the title *A Measurement of Molecular Velocity and a Test of the Cosine Law*.[17]

If there is reflection, the principle of detailed balancing[18] requires that the number of emitted molecules be equal to the number of incident molecules

638   P. Clausing

for every direction and every speed. The emitted molecules in general will follow another law than the cosine law. This, for example, will be the case if specular reflection occurs primarily for glancing incidence. Stern and Estermann's[19] fine diffraction experiments with molecular beams are recalled, where a much more complicated reflection is present than in the case of specular reflection.

In the case of partial reflection, the transmission probabilities have quite different values than those previously determined. In general these new values of W can be calculated only with great difficulty. Only in the case of a probability ρ of specular reflection, where the probability is the same for all angles, can the new W values be easily determined as

$$W_\rho = [(1+\rho)/(1-\rho)] \cdot W, \qquad (17)$$

where the probability ρ represents the coefficient of reflection.[3,4,5,7] If, therefore, in a flow experiment a different value of K is found, this must be interpreted as follows. If the Maxwellian speed distribution is assumed, then reflection is present, which *in relationship to the flow through the tube* can be characterized quantitatively by a ρ value using Eq. (17).[20] For practical purposes this may perhaps be of importance; however, it is of no fundamental significance.

Therefore: *A flow experiment can only inform us whether any reflection at all is present; the experiment tells nothing about the quantitative nature of the reflection. Otherwise, the flow experiment gives us only known information.*

In Sec. IV we will calculate the transmission probability for a round tube of any length; Sec. V is concerned with the special case of a tube which is long in comparison with its radius.

III. Derivation of the Dushman Formula[22]

We shall now show how Dushman's Eq. (9) can be derived. For this we give the essentials of the Knudsen proof of Eq. (1) for a long tube. Using Eqs. (10) and (11), Eq. (1) can be transformed into[23]

$$K = (16S/3sL) \cdot S\nu_1. \qquad (18)$$

Knudsen attributes to the gas in the tube a certain velocity c which, however, is not considered constant for the whole length of the tube. In dt sec, $S\,c\,dt\,n$ molecules pass any particular cross section and one finds that $c = K/nS$, because this number is also equal to $K\,dt$.

For the impulse B, on a square centimeter of wall per second, he finds in a somewhat objectionable manner (which we shall ignore here)

$$B = (3\pi/32) \cdot nmuc. \qquad (19)$$

There is therefore conveyed per second upon an annular element of wall, sdx (x varies from zero to L along the tube)

$$(3\pi/32) \cdot nmu \cdot (K/nS) \cdot sdx, \qquad (20)$$

and upon the whole tube

$$(3\pi/32) \cdot muK \cdot (sL/S). \qquad (21)$$

This value is equated to the force $p_1S = (\pi/8)n_1mu^2S$ exerted from the left end of the tube, which gives Eq. (18).

This equation must, however, be subject to a small correction, which has not been taken into account by Knudsen.

The force from the left end of the tube equals the impulse which is exerted per second from the left end of the tube. This impulse is made up of two parts, one is the impulse which the $S\nu_1$ molecules coming into the tube give to the tube and the second is the impulse which the $(S\nu_1 - K)$ molecules returning from the tube into the first volume convey. Both impulses are directed toward the right. If K were equal to zero, then $2S\nu_1$ molecules with their impulse would have contributed to the force acting from the left on the tube; if it is not desired to neglect K vs $S\nu_1$, then this amount is only $2S\nu_1 - K$. It is therefore obvious that the force on the left acting on the tube is not p_1S but only

$$[(2S\nu_1 - K)/2S\nu_1] \cdot p_1S, \qquad (22)$$

directed to the right. Analogously, the force on the right is expressed by

$$(K/2S\nu_1) \cdot p_1S, \qquad (23)$$

directed to the left. Again equating the total force exerted on the tube to the value of Eq. (21), with the help of

$$p_1 = (\pi/8) \cdot n_1mu^2 = (\pi/2) \cdot \nu_1mn,$$

the following equation for the determination of K is found:

$$\frac{3\pi}{32} \cdot muK \cdot \frac{sL}{S} = \frac{2S\nu_1 - 2K}{2S\nu_1} \cdot \frac{\pi}{2} \cdot \nu_1muS. \qquad (24)$$

The solution of this equation is

$$K = \frac{16S/3s}{L + 16S/3s} \cdot S\nu_1. \qquad (25)$$

Dushman's Eq. (9) produces [with the help of $p_2 = 0$ and Eqs. (6), (7), (10) and (11)]

$$K = (W \cdot S\nu_1) = \frac{8r/3}{L + 8r/3} \cdot \pi r^2\nu_1, \qquad (26)$$

thus the same value for K as Eq. (25) with $S = \pi r^2$ and $s = 2\pi r$.

We see therefore that instead of Eqs. (1) or (18) for the long tube, Knudsen should have found Eq. (25). If the incorrectness of Eq. (1) and the correctness of Eq. (2) or (12) is taken into consideration, it becomes apparent that Eq. (25) also will be incorrect, but that Eq. (26), on the other hand, offers a close second approximation for the problem of the long (but not infinitely long) circular cylindrical tube.

The fact that Eq. (26) for $L/r = 0$ does not give $K = \infty$, but rather Eq. (14), naturally does not prove its validity for short tubes, but only the possibility of (false) application. The proof of Eq. (25) shows us how the total resistance of the tube should be divided into resistances of the tube itself and of the openings. Both openings (the entrance as well as the exit opening) have the same resistance $\omega''/2$; interposition of the resistance ω' of the long tube itself produces the total resistance ω [see Eqs. (6)–(8)]. Even if a physical reason is bestowed on the addition of the resistances the author still does not wish to state that he feels this to be physical. The idea of the resistance of an exit opening, where the molecules enter freely into the vacuum, remains for him inexplicable. The author therefore believes that the introduction of this concept into *the problem* of the short tube is not desirable.

On the other hand, it is obvious that the author, too, regards the idea of the flow resistance and the possibility of the addition of such resistances to be of great importance for vacuum engineering. In this respect it is, for example, very useful to speak of the resistance of a short tube. At the end of this paper we shall revert to this subject (see Sec. VI).

IV. The Problem of the Short Round Cylindrical Tube[24]

A plane perpendicular to the axis of the cylinder cuts a disk-shaped cross section from the tube. Two cross sections closely adjacent to each other determine a "ring" on the wall of the tube.

Definitions are as follows:

(1) $w_{rr}(v)dv$ is the probability that a molecule, which in accordance with the cosine law leaves a ring $2\pi r \delta v$, strikes directly (i.e., without first colliding with the wall again) another ring $2\pi r dv$ located at a distance v from the first ring.

(2) $w_{rs}(v)$ is the probability that a molecule leaving a ring $2\pi r \delta v$ passes directly through a cross section at the distance v.

(3) $w_{sr}(v)dv$ is the probability averaged over a cross section πr^2 that a molecule leaving the section strikes directly a ring $2\pi r dv$ at a distance v.

(4) $w_{ss}(v)$ is the probability, averaged over a cross section πr^2, that a molecule leaving the section strikes directly a second cross section at a distance v.

It is easily seen that

$$w_{rr}(v) = -[dw_{rs}(v)/dv], \qquad (27)$$

and

$$w_{sr}(v) = -[dw_{ss}(v)/dv]. \qquad (28)$$

If the tube is thought of as being uniformly filled with molecules, according to the cosine law as many molecules go from a ring to a cross section as from the cross section to the ring in a certain period. Therefore

$$2\pi r \delta v w_{rs}(v) = \pi r^2 w_{sr}(v)\delta v, \qquad (29)$$

or

$$w_{rs}(v) = (r/2)w_{sr}(v). \qquad (30)$$

If, for example, $w_{ss}(v)$ is known, the functions w_{sr}, w_{rs}, and w_{rr} can immediately be calculated with the aid of Eqs. (27)–(30). According to Walsh[25] the function $w_{ss}(v)$ can be derived in a very simple manner.

First Auxiliary Theorem

A sphere is thought of as being filled with a gas at such low pressure that the molecules do not collide with each other. The cosine law then gives an even distribution of all molecules over the surface of the sphere which, in accordance with this law, have left a surface element of the sphere. If one integrates over many surface elements, it follows that every part of a spherical surface irradiates the entire spherical surface (and therefore also any particular part of this surface) evenly with molecules.

Second Auxiliary Theorem

In a volume of any shape the cosine law for incidence on, and escape from, any area which one considers in the vessel corresponds with the cosine law for the escape of molecules from the wall.

If we now consider the two sections S_A and S_B of the tube, we can visualize a (gas-filled) spherical surface passing through the edges of the sections (see Fig. 1). The sections divide this surface into three spherical zones A, B, and C.

If we confine out attention to the molecules which proceed upwards through S_A, it is seen that in accor-

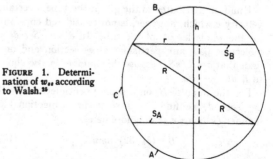

FIGURE 1. Determination of w_{ss} according to Walsh.[25]

640   P. Clausing

dance with the second auxiliary theorem they leave S_A in accordance with the cosine law. Further, these are the molecules going from the surface A to the surface B+C, and in accordance with the first auxiliary theorem these molecules fall upon B+C evenly, that is, they fall upon the two surfaces B and C in proportion to the surface areas $2\pi Rh$ and $2\pi Rv$. The fraction,

$$2\pi Rh/(2\pi Rh+2\pi Rv)=h/(h+v), \qquad (31)$$

of the molecules proceeding upward through S_A thus passes S_B, and Eq. (31) represents the probability $x_{ss}(v)$ being sought. A simple calculation from Fig. 1 gives

$$w_{ss}(v)=(1/2r^2)\{v^2-v(v^2+4r^2)^{\frac{1}{2}}+2r^2\}. \qquad (32)$$

According to Eqs. (28), (30), and (27) this gives

$$w_{sr}(v)dv=\frac{1}{2r^2}\left\{(v^2+4r^2)^{\frac{1}{2}}+\frac{v^2}{(v^2+4r^2)^{\frac{1}{2}}}-2v\right\}dv, \qquad (33)$$

$$w_{rs}(v)=\frac{1}{4r}\left\{(v^2+4r^2)^{\frac{1}{2}}+\frac{v^2}{(v^2+4r^2)^{\frac{1}{2}}}-2v\right\} \qquad (34)$$

and

$$w_{rr}(v)dv=\frac{1}{4r}\left\{2+\frac{v^3}{(v^2+4r^2)^{\frac{1}{2}}}-\frac{3v}{(v^2+4r^2)^{\frac{1}{2}}}\right\}dv. \qquad (35)$$

For $v>2r$ the following series applies[26]

$$w_{ss}(v)=\frac{1}{4}\left(\frac{2r}{v}\right)^2-\frac{1}{8}\left(\frac{2r}{v}\right)^4+\frac{5}{64}\left(\frac{2r}{v}\right)^6-\cdots, \qquad (36)$$

$w_{ss}(v)$, and therefore also $w_{sr}(v)$, $w_{rs}(v)$, and $w_{rr}(v)$ are functions which rapidly become smaller with an increase of v.

Up to now we have only considered the case $v>0$. For negative values of v the radicals of Eqs. (32)–(35) must be given a negative sign. Negative probabilities w_{rs} and w_{sr} will then indeed be encountered, but it will always turn out that these are probabilities which will have to be taken into account with the minus sign [see Eqs. (58) and (60)].

We now imagine an x axis running along the tube in such a way that x increases from zero to L from left to right, and we observe a molecule that is at the point x. The probability that the molecule will enter

the second volume without returning to the first we we shall call the escape probability and designate it $w(x)$. It is then clear that the transmission probability is determined by

$$W=\int_0^L w_{sr}(x)\cdot w(x)dx+w_{ss}(L), \qquad (37)$$

and $w(x)$ by the integral equation

$$w(x)=\int_0^L w_{rr}(\xi-x)d\xi\cdot w(\xi)+w_{rs}(L-x), \qquad (38)$$

or [using Eqs. (35) and (34)]

$$w(x)=\frac{1}{4r}\int_0^L\left\{2+\frac{(\xi-x)^3}{[(\xi-x)^2+4r^2]^{\frac{1}{2}}}\right.$$
$$\left.-\frac{3(\xi-x)}{[(\xi-x)^2+4r^2]^{\frac{1}{2}}}\right\}w(\xi)d\xi+\frac{1}{4r}\left\{[(L-x)^2+4r^2]^{\frac{1}{2}}\right.$$
$$\left.+\frac{(L-x)^2}{[(L-x)^2+4r^2]^{\frac{1}{2}}}-2(L-x)\right\}. \qquad (39)$$

From this integral equation two approximation solutions can be specified:

(1) For $r\ll L$, it is easy to show by substitution that for points of the tube at sufficient distance from the openings, $w(x)=x/L$ is a solution of Eq. (39). The substitution of $w(x)=x/L$ in Eq. (37), however, does not give the correct result $W=8r/3L$, but only $W=4r/3L$. Due to the rapid decrease of $w_{sr}(x)$, the integrand in Eq. (37) contributes to the value of the integral mainly for small values of x and in this case $w(x)=x/L$ is no solution.[27]

(2) If r/L is large (i.e., $r/L\geqslant 1$)

$$w(x)=\alpha+[(1-2\alpha)/L]\cdot x, \qquad (40)$$

(α is a constant) is a good approximation to the solution of Eq. (39) over the entire length of the tube. If Eq. (40) were the correct solution, the substitution of Eq. (40) in Eq. (39) would produce an equation, which upon solution for α would give a value independent of x. The actual substitution however gives

$$\alpha=\alpha\left(\frac{r}{L},\frac{x}{L}\right)=\alpha\left(\frac{r}{L},\frac{L-x}{L}\right)=\frac{\{(L-x)[(L-x)^2+4r^2]^{\frac{1}{2}}-(L-x)^2\}-\{x(x^2+4r^2)^{\frac{1}{2}}-x^2\}}{\dfrac{Lx^2-\{2x-L\}\{x^2+4r^2\}}{(x^2+4r^2)^{\frac{1}{2}}}-\dfrac{L(L-x)^2-\{2(L-x)-L\}\{(L-x)^2+4r^2\}}{[(L-x)^2+4r^2]^{\frac{1}{2}}}}. \qquad (41)$$

The remarkable thing about Eq. (41) is that, despite its dependence on x, for large values of r/L it is practically independent of x, as Table I demonstrates.

In order to obtain a definite result, an agreement should be reached on the choice of α. For example,

one might choose α averaged over the entire tube, or for every r/L take $\alpha=\alpha(r/L,0)$. For short tubes this selection is unimportant as is shown by Table I. If, however, it is considered important that the equation obtained by the substitution of Eqs. (33),

TABLE I.

$$\alpha\left(\frac{r}{L},\frac{x}{L}\right)=\alpha\left(\frac{r}{L},\frac{L-x}{L}\right)$$

x/L \\ r/L	10	2	1	0,25	0
0	0.4759091	0.3963046	0.3262380	0.1631190	r/L
0.2	0.4759167	0.3959445	0.3247135	0.1649489	2/25
0.4	0.4759141	0.3957635	0.3239510	0.1705989	3/25
0.5	0.4759143	0.3957406	0.3238563	0.171574	1/8

(40), and (32) in Eq. (37)

$$W=\frac{1-2\alpha}{3r^2L}\{4r^3+(L^2-2r^2)(L^2+4r^2)^{\frac{1}{2}}-L^3\}+\alpha$$
$$+\frac{1-\alpha}{2r^2}\{L^2-L(L^2+4r^2)^{\frac{1}{2}}+2r^2\},\quad(42)$$

gives the correct value $W=8r/3L$ also for long tubes, α should be chosen in such a way that Eq. (40) for low values of r/L approaches $w(x)=x/L$. The averaged α is completely unsuitable for this purpose, whereas $\alpha=\alpha(r/L,0)$ gives $\alpha=r/L\approx0$, and thus Eq. (40)$\equiv w(x)=x/L$. Using $\alpha=\alpha(r/L,0)$, Eq. (42) gives $W=7r/3L$ for small r/L, a result that is about 12% too small. The simplest good selection for α (one choice of many, however) is

$$\alpha=\alpha\left[\frac{r}{L},\frac{2r(7)^{\frac{1}{2}}}{3L+2r(7)^{\frac{1}{2}}}\right].\quad(43)$$

For very small r/L, then $\alpha=4r/3L\approx0$ and Eq. (42) gives $W=8r/3L$.

Thus we have in the expression

"Eq. (42) with α from Eq. (43)," (44)

a result which is correct for very long tubes and for very short tubes with arbitrary approximation. In Table II numerical values of Eq. (44) have been complied to which we have added for comparison the Dushman values calculated from Eq. (26).

To what extent Eq. (44) gives a good approximation cannot be stated. The author was able to convince himself that the formula is very strictly valid for tubes with $r/L\gtrless1$. This resulted from numerical calculation for the case when $r=L$, whereby the integral equation Eq. (39) was replaced by a system of 11 linear equations with 11 unknowns, a system which, thanks to the relationships $w(L/2)=\frac{1}{2}$, and $w(x)+w(L-x)=1$, can be reduced to five equations with five unknowns. The solution of this system was substituted in the integral from Eq. (37) which had been transformed into a sum and gave $W=0.6718$, a result which differs only slightly from the tabulated value of 0.6720.

Such a calculation is too complicated for smaller values of r/L and the author did only a graphical interpolation between the two extreme cases, (a) $\log W$

TABLE II.

L/r	W from author Eq. (44)	W from Dushman Eq. (26)	L/r	W from author Eq. (44)	W from Dushman Eq. (26)
0	1	1	3.2	0.4062	0.454
0.1	0.9524	0.965	3.4	0.3931	0.439
0.2	0.9092	0.931	3.6	0.3809	0.426
0.3	0.8699	0.899	3.8	0.3695	0.412
0.4	0.8341	0.870	4.0	0.3589	0.400
0.5	0.8013	0.842	5	0.3146	0.348
0.6	0.7711	0.816	6	0.2807	0.307
0.7	0.7434	0.792	7	0.2537	0.276
0.8	0.7177	0.769	8	0.2316	0.250
0.9	0.6940	0.747	9	0.2131	0.229
1.0	0.6720	0.727	10	0.1973	0.210
1.1	0.6514	0.708	12	0.1719	0.182
1.2	0.6320	0.690	14	0.1523	0.160
1.3	0.6139	0.672	16	0.1367	0.143
1.4	0.5970	0.656	18	0.1240	0.129
1.5	0.5810	0.640	20	0.1135	0.117
1.6	0.5659	0.625	30	0.0797	0.0817
1.7	0.5518	0.611	40	0.0613	0.0625
1.8	0.5384	0.597	50	0.0499	0.0506
1.9	0.5256	0.584	60	0.0420	0.0425
2.0	0.5136	0.572	70	0.0363	0.0367
2.2	0.4914	0.548	80	0.0319	0.0322
2.4	0.4711	0.526	90	0.0285	0.0288
2.6	0.4527	0.506	100	0.0258	0.0260
2.8	0.4359	0.488	1000	0.002658	0.002660
3.0	0.4205	0.470	∞	$8r/(3L)$	$8r/(3L)$

as a function of $\log(r/L)$ from Eq. (44) for $r/L\gtrless1$, and (b) $\log W$ as a function of $\log(r/L)$ from $W=8r/3L$. This interpolation gave values which differed less than 2% from the values in Table II, and the author believes therefore that the errors of the tabulated values are not greater than 2%.

The author therefore regards Eq. (44) as the best-known formula for the molecular flow through a round tube of any length.

Perhaps, particularly in the range between $(r/L\gtrless1)$ and $(r/L\ll1)$, an even better approximation could be obtained by such a selection of α, that Eq. (42) for $r\ll L$ gives

$$W=\frac{8r/3}{L+8r/3}$$

[see the remark about Eq. (26)]. It is easy to determine such a solution; but the calculation of W remains very complicated and the result of this calculation would be just as difficult to check as the result of Eq. (44).

A first application of this theory of molecular flow through a short tube is made in the author's work on beam formation in molecular flow.[28] The calculated spatial distribution of the molecules emanating from a short tube apparently has been verified by Ellett.[29]

Similarly to the calculation of the short round tube the author has solved the problem of molecular flow through an aperture with rectangular cross section. If a is the length of the aperture, b the width, and L the depth (that is, the length of the tube), then these magnitudes should satisfy the inequalities

$$b\ll a \quad\text{and}\quad L\ll a.\quad(45)$$

642   P. Clausing

Under this condition the following is found

$$W = \alpha\left[1 - \frac{b}{L}\ln\frac{L+(L^2+b^2)^{\frac{1}{2}}}{b}\right] - \frac{L}{2b} + \frac{(L^2+b^2)^{\frac{1}{2}}}{2b} + \frac{b}{2L}\ln\frac{L+(L^2+b^2)^{\frac{1}{2}}}{b},$$

for $b \leqq L$ the following is true

$$\alpha = \alpha\left[\frac{b}{L}, \frac{b\ln\left(\frac{L}{b}\right)}{2L+b\ln\left(\frac{L}{b}\right)}\right], \text{ where } \alpha\left(\frac{b}{L}, \frac{x}{L}\right) = \alpha\left(\frac{b}{L}, \frac{L-x}{L}\right)$$

$$= \frac{\{(L-x)-[(L-x)^2+b^2]^{\frac{1}{2}}\}-\{x-(x^2+b^2)^{\frac{1}{2}}\}}{\left\{2(L-x)-2[(L-x)^2+b^2]^{\frac{1}{2}}+\dfrac{L(L-x)}{[(L-x)^2+b^2]^{\frac{1}{2}}}\right\}-\left\{2x-2(x^2+b^2)^{\frac{1}{2}}+\dfrac{Lx}{(x^2+b^2)^{\frac{1}{2}}}\right\}}$$ (46)

for $b \geqq L$ the following is true

$$\alpha = \alpha\left(\frac{b}{L}, 0\right) = \frac{L+b-(L^2+b^2)^{\frac{1}{2}}}{2[L+b-(L^2+b^2)^{\frac{1}{2}}]+\dfrac{L^2}{(L^2+b^2)^{\frac{1}{2}}}}.$$

The numerical values are given in Table III.

It has again been shown by a numerical solution of the pertinent integral equation for $b = L$ that Eq. (46) for thin apertures offers a very good approximation to reality. This solution (of a system of 21 linear equations capable of being reduced to 10) gave $W = 0.6843$, thus in very good agreement with the value of $W = 0.6848$ given in Table III.

For thick apertures which satisfy not only the conditions of Eq. (45) but also the condition,

$$b \ll L,$$ (47)

the author derived[30] the formula

$$W = b/L(\tfrac{1}{2}+\ln L/b),$$ (48)

which for a very small b/L becomes

$$W = (b/L)\ln L/b.$$ (49)

TABLE III.

L/b	W from Eq. (46)	L/b	W from Eq. (46)	L/b	W from Eq. (46)
0	1	1.3	0.6321	3.2	0.4439
0.1	0.9525	1.4	0.6168	3.4	0.4318
0.2	0.9096	1.5	0.6024	3.6	0.4205
0.3	0.8710	1.6	0.5888	3.8	0.4099
0.4	0.8362	1.7	0.5760	4	0.3999
0.5	0.8048	1.8	0.5640	5	0.3582
0.6	0.7763	1.9	0.5525	6	0.3260
0.7	0.7503	2.0	0.5417	7	0.3001
0.8	0.7266	2.2	0.5215	8	0.2789
0.9	0.7049	2.4	0.5032	9	0.2610
1.0	0.6848	2.6	0.4865	10	0.2457
1.1	0.6660	2.8	0.4712		
1.2	0.6485	3.0	0.4570	∞	(b/L) ln (L/b)

V. The Problem of the Long, Round Cylindrical Tube

In Sec. IV it is shown that Eq. (39) has a solution for $w(x) = x/L$ for $r \ll L$ if one disregards the two ends of the tube. This solution, however, does not permit calculation of the transmission probability according to Eq. (37). In order to determine this probability, we shall introduce $g(x)$, the incidence rate of molecules on the wall of the tube. At the ends of the tube the $g(x)$ values, $g(0)$ and $g(L)$, connect in some manner with the incidence rates ν_1 and $\nu_2 = 0$ in the two volumes.

If we further define a *relative incidence rate* by means of the equation

$$h(x) = g(x)/\nu_1,$$ (50)

the identity

$$h(x) \equiv w(L-x),$$ (51)

can be proved.

It is easily shown that $g(x)$ satisfies the integral equation

$$2\pi r dx \cdot g(x) = \int_0^L 2\pi r d\xi \cdot g(\xi) \cdot w_{rr}(\xi-x)dx + \pi r^2 \nu_1 w_{sr}(x)dx.$$ (52)

However, the number of molecules per second leaving the ring $2\pi r dx$ is equal to the number which the ring receives from anywhere in the same time. From Eqs. (50) and (30) we obtain, in view of Eq. (52)

$$h(x) = \int_0^L w_{rr}(\xi-x) \cdot h(\xi)d\xi + w_{rs}(x).$$ (53)

Thus

$$h(L-x) = \int_0^L w_{rr}(\xi - (L-x)) \cdot h(\xi) d\xi$$
$$+ w_{rs}(L-x), \quad (54)$$

and with the aid of the substitution $\xi = L - \eta$, we find

$$h(L-x) = -\int_L^0 w_{rr}(x-\eta) \cdot h(L-\eta) d\eta$$
$$+ w_{rs}(L-x), \quad (55)$$

or

$$h(L-x) = \int_0^L w_{rr}(\eta-x) \cdot h(L-\eta) d\eta + w_{rs}(L-x). \quad (56)$$

If one compares Eq. (56) with Eq. (38), one sees that $h(L-x)$, regarded as a function of x, satisfies the same integral equation as $w(x)$, whereby Eq. (51) is proved.[31]

The solution $w(x) = x/L$ is therefore equivalent to

$$g(x) = (1-x/L)\nu_1. \quad (57)$$

We now consider an arbitrary cross section in a practically infinitely long tube and calculate how many more molecules per second pass this section to the right than to the left. From Eq. (16) we have already identified this excess with $K = WS\nu_1$, and a comparison with this K will thus give directly the W being sought. It is easily seen that this excess is determined by[32]

$$K = \pi r^2 \nu_1 w_{ss}(x) - \int_0^L 2\pi r d\xi \cdot g(\xi) w_{rs}(\xi-x). \quad (58)$$

The substitution of Eq. (57) in Eq. (58) and performance of the integration results in

$$K = \frac{8r}{3L}\left\{1 - \frac{3}{8}\cdot\frac{r}{x} - \frac{3}{8}\cdot\frac{r}{L-x}\right.$$
$$\left. + \frac{1}{4}\cdot\frac{r^3}{x^3} + \frac{1}{4}\cdot\frac{r^3}{(L-x)^3} + \cdots\right\}\cdot\pi r^2 \nu_1. \quad (59)$$

If it is remembered that r/x and $r/(L-x)$ are very small quantities when the observed section is thought of as being far removed from the ends of the tube, then the desired Eq. (12) independent of x is obtained from Eq. (59).

It is worthy of note that in the previous sections probabilities only were discussed, whereas here, for the problem of the long tube, the escape probability is replaced by the rate of incidence. This is not absolutely necessary. If one substitutes in Eq. (58) K from Eq. (16), $g(\xi)$ according to Eqs. (50) and (51) and $w_{rs}(\xi-x)$ from Eq. (30), the following is obtained[32]

$$W = -\int_0^L w_{sr}(\xi-x) \cdot w(L-\xi) d\xi + w_{ss}(x). \quad (60)$$

This equation, which is valid for both short and long tubes, independent of the selection of (x) is quite analogous to Eq. (37). But up to now the author has not been able to understand *a priori* this equation, as it was so easy for him to do for Eq. (37).

From the derivation of Eq. (60) it is clear, that the substitution of $w(x) = x/L$, the carrying out of the integration and the neglect of the powers of r/x and $r/(L-x)$ higher than the first, will again give the correct value $W = 8r/3L$.

The complete determination by probability calculation of W for a long round tube should therefore be made in the following manner: Equations (32)–(35) are determined, the integral Eq. (39) is written down, and it is shown that for the long tube $w(x) = x/L$ is a solution. Equation (60), which can be understood *a priori*, is written down, the value of $w(x) = x/L$ obtained is substituted, and by neglecting the small values r/x, etc., $W = 8r/3L$ is obtained.

VI. Application of the Flow Equations in High-Vacuum Engineering

One of the most important applications of the equations for molecular flow is certainly an evaluation of an apparatus for obtaining a high vacuum in technology. In the manufacturing of incandescent lamps, of radio receiving and transmitting tubes, of x-ray tubes, etc., the problem is always present of pumping out, as quickly as possible, the gas developed in the tube. It is therefore also understandable that almost all authors devote a short discussion to Knudsen molecular flow in their presentations of high-vacuum engineering devices. The well-known prescription is always stated: Choose the diameter of the tubes as large, but the lengths as short as possible. The flow resistance is proportional to the length, and inversely proportional to the cube of the diameter.

To this, little of essential importance can be added. The author would like only to elucidate this prescription with a table, from which the time required for the evacuation of a vessel through a given piping system can be extracted almost immediately.

We consider a vessel with volume V, which is connected through pipes to a high-vacuum pump. At the time $t=0$, the gas in the vessel has the pressure $p_1 = p_{10}$ and is removed by the pump. The instantaneous pressure p_1 satisfies the equation

$$-V dp_1 = J dt, \quad (61)$$

or from Eqs. (9) and (8)

$$-V dp_1 = \frac{p_1 - p_2}{\omega} dt, \quad (62)$$

if p_2 is the pressure above the pump. On the other

644   P. Clausing

hand, p_2 satisfies the equation

$$-Vdp_1 = Gp_2dt, \qquad (63)$$

where G represents the speed of the pump. The elimination of p_2 from Eqs. (62) and (63) produces

$$-Vdp_1 = \left[p_1 \Big/ \left(\frac{1}{G}+\omega\right)\right]dt, \qquad (64)$$

and the integration of this equation gives

$$t = V[(1/G)+\omega]\ln(p_{10}/p_1). \qquad (65)$$

The pipes which connect the vessel with the pump consist of various successively connected tubes with different diameters. Also, a valve inserted in the line is equivalent to a short tube. These various tubes represent various resistances $\omega', \omega'', \omega'''\ldots$. Therefore Eq. (65) can be thought to be written as follows:

$$t = V\left(\frac{1}{G}+\omega'+\omega''+\omega'''+\cdots\right)\ln(p_{10}/p_1). \quad (66)$$

As in the case of Eq. (65), Eq. (66) shows that the reciprocal speed of the pump can be regarded as a resistance.[33] We now imagine that the following series of experiments have been made.

(1) The vessel is connected directly to the pump without the interposition of any tube. As previously, it is evacuated from the pressure p_{10} to the pressure p_1. According to Eq. (65) the time necessary for this is given by

$$t_g = V\cdot(1/G)\ln(p_{10}/p_1). \qquad (67)$$

(2) The vessel is attached through a tube to an infinitely large volume, which is completely empty. Again the time for the reduction from p_{10} to p_1 is to be determined. For this time the following results:

$$t' = V\cdot\omega'\ln(p_{10}/p_1). \qquad (68)$$

(3) Experiments as under (2), but with second, third, and further tubes:

$$t'' = V\cdot\omega''\ln(p_{10}/p_1), \quad \text{etc.} \qquad (69)$$

Using Eqs. (67)–(69), Eq. (66) can thus be written in the form:

$$t = t_g+t'+t''+t'''+\cdots. \qquad (70)$$

In a somewhat arbitrary manner we now want to designate those values of t, t_g, t', t'', $\ldots$ which correspond to the relationship

$$p_{10}/p_1 = 10^3 \quad \text{or} \quad \ln(p_{10}/p_1) = 6.9078,$$

as *pumping time* or *partial pumping time*. Further, according to Eq. (67) we write

$$t_g = V\cdot(6.9078/G) = V\cdot\tau_g, \qquad (71)$$

and for the valves in the system we write, in accordance with Eq. (69), equations of the form

$$t^{(n)} = V\cdot\omega^{(n)}\cdot6.9078 = V\cdot\tau^{(n)}. \qquad (72)$$

The magnitudes τ are designated as *partial pumping times per unit volume*.

For the tubes of the system, which can be regarded almost always as "long tubes," with a resistance proportional to their length [see Eq. (6)], we define *partial pumping times σ per unit volume and per unit length* through the equations corresponding to Eq. (69)

$$t^{(m)} = V\cdot L^{(m)}\cdot(\omega^{(m)}/L^{(m)})\cdot6.9078 = V\cdot L^{(m)}\cdot\sigma^{(m)}. \quad (73)$$

It is now clear that, with the help of the formula

$$t = V(\tau_g+\sum\tau^{(n)}+\sum L^{(m)}\cdot\sigma^{(m)}), \qquad (74)$$

the pumping time for every tube system can be easily calculated, if a table containing the values τ and σ for the usual valves and tubes is available. Table IV is such a compilation assuming that the gas to be evacuated is nitrogen at 18°C. To use this table, the lengths $L^{(m)}$ and the volume V must be expressed in

TABLE IV. D—diameter of the tube (mm); σ—partial pumping time for nitrogen at 18°C (sec/liter/m).

D	σ	D	σ	D	σ
0.2	7032000	4.5	617	16	13.7
0.4	879000	5.0	450	17	11.5
0.6	260500	5.5	338	18	9.65
0.8	109900	6.0	260	19	8.20
1.0	56260	6.5	205	20	7.03
1.2	32560	7.0	164	22	5.28
1.4	20500	7.5	133	24	4.07
1.6	13730	8.0	110	26	3.20
1.8	9646	8.5	91.6	28	2.56
2.0	7032	9.0	77.2	30	2.08
2.2	5283	9.5	65.6	32	1.72
2.4	4070	10	56.3	34	1.43
2.6	3201	11	42.3	36	1.21
2.8	2563	12	32.6	38	1.04
3.0	2084	13	25.6	40	0.88
3.5	1312	14	20.5		
4.0	879	15	16.7		

D—diameter of the stopcock bore (mm); L—depth of the stopcock bore (mm); τ—partial pumping time for nitrogen at 18°C (sec/liter).

Description of the stopcock	D	L	τ
No. 106, 3-way cock, 1 hole in hollow tap	6	1.5	2.60
No. 106, 2-way cock, 2 holes in hollow tap	6	2×1.5	5.20
No. 107, 2-way cock, 2 holes in hollow tap	10	2×2	1.80
No. 110, 3-way cock, 1 hole in hollow tap	5	1.5	3.89
No. 111, 2-way cock	2	9	88.0
No. 117, 2-way cock	6	15	6.62

G—speed of the pump (liters/sec); τ_g—partial pumping time (sec/liter).

Pump	G	τ_g
Langmuir metal condensation pump[34]	3	2.3
Newer metal diffusion pump[35]	5.5	1.3
Gaede three-stage pump[35]	15	0.46
Kurth–Ruggles glass pump[35]	3.5	2.0
Payne two-stage pump[35]	60	0.11
Gaede–Keesom large diffusion pump[36]		
(a) for air	130	0.053
(b) for helium	420	0.0164

meters and liters, respectively, in Eq. (74) to obtain the pumping time in seconds.

The magnitude of σ used in the table is calculated from

$$\sigma = 3(6.9078/2\pi r^3 u)\cdot 10^5, \qquad (75)$$

which results from Eq. (73), the second Eq. (6), and the second Eq. (10), multiplying by 10^5 to convert from cgs units to liters and meters. It is well known that the mean molecular velocity is given by

$$u = (8RT/\pi M)^{\frac{1}{2}} = 14551(T/M)^{\frac{1}{2}}\ \text{cm}\cdot\text{sec}^{-1}, \qquad (76)$$

(where R is the gas constant, T the absolute temperature, and M the gram molecular weight).

The τ of the table is determined by the formula

$$\tau = 4(6.9078/WSu)\cdot 10^3, \qquad (77)$$

which is derived from Eq. (72), the first equation of Eq. (6), $p_2=0$, Eqs. (10), (11), and (16), by multiplying by 10^3 in order to convert from cgs units to liters. W is taken from the second column of Table II. Example: A 3-liter vessel is to be pumped out through a system composed of a tube of 0.8-m length and 17-mm diameter, a No. 117 stopcock, and a tube of 0.5-m length and 11-mm inside diameter. The pump speed is 3 liters/sec. From Eq. (74) and Table IV the pumping time is calculated as $t = 3(2.3+6.6+0.8 \times 11.5+0.5\times 42.3) = 118$ sec.

It is clear from Eqs. (75)–(77) how the pumping times for gases other than nitrogen are obtained by multiplying by $(M/M_{\text{nitrogen}})^{\frac{1}{2}}$. For hydrogen, for example, the pumping times are about four times smaller. For gas mixtures (such as air), the pumping times for the individual components must each be considered separately for an exact calculation; partial separation occurs.

Dependence of the pumping speed on the nature of the gas has not been in general taken into consideration here.

The times required to reduce the pressure by a ratio other than 1000 can be easily determined. Equation (65) shows that these times are proportional to the logarithm of this ratio; the pumping time must therefore be multiplied by $q/3$, when the ratio is 10^q.

Naturally the calculated pumping times are only correct for pressures which lie entirely in a range for which the condition "free path length$\gg$the diameter of the tube" prevails. For example, it requires no further proof that it will take a long time before this condition is achieved in the case of a long capillary tube used to evacuate a large, gas-filled vessel. For orientation we give in Table V the mean free path lengths λ of various gases at 25°C and 1 bar $(7.5\times 10^{-4}$ Torr) and recall that λ is inversely proportional to the pressure.

The intention of this calculation of the pumping

TABLE V.[a]

Gas	λ (cm)
H_2	19.2
He	29.6
N_2	10.0
O_2	10.7
A	10.6
CO	9.92
CO_2	6.68

[a] From Dushman's book,[10] p. 237.

time is merely to provide such a simple method for the approximate evaluation of a high-vacuum system, that the quantitative evaluation of the pumping speed of this system can also be entrusted to persons not trained in physics.

Finally, a second possibility of application of Table IV should be mentioned which, in line with what was stated at the beginning of this paragraph, has very great significance for high-vacuum engineering.

In a vessel there is considered to be a *continuous evolution of gas*, for example, through outgassing of electrodes. If this is not pumped out, then the pressure p_1 rises every second by the increment Δ. What pumping time must be selected so that p_1 during the pumping maintains a prescribed small value.

In agreement with Eq. (64) the following is true

$$V\cdot\Delta = \frac{p_1}{1/G+\omega}, \qquad (78)$$

or using Eq. (65) and $p_{10}/p_1 = 10^3$,

$$p_1/\Delta = t/6.9078 \approx t/7. \qquad (79)$$

If the rate of gas development is known, in accordance with Eqs. (79) and (74), it can be found from the table how the pumping equipment is to be designed in order to obtain the desired low pressure p_1 in the vessel.

It is clear that for modern high-vacuum pumps and the low pressures which can be achieved with them, this second application of Table IV produces very reliable results for practical purposes.

Summary

Knudsen and v. Smoluchowski have derived expressions for the steady-state flow of highly rarefied gases through cylindrical tubes with a length long in comparison to the transverse dimensions. Dushman has presented a similar expression for round tubes of any length. In this paper it is shown that the Dushman expression, which can be made very plausible for long tubes (Sec. III), may only be regarded, at most, as a rough approximation for short tubes, and that the rigorous solution for a tube of any length results from an integral equation, which refers to geometrical probabilities. For the case in which the radius of the tube r is larger than the length L or smaller but of the same magnitude as L ($r \gtrless L$), and for the case where r is very small in comparison to L, solutions of the integral equation can be found. The solution for $r \ll L$ (Secs. IV and V) leads to the Knudsen formula for the long round tube, while the solution for $r \gtrsim L$ (Sec. IV) gives such a good approximation for short tubes that the accuracy is greater than 1 part in a thousand even for the case of $r = L$, as has been indicated by a numerical solution of the integral equation for this case.

In Sec. II the common expressions for molecular flow are transformed to kinetic variables and in this way take on the quite general form

$$K = WS\nu_1.$$

K is the number of molecules per second going from a first vessel through the observed tube to a second vessel; S is the surface area of the cross section of the tube; and ν_1 is the number of molecules which in one second strike 1 cm² of the wall of the first vessel. The corresponding number ν_2 for the second vessel is thought of as being zero. The dimensionless magnitude W therefore represents nothing other than the transmission probability. W has been calculated in this article for short and long round tubes. For long, thin apertures and for long, thick apertures only, the corresponding results are given (Sec. IV).

The physical significance of molecular flow experiments is particularly emphasized in Sec. II. If a Maxwellian velocity distribution of the molecules is assumed, the flow experiment only informs us whether any reflection is present or not; the experiment does not tell us anything quantitative about this reflection.

In Sec. VI a table is given for practical use, which enables, in a very simple manner, a quantitative evaluation of the pump speed for any high-vacuum pumping system. For the case of continuous gas evolution in a vessel connected to the pump, the steady-state final pressure in the vessel can be easily determined with the aid of this table when the rate of gas evolution and the design of the piping between vessel and pump are given.

References

1. M. Knudsen, Ann. Physik (4)**28**, 75 (1909).
2. For the significance of the cosine law, see Ref. 3; for a recent test of this law see Refs. 4 and 5.
3. P. Clausing, Ann. Physik (5)**4**, 533 (1930).
4. P. Clausing, Ann. Physik (5)**7**, 489 (1930).
5. P. Clausing, Ann. Physik (5)**7**, 569 (1930).
6. M. Knudsen, Ann. Physik (4)**28**, 999 (1909).
7. M.v. Smoluchowski, Ann. Physik (4)**33**, 1559 (1910).
8. For the special case of Eq. (5) calculated to date, see M.v. Smoluchowski, Ref. 7 (tube with rectangular cross section) and the author, Ref. 9: round tube with eccentric round axis (Sec. 3, c_2); round tube with coaxial round axis (Sec. 3, c_3).
9. P. Clausing, Physica **9**, 65 (1929).
10. S. Dushman, "Production and Measurement of High Vacuum," p. 32, Schenectady (1922); Int. Crit. Tabl., Vol. I, p. 91, New York (1926); J. Franklin Inst. **211**, 689 (1931).
11. P. Clausing, Verslag. Amsterdam **35**, 1023 (1926).
12. P. Clausing, dissertation in Leyden, Amsterdam (1928).
13. For a simple proof of this equation without any integration, see Ref. 12, p. 156, note 2.
14. By the word *reflection* we mean that type of *rebounding* from the wall where for each individual molecule the characteristics of emergence from the wall are dependent on the characteristic of incidence. The opposite case we call scatter. Scatter, according to the cosine law, will be called *diffuse*. Dispersion includes *adsorption* and *emission*. (See Sec. 1 of Ref. 3.)
15. It cannot be so proved that the scatter is diffuse. It can be easily seen that there are an infinite number of scattering laws, which for a given tube lead to the same W as the cosine law of scattering. Only agreement of the results from many flow experiments with tubes of greatly different form (not solely with a number of long round tubes with different r/L ratios) with the results of calculations according to the cosine law for these tubes would make another scattering law highly improbable. This remark is more one of principle than a practical one, but it does indeed demonstrate the inadequacy of the proof of the cosine law in previous flow experiments.
16. W. Gaede, Ann. Physik (4)**41**, 289 (1913).
17. See Ref. 5. It is only of minor concern here that in Ref. 5 not K but rather a mean transit time $\bar{t} = L^2/4ru$ plays the principal role. This is because the author observed a nonsteady flow while all other experimenters observed flows which were "steady for short periods of time."
18. See Sec. 1 of Ref. 3.
19. O. Stern, Naturwiss. **17**, 1391 (1929); I. Estermann and O. Stern, Z. Physik **61**, 95 (1930).
20. W. Gaede (Ref. 16) even found a negative value, $\rho = -z$. He considered the magnitude z to be the probability that a molecule striking the wall rebounds in the incident direction. (See also Sec. 8 of Ref. 3, and Ref. 21.)
21. P. Clausing, Ann. Physik (5)**4**, 567 (1930).
22. See Sec. 22, Ref. 12.
23. It is noteworthy that Eq. (18) is also obtained if the flow is regarded as diffusive with diffusion constant $(1/3)\lambda u$. λ is here the mean free path of the molecules in the tube, which according to R. Clausius (Die mechanische Wärmetheorie, Vol. 3, Sec. II, Braunschweig 1889–1891) equals $4\,S/s$. No theoretical value, however, can be ascribed to this finding.
24. See Refs. 9 and 11. Appendix III of Ref. 12 gives many details.
25. J. W. T. Walsh, Proc. Phys. Soc. (London) **32**, 59 (1919/1920); Phil. Mag. (7)**7**, 1092 (1929). Walsh actually treats a light radiation problem which, by virtue of the cosine law applying to it, coincides in the mathematical description completely with our problem.
26. See Eq. (24), Ref. 9, which unfortunately contains two erroneous numerical factors.
27. How the correct result $W = 8r/3L$ can be obtained will be shown in Sec. V.
28. P. Clausing, Z. Physik **66**, 471 (1930).
29. A. Ellett, Phys. Rev. (2)**37**, 1699 (1931).
30. Section 29 of Ref. 12.
31. The author has shown in Sec. 26 of Ref. 12 that Eq. (51) is valid for tubes of any shape.
32. For negative values of $(\xi - x)$ see the second note following Eq. (36).
33. This interpretation of the reciprocal of the pump speed can be found on p. 40 of Dushman's book (Ref. 10).
34. See Dushman's book, p. 66, Ref. 10.
35. S. Dushman, J. Franklin Inst. **211**, 689 (1931).
36. W. Gaede and W. H. Keesom, Comm. Leiden No. 195, 1 (1929).

HIGH-VACUUM GAUGES

by F. M. PENNING.

Summary. The various principles for measuring high vacua are briefly discussed. Some new designs of gauge are described in detail, particularly those which are based on the characteristics of electric gas discharges.

Definition of Gas Pressure; Units

The pressure of a gas is defined as the force applied by the impact of the gaseous molecules against unit surface of the walls of the enclosure containing the gas. The C.G.S. unit in which gas pressures are measured is 1 dyne per sq cm and is termed a microbar [1]). Low pressures are usually expressed in units of millimetres of mercury, a pressure of 1 mm. of Hg being termed the Torricelli or sometimes the Tor. The relationship between these units is:

1 mm Hg = 1 Tor = 1333 microbar = 1333 dynes per sq cm = 1/760 atmosphere.

In this paper, discussion will be confined to gas pressures below one atmosphere, and particularly to those below 1 mm Hg.

Principles of Pressure Measurements

Apart from a direct measurement of the gas pressure, every other property of a gas which varies with the pressure can theoretically be employed as a measure of that pressure. The associated properties which can be used for this purpose are:

1. Gas pressure,
2. Thermal conductivity of the gas,
3. Internal friction or frictional drag,
4. Radiometer effect,
5. Dependent gas discharge, and
6. Independent gas discharge.

Where high-vacuum gauges based on these principles have already been described in the literature (see bibliography), brief mention only will be made to them in the present article; while a more detailed description will be given of any individual new types which have been devised.

Gauges depending on a measurement of the Gas Pressure per se

Manometers or pressure gauges based on this principle obtain a pressure reading by measuring the displacement of the surface of a liquid (liquid

gauges) or of an elastic (mechanical gauges). To magnify the sensitivity of measurement, the pressure can be multiplied before measurement, viz, by compression in a known ratio (Mac Leod gauge). The surface, which is displaced under the action of the pressure, need not form part of the wall of the actual enclosure, but can be freely suspended. This method is adopted in the

Molecular-Jet gauge (fig. 1), which was

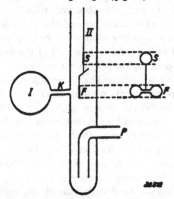

Fig. 1. Molecular-jet gauge for measuring vapour pressures (diagrammatic). The space II is connected to a pump at P; the vapour pressure of the substance in I deflects the vane F from its equilibrium position.

designed by Mayer for determining very low vapour pressures [2]). The substance whose vapour pressure is to be determined is contained in the bulb I, which is connected to the tubular vessel II through a capillary; II is highly evacuated through the wide connecting tube P. In vessel II a small, very light vane with two aluminium discs, 0.01 mm thick and 4 mm in diameter, is suspended; one of the discs is situated directly in front of the opening of K. The stream of molecules issuing from K displaces the small vane from its position of equilibrium, the magnitude of the angle of deflection being measured by means of a small mirror S. In the gauge constructed, this angle of deflection was found to be proportional to the pressure below a limit of 10^{-3} mm. The gauge must be calibrated with a substance of known vapour pressure.

[1]) A considerable divergence of opinion is found in the literature in regard to the units bar and microbar (barye). In Germany and the United States the C.G.S. unit has sometimes been termed the bar. At present the universal definition of these units is: 1 bar = 10^6 dynes per sq cm = 0.99 atmosphere and 1 barye = 1 microbar = 1 dyne per sq cm.

[2]) H. Mayer, Z. Physik, 67, 240, 1931.

Editor's note: Penning published his first paper on the Penning (or Phillips) gauge in German in Physica 1937; 4:71. The paper reproduced here contains the first description in English of the Penning gauge, or as it is called here, the Phillips vacuum-meter; it appeared in Philips Tech Rev 1937; 2:201.

Mayer obtained a sensitivity of 10^{-6} mm Hg per mm deflection. A serious drawback of this gauge is that it is very susceptible to tremor, and it must, therefore, be carefully set up to protect it against extraneous vibrations. By virtue of its construction this gauge cannot be used for measuring constant gas pressures.

Gauges depending on the Variation of Thermal Conductivity with Pressure

One of the most remarkable achievements of the kinetic theory of gases was to account for the fact that the thermal conductivity of a gas is independent of its pressure. From this it appears impossible to construct a pressure gauge depending for its action on variations in the thermal conductivity. But this conductivity is independent of the pressure only within a narrow range of pressures, for at high pressures the transmission of heat through a gas is a function of the pressure, since in addition to conduction convection of heat also takes place. Theoretically, therefore, a thermal-conductivity gauge is feasible within this range of pressures; a gauge of this type has indeed been described.

At low pressures, the thermal conductivity is independent of the pressure only for as long as the free path λ of the gaseous molecules is small compared with the distance d between the hot and the cold masses ($\lambda \ll d$). It is evident that at zero pressure ($\lambda \gg d$) the thermal conductivity also must be zero and that there must hence be a transition region in which the thermal conductivity varies with the pressure. If $d = 1$ cm, then $\lambda = d$ for a pressure of approximately 0.01 mm Hg and a transition stage becomes feasible between 10^{-1} and 10^{-3} mm Hg.

In this transition region the pressure can be measured as follows: A metal radiator to which a constant quantity of heat is supplied per second is fixed along the centre of a glass tube. The heat loss to the glass wall determines the temperature of the metal, which is hence a measure of the pressure. Gauges of this type require calibration before use, e.g. against a MacLeod gauge. The more common forms of this gauge are:

1) The Pirani Gauge.

In this gauge, a thin wire which is heated electrically and whose resistance is a measure of the temperature is generally used as a metal radiator (*fig. 2a*).

2) The thermo-couple gauge.

In addition to the resistance, other properties

of a metal, such as the thermo-electric force, can also be employed for measuring the temperature. In this case a heated wire is fitted with a thermo-couple (*fig. 2b*), and instead of passing an electric

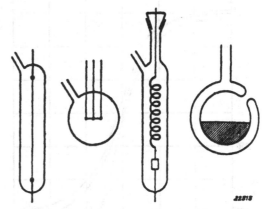

Fig. 2. Principle of the thermal-conductivity gauge.
  a) Resistance (Pirani)
  b) Thermo-couple (Voege and Rohn).
  c) Bimetallic (Klumb and Haase)
  d) Dewar flask (Herzog and Scherrer).

current through the wire, heat is supplied to it by irradiation of the thermo-couple with a light-source of constant intensity. In both arrangements the temperature to which the thermo-couple is raised is a measure of the pressure.

3) Bimetallic manometer.

Recently Klumb and Haase[4] made use of the change in shape of a bimetallic strip (straight strip or helix) to measure the temperature. The application of the principle of a bimetallic helix to the construction of a gauge is shown in *fig. 2c*. The helix is here again heated by the electric current; the rise in temperature causes the small mirror attached to the helix to turn through a specific angle, such rotation being compensated by a torsion head. The number of degrees which the torsion head has to be turned is a measure of the pressure. Results obtained with this gauge for a number of gases are shown in *fig. 3*. The time taken to obtain a reading with the gauge is comparatively long, being about 1 minute.

4) Thermal-conductivity gauge depending on the Rate of Volatilisation of Solid or Liquid Substances.

A gauge constructed on this principle has been described by Herzog and Scherrer[5]. The vessel in which the pressure has to be measured communicates with the space between the walls of a

[3]) Th. Haase, G. Klages and H. Klumb, Phys. Z., 37, 441, 1936.

[4]) H. Klumb and Th. Haase, Phys. Z., 37, 27, 1936.
[5]) G. Herzog and P. Scherrer, Helv. Phys. Acta, 6, 277, 1933.

double-walled sphere which is filled with solid carbon dioxide (*fig. 2d*). The sphere is thus a type

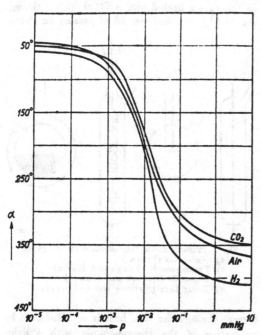

Fig. 3. Angle α through which the torsion head of the bimetallic manometer must be turned to compensate the rotation of the mirror.

of Dewar flask in which the carbon dioxide volatilises the slower the lower the pressure of the gas between the walls of the vessel, so that the rate of volatilisation of the carbon dioxide is a direct measure of the pressure. This gauge is not intended for accurate pressure measurements.

Gauges depending on the variation of Internal Friction with Pressure

Similarly to the thermal conductivity, the internal friction of a gas is also independent of the pressure as long as $\lambda \ll d$, but over a specific range of pressures this property also can be used for measuring high

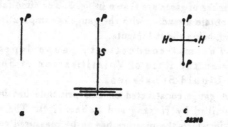

Fig. 4. Principles of the damping-type gauges.
a) Oscillating quartz fibre.
b) Rotatable disc with small mirror S.
c) Rotatable quartz cross HH; the surface of H is used as a convex mirror. The point of suspension is marked P in each case.

vacua. Gauges operating on this principle again require special calibration, while the quantitative readings obtained depend on the type of gas under measurement. There are two types of this gauge, viz:
1) Damping gauges, in which the damping of linear or circular oscillations is measured. Of the three forms which have been devised, the first has a thin fibre of glass or other material which is set in vibration by an impulse (*fig. 4a*), while in the second and third forms circular oscillations are imparted by small magnets to an oscillator, either a disc (*fig. 4b*) or a small quartz cross (*fig. 4c*), suspended by a fibre; in the latter case one of the small spheres on the cross acts as a mirror. In all three forms the damping of the motion is a measure of the pressure.
2) Molecular gauge of Langmuir and Dushman in which a rapidly-rotating aluminium disc A applies a torque to a thin mica disc M (*fig. 5*), this torque being directly proportional to the pressure of gas in the gauge.

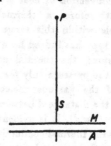

Fig. 5. Molecular gauge of Langmuir and Dushman. A - Rapidly-rotating aluminium disc; M - mica disc, whose deflection from the equilibrium position is read off by means of the small mirror S.

Gauges depending on the Radiometer Effect

These gauges are based on the principle that in very highly-rarefied gases (free path ≫ dimensions of the containing vessel) two surfaces at different temperatures exert a mutual mechanical force. The rotation of the vanes in the Crookes' radiometer also is due to this force. Consider a flat plate B (*fig. 6*) at the absolute temperature T

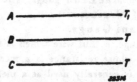

Fig. 6. Mechanical force applied to a plate B at the temperature T situated between a second plate at the same temperature and a third plate at a higher temperature T_1. Owing to the higher velocity of the molecules arriving from above, the pressure applied to the upper side of B is greater than that applied to the under side.

which is located between two other plates A and C at temperatures of T_1 and T respectively. At equilibrium the number of molecules striking against each sq cm of the three plates will be the same for each plate, viz. n. At the low gas pressures in question here nearly every molecule which strikes against B, will have registered its last impact at the surface of the opposite plate and will therefore be at the temperature of this plate. Molecules at a temperature of T_1 thus strike against the upper surface of B and molecules at a temperature of T against the lower surface; the molecules arriving from B are also at a temperature of T. The force applied to B per sq cm is, according to the kinetic theory of gases, $nm(v_a + v_b)$, where m is the molecular weight of the gas in question, v_a and v_b the average velocity components perpendicular to the surface with which the gaseous molecules either arrive at or leave the surface respectively, and n is the number of collisions per sq cm per sec. Since v is proportional to $\sqrt{T}$, i.e. $v = c\sqrt{T}$, the difference in pressure p between the top and bottom surface of B is:

$$\Delta p = mnc\,(\sqrt{T_1} - \sqrt{T}).$$

If the external wall of the enclosure is at the temperature T the pressure in the rest of the enclosure will be

$$p = 2\,mnc\,\sqrt{T}$$

and hence

$$\Delta p = \frac{p}{2}\left(\frac{\sqrt{T_1}}{\sqrt{T}} - 1\right).$$

The resultant force is, therefore, independent of the type of gas in the gauge.

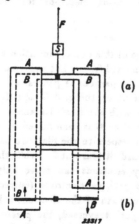

Fig. 7. Absolute gauge of K n u d s e n. Front view (*a*) and plan (*b*). The fixed plates are maintained at a higher temperature than the rest of the enclosure and thus apply a force on the vanes B in the direction of the arrow. The rotation of B is read off by means of the small mirror S.

1) Absolute gauge of Knudsen.

The absolute gauge of K n u d s e n consists of the type in fig. 6, but with a glass wall in place of the plate C. The two strips B (*fig.* 7) have been combined to form a small frame which is suspended by a fibre F. The fixed plates A are electrically heated to a temperature T_1, which is higher than the rest of the enclosure, and thus apply a pressure on B in the direction of arrow; the rotation of B which is read off by means of a small mirror attached to F, is a measure of the pressure. The absolute value of the pressure is obtained from the dimensions and the period of oscillation, etc., of the system using the equation given above. The K n u d s e n gauge is the only gauge which always give a direct reading of the absolute pressure of a gas; this is only possible with the M a c L e o d gauge when the gases and vapours present obey B o y l e's law.

A short time ago D u M o n d and P i c k e l s [6] described a new form of this gauge, which is suitable for general purposes. In this design a scale is provided on the gauge, which permits a direct reading of the pressure between 10^{-4} and 10^{-6} mm Hg. This particular gauge also requires calibration, and still shares the same disadvantages as the original Knudsen m a n o m e t e r in that it needs careful handling, is difficult to make and can only be used when suitably protected against vibration.

2) Aluminium-Foil gauge of Knudsen.

This gauge is based still more closely on the principle illustrated in fig. 6. A vertically-suspended aluminium foil of the type used in electroscopes here serves as the plate B, its deflection from the equilibrium position being a measure of the pressure.

3) The Molecular vacuum-meter of Gaede [7]

A new gauge of this name and based on the principle under discussion here was recently placed on the market. Two strips HH mounted along the glass wall (*fig.* 8) which are raised to a suitable temperature by electrical heating elements here act as the hot plates. A small frame with two thin strips BB is suspended from a fibre. The equilibrium position which BB assumes under the action of the gas pressure lies in the plane EE and is made to coincide with the elastic equilibrium position. To displace the frame BB, e.g. to the position shown in fig. 8, a specific force which is directly proportional to the pressure must be applied to it. This deflection is obtained by passing a current i through

[6] J. W. M. D u M o n d and W. M. P i c k e l s, Rev. Sci. Instr., 6, 362, 1935.

[7] W. G a e d e, Z. techn. Phys., 15, 664, 1934.

the coils *S*, thus causing a rotation of the small magnet *M* suspended from the fibre. The magnitude of *i* for a particular position of *BB* is, therefore, a measure of the pressure, which latter is read

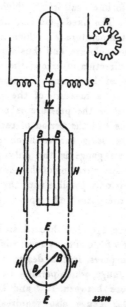

Fig. 8. Molecular vacuum-meter of G a e d e. The strips *H* on the glass wall are raised to a temperature higher than the rest of the enclosure by means of heating elements placed in contact with them. A thermal directive force is applied to the small frame *BB* provided it is not kept in its equilibrium position *EE* by an elastic opposing force. By means of the small magnet *M* and the current *i* in the coils *S*, the frame is brought back to its initial position. *i* is proportional to the pressure in the tube. The magnitude of *i* which is read off from the adjustment of the variable resistance is a measure of the pressure.

off directly on the control knob of the variable *R*. The position of *BB* is adjusted by means of the pointer *W*. This gauge is calibrated empirically, although the deflection is independent of the type of gas used. Since the "thermal" directive force determines the period of oscillation of the system as well, the latter can also be taken as a measure of the pressure; the relative accuracy of this method is stated to be one per cent. The gauge can be used as a damping manometer also (see above), in which case readings are governed by the internal friction, i.e. to a first approximation by the molecular weight of the gas present. By combining the two methods, the molecular weight can be determined in addition to the pressure.

Gauges based on a Dependent Gas Discharge

In the case of a dependent gas discharge the primary carriers of electricity are not produced by the discharge itself, but derived from an external source, e.g. from a hot wire or by an actinic action.

If the electrons, ions or radiation quanta of sufficient energy generated in this way pass through a gas the latter will be ionised; if the pressure of the gas is sufficiently low the number of ions and electrons produced will be proportional to the pressure and the path traversed by the ionised particles. *Fig. 9* shows the number of ionisations *N*

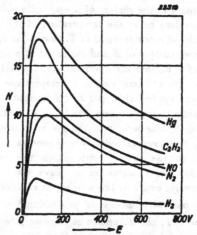

Fig. 9. Ionisation by electrons plotted against the accelerating potential *E* in volts. *N* = number of ionisation reactions per cm of path and per mm pressure.

per cm of path and per mm of pressure of the electrons in a variety of gases as a function of the electron energy. It is seen that for most gases *N* is a maximum at an energy of the order of 100 electron-volts, and that therefore for measuring low pressures a mean electron energy of this magnitude should be used. Since a much smaller number of ionisation reactions is produced by positive ions and radiation quanta of moderate energy, the importance of using electrons for ionisation is evident.

Ionisation Gauge.

In the gauge of this name, ionisation is produced by electrons emitted from a hot tungsten filament *F* (*fig. 10*). The apparatus is generally in the form of a triode, in which the grid *G* and the anode *A* are concentric with the heated filament *F*. One of the electrodes *A* or *G* is maintained at a negative potential with respect to the filament, so that ions alone flow towards it (current i_+) while the primary and secondary electrons can move only towards the third and positive electrode (current i_-). As long as the free path of the electrons is large compared to the path traversed, the pressure is then proportional to i_+/i_-. If i_- is made large, e.g. 10 milliamps, and a sensitive galvanometer is used for measuring i_+, pressures down to 10^{-6} mm can be measured. If desirable, the ion current also can be amplified to facilitate measurements.

The maximum sensitivity with the method is obtained when G is made positive, for part of the

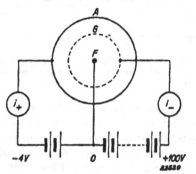

Fig. 10. Ionisation gauge. The ion current i_+ flows to A, and the electron current i_- to G. The pressure is proportional to i_+/i_-.

electrons will then shoot through the mesh of the grid G' and describe longer paths before coming back to G, thus having more chance to cause ionisation. By means of a magnetic field the path of the electrons can be still further lengthened [7].

It follows from fig. 9 that the readings of the ionisation gauge depend on the type of gas used. One of the first observations of Found and Dushmann was that the ionisation was proportional to the molecular weight of the gas used, but it is seen from fig. 9 that this is not the case, and more accurate measurements carried out later by Reynolds [8] with ionisation gauges showed that this proportionality is not generally valid.

Instead of by direct measurement of the ion current, the pressure of a gas can be determined in still another way with the ionisation gauge. If A is taken as the anode and G as the negative electrode, the insertion of a resistance in the grid circuit will not affect the anode current in any way if the electrodes are located in a vacuum, since no grid current will flow. When gas is admitted to the gauge a positive ion current will flow towards the grid, and the insertion of a resistance in the path of this current will alter the grid voltage and in consequence the anode current also. The alteration in current may be used as a measure of the pressure [9]. Another feasible method is to use the ion current for charging a condenser which is connected either to the grid of another vacuum triode or to a relay-amplifying valve so that the anode current of the second valve is dependent on the condenser volt-

age [10]). The total rate of change of anode current of this second valve is then a measure of the pressure.

Finally, instead of deriving an electron saturation current from the cathode, the anode voltage can be made so small that the current in vacuo is limited by the electron space charge. If positive ions are then generated at low gas pressures, these are able to neutralise the space charge due to a larger number of electrons and thus allow the cathode current to increase considerably. In this method of measurement proposed by Spiwak and Ignatow[11] the anode current itself is a measure of the pressure. Theoretically this method was already employed some years previously by G. Hertz [12] for measuring the ionisation potential of gases.

Gauges based on an Independent Gas Discharge.

In an independent gas discharge, the particles responsible for carrying the charge are generated in the gas or at the electrodes by the discharge itself and without any external agency other than the applied voltage. Although in this case a number of factors, such as the current, current density, and the dimensions of the various components of the discharge, are dependent on the pressure, these have hitherto been used rarely for quantitative pressure measurements, since in general they are difficult to reproduce and also vary within wide limits with the characteristics of the gas present [13]. In order to obtain a rough idea of the vacuum in a glass vessel, a common method used in high-vacuum technique is to impose a high-frequency electrical field and observe the luminous effect resulting therefrom. Down to about 10^{-3} mm Hg the gas itself then appears luminous, while at lower pressures the glass walls alone exhibit a green fluorescence.

This form of gas discharge ceases at a gas pressure of approximately 10^{-3} mm, when using a direct voltage or low-frequency alternating voltage in a valve of moderate dimensions and at pressures of the order of 1000 volts, and for this reason will be quite useless for making measurements at low pressures. But by using a magnetic field and suitably disposing the electrodes, measurements

[8]) N. B. Reynolds, Physics, 1, 182, 1931.
[9]) H. Teichmann, Z. techn. Phys., 9, 22, 1928.

[10]) R. Sewig, Z. techn. Phys., 12, 218, 1931.
     A. Butschinsky, Techn. Phys. U.S.S.R., 3, 223, 1936.
[11]) A. Spiwak and A. S. Ignatow, Sow. Phys., 6, 53, 1934.
[12]) G. Hertz, Z. Phys., 18, 307, 1923.
[13]) By using high alternating voltages (7000 volts eff.) Wellauer obtained good results in a range of pressures from 5·10⁻³ to 10⁻¹ mm Hg (Arch. f. Elektrot., 24, 4, 1930).

can be made at much lower pressures and on this principle a simple gauge can be constructed for pressures down to 10^{-5} mm.

Philips' Vacuum-meter

The Philips' vacuum-meter is based on the principle just discussed and will be described in greater detail by reference to *fig. 11.* If two plates

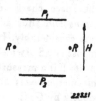

Fig. 11. Principle of the Philips' Vacuum-meter. The electrons at the plates P_1 and P_2 (connected together as a common cathode) do not move along straight lines to the ring RR (anode), but oscillate in spiral paths between P_1 and P_2 under the influence of the magnetic field.

P_1 and P_2 situated opposite to each other form the cathode for the discharge and a ring R acts as a anode, then in the absence of an applied magnetic field the electrons emitted from P_1 and P_2 will travel along curved paths to the ring R. But if a sufficiently powerful magnetic field is applied in the direction perpendicular to P_1, the electrons emitted from P_1 will travel in narrow spirals round the magnetic lines of force towards P_2, will be repelled towards P_1 by the retarding electric field and will thus oscillate many times between P_1 and P_2 before reaching the anode. In consequence. even at very low gas pressures the electrons will be able to collide with a sufficient number of gas molecules producing an adequate ionisation to permit an independent discharge through the gas.

The gauge constructed on this principle is shown in *fig. 12.* The plates P and the ring R are located

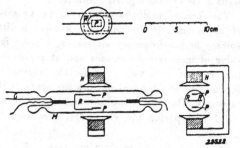

Fig. 12. Philips' Vacuum-meter with two plate-type cathodes P_1 and P_2 and a ring anode R in the field of a permanent magnet H.

in the field of a permanent magnet H, which midway under the pole pieces has a field intensity of approximately 370 oersted. The electrodes are connected

up according to the circuit shown in *fig. 13,* viz, through a resistance of 1 megohm to a low current source of 2000 volts. The current i through the gauge

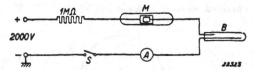

Fig. 13. Circuit of the Philips' Vacuum-meter with the tuning lamp B connected as a current measurer, and the microammeter A.

valve M is hence a measure of the gas pressure, and can be read off on a micro-ammeter with resistances connected in parallel. In qualitative measurements, the instrument can be replaced by a small glow-discharge lamp, such as the Philips 4662 tuning lamp in which the length l of the glow discharge is a measure of the current flowing and hence also of the pressure in the valve M.

The relationship between i or l and the pressure p is shown in *fig. 14* for a pressure range from $2 \cdot 10^{-3}$ to 10^{-5} mm Hg. The curves are based on mean

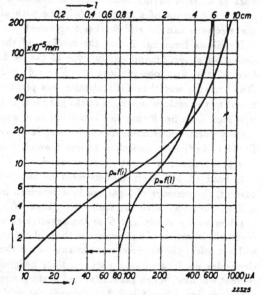

Fig. 14. Gas pressure p in the Philips' Vacuum-meter plotted as a function of the discharge current i or the length l of the glow-discharge in the tuning lamp (mean values for H_2, CO, A and air).

values for air, hydrogen, carbon oxide and argon, while the values of p are in agreement within a factor of 2. This gauge [14]) which has been described in detail in another paper [15]) has the advantage of being extremely simple in construction and does

[14]) Marketed by E. Leybold's Nachf. A.G., Cologne-Bayental.

[15]) F. M. Penning, Physica, 4, 71, 1937.

208   PHILIPS TECHNICAL REVIEW   Vol. 2, No. 7

not require protection against tremor, while one reading made without preliminary adjustment is sufficient to measure a pressure value. It is therefore extremely suitable for industrial use and permits simultaneous and continuous supervision of a multiplicity of pumps. Moreover, it can be sealed to the apparatus of vessels being evacuated or in close proximity to them. Compared with the Mac Leod gauge, it has the advantage of giving continuous and instantaneous readings which are easily visible at a distance; it can be used also for measuring the pressure of condensible vapours and does not itself introduce a supplementary vapour pressure. The latter feature is important when the gauge is used for measurements in conjunction with oil high-vacuum pumps. In conclusion, it should be stated that in the first models of this gauge made the cathode plates P were of iron, but it was shown by experience that the sensitivity of the gauge was increased when the cathode was made of thorium or zirconium; moreover with these metals cathode sputtering was less than with iron.

Tube connecting Gauge and Gas Chamber

In gauges in which gas is continuously consumed, as for instance in those instruments designed on the electrical-conductivity principle, a pressure gradient is produced in the tube connecting the gauge and the gas chamber; this tube must therefore not be made too narrow.

Other potential errors in pressure measurement may be caused by inserting a liquid-air trap between the gauge and the gas chamber, when, in particular, the pressure of condensible gases and vapours, which are given off in the gas chamber, escapes measurement entirely or is only partially measured. A second source of error may be caused by the bore of the trap *not being the same where it enters or leaves the liquid air* [16]). In *fig. 15* the section between A and B represents a trap at a temperature T', with one end of diameter d and the other of diameter D; to the left of A and the right of B the temperature is T. The pressure to the left of A is p, and that to the right of B is P. Over the pressure range in which the conditions of flow conform with the Poiseuille formula (free path $\lambda \ll$ diameters of tubes d and D) we have:

$$\lambda \ll d, \quad p = p' = P.$$

[16]) M. Rusch and O. Bunge, Z. techn. Phys., **13**, 77, 1932.

In the range covered by the laws of Knudsen ($\lambda \gg D$) we have:

$$\lambda \gg D, \frac{p}{p'} = \sqrt{\frac{T}{T'}} \text{ and } \frac{P}{p'} = \sqrt{\frac{T}{T'}}, \text{ also } p = P.$$

Although d and D are different and no errors are likely to accrue in the two boundary pressure ranges, this is no longer true for the transition range in which λ is approximately equal

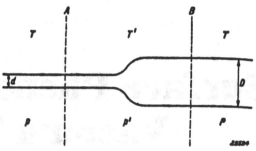

Fig. 15. Gas pressure in a tube of variable diameter at different temperatures.

to d. In this range p/p' and P/p' are determined by d and D respectively, and for $d \neq D$ we have also $p/P \neq 1$; errors of 10 per cent can readily accrue here. If a trap is necessary, then if accurate measurements are required one with a uniform cross-section should be used.

Synopsis

A synopsis of the more common high-vacuum gauges in use is given in Table I, together with the principal ranges of pressures for which they are suitable.

BIBLIOGRAPHY.

S. Dushman. Production and measurement of high vacua. Schenectady, 1922.

A. Goetz: Physik und Technik des Hochvakuums. Brunswick, 1926.

F. H. Newman: The production and measurement of low pressures. London, 1925.

G. W. C. Kaye: High vacua. London, 1927.

S. Dushman: Recent advances in the production and measurement of high vacua. Journ. Franklin Inst. 211, 689, 1931.

Table I.

Type of Gauge	Range (mm Hg)										Pressure reading dependent on type of gas	Principle
	10^2	10^1	10^0	10^{-1}	10^{-2}	10^{-3}	10^{-4}	10^{-5}	10^{-6}	10^{-7}		
Hg gauge											No.	Pressure
Macleod gauge											Partially*)	Press. after preliminary compr.
Pirani gauge (fig. 2a)											Yes	Thermal conductivity.
Knudsen gauge (fig. 7)											No	Radiometer effect.
Molecular vacuum-meter (fig. 8)											No	Radiometer effect
Ionisation gauge (fig. 10)											Yes	Ionisation by electrons
Philips' vacuum-meter (fig. 11)											Yes	Gas discharge in magnetic field

*) In the presence of condensable vapours.

846   INDUSTRIAL AND ENGINEERING CHEMISTRY   Vol. 40, No. 5

Surface Phenomena Useful in Vacuum Technique

Le Roy Apker

GENERAL ELECTRIC COMPANY, SCHENECTADY, N. Y.

Studies of surface phenomena can give valuable information at pressures so low as to be unmeasurable by the usual methods. Previous work on thermionic emission from wires, and on field emission from single crystals, is mentioned briefly. The photoelectric emission from tungsten is very sensitive to residual gas in vacuum systems and can be used to estimate partial pressures of active gases. A simpler method involving a sudden burst of adsorbed gas into an ionization gage is also described.

MODERN vacuum techniques can produce pressures so low that they cannot be detected with available gages. Experiments in Nottingham's laboratory have shown that the common form of ionization gage fails at about 10^{-5} micron. This effect is due to x-rays generated by electrons striking the grid (6). These rays eject photoelectrons from the collector and produce the same effect as positive ions flowing to this electrode. Thus the gage indicates 10^{-5} micron, even though the actual pressure is lower. Obviously, other methods are needed for measuring the ultimate vacua attainable today.

At 10^{-5} micron, enough molecules strike a clean surface to cover it with a monolayer in a few minutes. The adsorbed gas may produce radical changes in the behavior of the material. Evidently, experiments must be done quickly. For this reason, little is known about clean surfaces, except for a few refractory metals and several materials that are easily evaporated. For this reason, also, adsorption phenomena may be used as sensitive detectors of certain gases.

Thermionic Emission from Tungsten Wires

Langmuir's experiments on adsorption by tungsten filaments (3) showed that the thermionic emission from these wires was extraordinarily sensitive to surface contamination—for example, cesium increased the current at 800° K. by a factor of 10^{20}. At 1500° K., oxygen decreased it by a factor of 10^5. This decrease is used in one modern form of leak detector (5).

Field Emission from Single Crystals

In Nottingham's laboratory, a different method has proved successful (2). A phosphor screen collects the field emission from a very small single crystal of tungsten. After the metal is cleaned by flashing at a high temperature, it gives a characteristic emission pattern. Residual gases are adsorbed selectively by the various exposed crystal faces, and the pattern changes radically as the surface becomes contaminated. Because the gas condenses on a relatively cool substrate, this technique gives information that cannot be obtained from thermionic emission measurements. The tube walls must be very clean, since the electrons are accelerated by several kilovolts. Thus, gas can be dislodged by impinging electrons. This field emission method has been used in a valuable study of techniques for reducing active gas pressures (4).

Photoelectric Emission from Tungsten

In the laboratory at Schenectady, a procedure combining features of both thermionic and field emission methods has been used which involves measuring the photoelectric emission from a tungsten ribbon as a function of time (1). The photocurrent from a metal increases approximately as the square of the quantity, $h\nu - \varphi$. Here $h\nu$ is the energy of the incident photons, and φ is the work function of the surface. The primary effect of contamination is to change the latter quantity. Therefore, the photoelectric method is not so sensitive as the thermionic, which involves currents varying as $\exp(-\varphi/kT)$. Both techniques measure an average effect for a polycrystalline surface and cannot detect highly selective adsorption by preferred crystal faces. Despite these drawbacks, the photoelectric method is convenient because it can be used at low temperatures and because little or no accelerating voltage is required.

A typical photocell described in (1) was sealed off and was gettered with tungsten vapor. After the tube had been standing for a week at 300° K., no photocurrent was ejected from the tungsten cathode by radiation of wave length 2537 Å. ($h\nu = 4.89$ electron volts). The ribbon was then flashed for 10-second periods at

May 1948   INDUSTRIAL AND ENGINEERING CHEMISTRY   847

temperatures that were increased successively in 100° steps. Nothing happened until 1000° K. was reached. At this point, the ionisation gage indicated a burst of gas. The peak pressure was 5×10^{-5} micron and the reading quickly returned to 2×10^{-6} as the gas was taken up by the walls of the tube.

After this flash, photocurrent was observed. The work function had therefore decreased to a value below 4.89 electron volts; Flashing at temperatures up to 2000° K. did not affect the emission, but at 2100° K. it increased by about 5%. The work function was then 4.5 volts. It was assumed that this treatment produced a clean surface, as the current was not changed by flashing the tungsten even at temperatures just below its melting point.

When this clean surface remained at 300° K., the emission slowly decreased. It fell exponentially at a rate of about 0.1% per minute, and reached immeasurably small values after several days. Flashing another filament or heating the glass wall of the tube contaminated the ribbon very quickly. Contamination rates in other tubes varied by several factors of 10, although the ionization gages indicated identical pressures. The exhaust schedule, the getter, and the collector material all affected these rates, but they were uninfluenced by the presence of the radiation. Changing the magnitude of the photocurrent (normally about 10^{-11} ampere) had no effect. However, with an accelerating potential of 50 volts, a few milliamperes of thermionic current from the tungsten dislodged enough gas from the collector to be detected even by the ionisation gage.

Although these observations are not understood in detail, they are valuable for detecting active gases. The sensitivity of such adsorption methods increases with the time that one is willing to wait for contaminants to condense. Although the results are not quantitative on an absolute scale, relative contamination rates can be determined with considerable precision.

Let us assume for the moment that the work function, φ, increases linearly with the amount of gas adsorbed. Then the rate at which φ changes can be used to determine the rate at which molecules condense on the surface. This in turn gives the partial pressure of active gases if one knows the reflection coefficient for the incident molecules. In the absence of other information, it may be assumed that reflection is negligible. For the photocell mentioned above, the partial pressure thus computed was about 10^{-9} micron.

Sudden Evaporation of Adsorbed Gas

Tungsten is useful also for detecting active gas by a simpler method.

When the pressure in a system has fallen below the usable range of an ionization gage, flashing a small tungsten filament produces a burst of gas that is easily measurable (say, 10^{-5} micron). If the filament is allowed to stand for 2 or 3 minutes, it will adsorb enough gas to give another readable burst (10^{-4} micron). The peak pressure increases at first linearly with the time during which the filament remains cool. It finally saturates at the value first mentioned. This reading agrees, in order of magnitude, with that computed for the evaporation of a monolayer from the wire.

Tungsten is highly sensitive to the gases from glass; hence this technique is convenient in determining when a tube or constriction is sufficiently degassed to permit a clean seal-off. For a system suitably prepared, the contamination rate decreases by a factor of several thousand, a short time after the constriction is closed. A similar technique is useful for measuring flow of active gas during an exhaust. Two detectors are connected to opposite ends of a constriction—the seal-off constriction, for instance. From the contamination rates it is a simple matter to determine the direction and relative rate of flow.

Nature of Adsorbed Gas

Well known researches on the composition of the active gases mentioned have been carried out at higher pressures. From Langmuir's data, the author has concluded that the component which evaporated in his experiments in 10 seconds at 2100° K. was probably oxygen. The gas that came off at 1000° K. has not been definitely identified. Work on this subject and related ones is in progress in several laboratories, with the mass spectrometer playing an invaluable role. Although adsorption phenomena may be complex in mixtures such as the residual gases in vacuum systems, one may anticipate that methods of the type discussed here will be applied to analysis as well as detection. Even at present, they are extremely valuable for improving vacuum techniques.

Acknowledgment

It is a pleasure to acknowledge indebtedness to Miss J. E. Dickey, E. A. Taft, and R. L. Watters for their generous help and to S. Dushman and A. W. Hull for valuable discussions.

Literature Cited

(1) Apker, L., Taft, E., and Dickey, J., *Phys. Rev.*, 73, 46 (1948).
(2) Daniel, J. H., *Ibid.*, 61, 658 (1942); see also earlier papers cited in this reference.
(3) Langmuir, I., IND. ENG. CHEM., 22, 390 (1930).
(4) Moore, N. H., thesis, Massachusetts Institute of Technology, 1941.
(5) Nelson, R. B., *Rev. Sci. Instruments*, 16, 55 (1945).
(6) Nottingham, W. B., paper presented at M.I.T. Electronics Conference, 1947.

RECEIVED November 12, 1947.

SOME RECENT DEVELOPMENTS IN VACUUM TECHNIQUES

D. Alpert and R. T. Bayard   Westinghouse Research Laboratories

It is a relatively easy matter to evacuate a given vessel to pressures below 10^{-8} mm Hg. If there is a problem in the field of very low pressures, it lies in the development of direct methods for measuring them. The first and most important new development to be described is a new ionization gauge of very simple design, which is capable of indicating pressures below 10^{-10} mm Hg.

In the field of gaseous electronics, the evacuation of the system is just a preliminary to the usual experiment in which one then introduces a sample of gas under investigation. The degree to which a given system can be evacuated obviously sets one limit on the degree of gas purity which is inherently possible. Under the present circumstances in which these very low residual pressures are possible, the limits on gas purity are set predominantly by the conventional techniques used: (1) in the transferring of gases from one place to another and, (2) in the measurement of higher gas pressures by McLeod gauges or liquid manometers. Two other new developments have been carried out to remove the above difficulties: first, an all-metal vacuum valve which can be baked out at high temperature; and second, a manometer for reading pressures in the millimeter range which can also be baked out. Finally, all of these components have been used in a simple vacuum and gas handling system which can easily be evacuated by means of an oil diffusion pump to 10^{-10} mm Hg.

It is by now well known that the conventional ionization gauge gives indications of pressures no lower than an equivalent nitrogen pressure of 10^{-8} mm Hg. On the other hand, there is good evidence that considerably lower pressures than this have actually been attained. This suggests the existence of a

Originally published in *Report on 10th Annual Conference on Physical Electronics* (MIT, Cambridge, 1950), p. 88.

-89-

residual current to the ion collector which is independent of pressure. Professor Nottingham some time ago suggested that this residual current is due to soft x-rays which release photo-electrons from the ion collector, the x-rays being created at the grid by the incidence of electrons from the cathode.

Evidence for the existence of a residual current comes from the observed characteristics of the ionization gauge itself. If for a given pressure and electron current (10 ma), the ion collector current i_c is measured as a function of grid potential V_g, the plot of i_c vs. V_g, shown in Fig. 1a, has a characteristic shape for pressures above 10^{-7} mm Hg. When the pressure is considerably lowered (ion gauge reading 10^{-8} mm Hg), the curve is radically different. This residual (lower) curve continues to rise over the entire range of voltage, the slope on a log-log plot being between 1.5 and 2. In the intermediate range of pressures ($10^{-7} - 10^{-8}$ mm Hg), the characteristic corresponds to a superposition of the "gas ionization" curve and the "residual" curve.

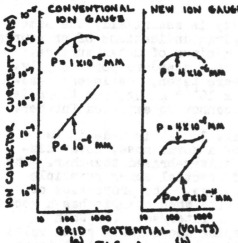

FIG. 1

With the explicit purpose of further investigating the x-ray effect, a design was sought for an ion gauge which would operate in the usual manner, but whose ion collector would intercept only a small fraction of the x-rays produced at the grid. The ionization gauge which evolved from these considerations is shown in Fig. 2. The new gauge has the usual three elements, but is inverted from the conventional arrangement; the filament A is outside the cylindrical grid B, while the ion collector C, consisting of a fine wire, is suspended within the grid. The positive potential on the grid forms a barrier to the ions formed inside the grid enclosed volume so that they are eventually collected at the center wire. The ion collection efficiency is thus comparable to that of the conventional ionization gauge. On the other hand, the geometrical cross section of the ion collector to radiation from the grid is approximately one hundred times smaller than that of the conventional cylindrical collector.

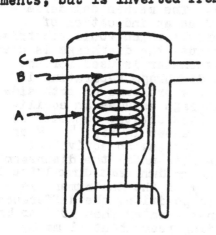

FIG. 2

-40-

The properties of the new ion gauge are shown by the curves of Figure 1b. At higher pressures the characteristics of the new gauge are very similar to those of the standard (RCA 1949) gauge, but it is evident that the new tube continues to have typical "gas ionization" characteristics at much lower pressures; it is not until the pressure is less than 10^{-10} mm Hg that the residual current predominates. The ratio of the residual currents for the two types of ion gauges is the same as the ratio of their geometrical cross sections to radiation. These general characteristics of the new gauge clearly substantiate the x-ray hypothesis and indicate that ionization gauges can be built to measure even lower pressures.

The new gauge, whose sensitivity is essentially the same as that of the conventional gauge, gives indications of pressures at least 100 times lower. It has a minimum of metal surface, is easily outgassed, and is operated on a standard power supply. In a gauge made since these curves were taken, the lowest reading corresponded to a pressure of 5×10^{-11} mm Hg, and there is no reason to believe that the range cannot be extended further.

The vacuum valve is shown in Fig. 3. It is made in two parts; the vacuum portion consists of a kovar nose, kovar diaphragm and copper body and seat all silver-brazed together. The actuating member operates on the differential screw principle and has a very large mechanical advantage. The properties of the valve are summarized as follows: (1) When open, it has a good pumping speed or conductance - about 1 liter/sec. (2) When closed, it is really closed; the pumping speed for a given valve when shut is estimated to be 10^{-12} liters per sec. (3) It is an excellent adjustable leak, and (4) probably most important is the property that opening and closing the valve introduces no contamination. At a pressure of 10^{-7} mm Hg, opening or closing the valve showed no significant fluctuation in the reading of an adjacent ion gauge.

The manometer is shown in Fig. 4. The mechanical deflection of a thin metal diaphragm is used as an indication of pressure. To eliminate calibration errors which would otherwise be introduced by high temperature bakeout, the diaphragm is used in a null reading device. A liquid manometer is used for the absolute measurement of pressure, the diaphragm separating it as shown from the clean system. When the pressure on both sides of the diaphragm is the same, the diaphragm assumes an equilibrium value which is measured by the electrostatic capacitance between the diaphragm and a "probe", as shown in Fig. 4. When a sample of gas is introduced into the system, an equivalent amount of air is let into the opposite side until the diaphragm assumes its equilibrium reading. The liquid manometer then reads directly the true pressure in the system. The sensitivity of the instrument is such that when the diaphragm is at zero position, the difference in pressures on either side of it is not greater than 10^{-2} mm Hg. The instrument is thus capable of reading from about .1 mm Hg up. Since no refrigerants are used, it may be used for the pressure measurement of condensible vapors.

-91-

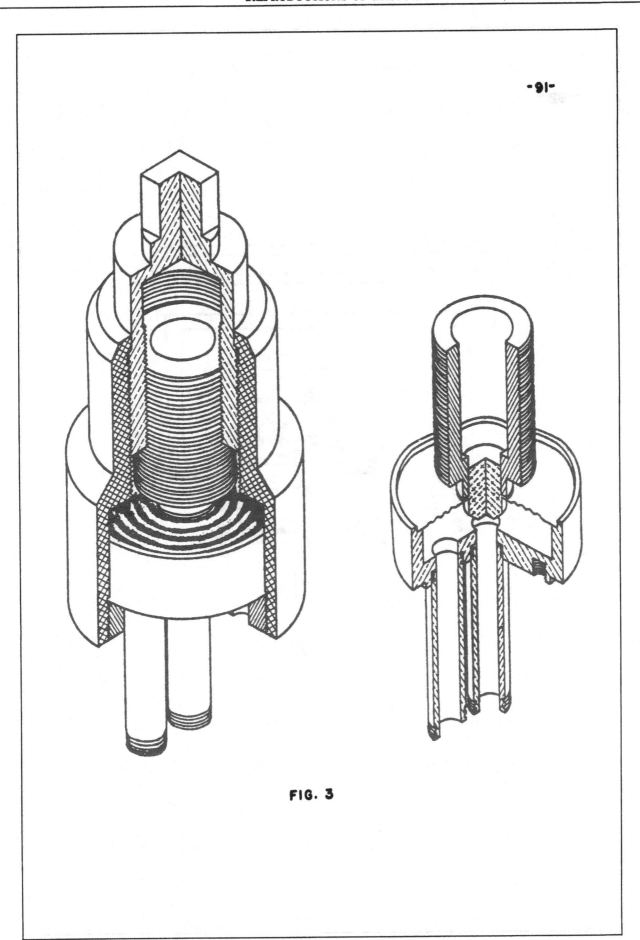

FIG. 3

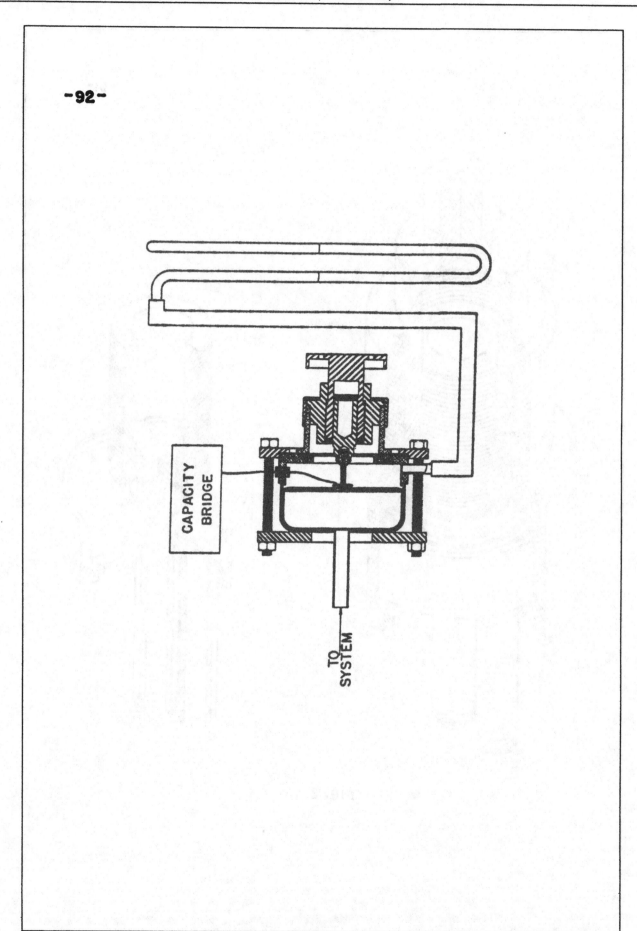

-93-

A simple compact vacuum system has been built incorporating all of these components. This has not only proved to be an excellent system for the handling of gases, but has been ideally suited for the further investigation of the properties of ionization gauges. Preliminary experiments have been carried out to measure the pumping speed of ionization gauges, and to make an absolute calibration of the ionization gauge at very low pressures. From these experiments we feel we understand the mechanisms which take place in an ionization gauge and have confidence in our ability to use them to read very low pressures.

Nottingham remarked that he considered this to be an epoch-making presentation, and added that this new ion gauge looked like such a simple and obvious design, once it had been built, that it was a wonder nobody had ever thought of it before. He asked how the metal parts were outgassed; Alpert replied that the collector and cage were connected to a high voltage source, and outgassed to 1400-1500C by electron bombardment from the gauge filament itself.

Vacuum Factor of the Oxide-Cathode Valve

By G. H. METSON, M.C., Ph.D., M.Sc., A.M.I.E.E., Electronics Division,
Post Office Research Station, Dollis Hill, London, N.W.2

[Paper received 9 September, 1949]

The vacuum factor k of an oxide-cathode triode valve is defined as the positive ion current measured in a negative control-grid per milliampere of ionizing electron current to the anode. Measurements on a variety of new valves show wide variations of k, but these values fall, after a period of operation, to an approximately constant value k_0 in the range 300 to 900 $\mu\mu$A/mA. The characteristics of this residual vacuum factor k_0 are examined in some detail and a theory proposed to explain the anomalous form of variation of k_0 with anode voltage. This theory has a bearing on the interpretation of ionization gauge measurements at pressures below 1×10^{-6} mm. of mercury.

It is commonly supposed that the life of the high-vacuum oxide-cathode valve is a function of the quantity of residual gas left in the tube after production processing. Measurement of the residual gas pressure is therefore a matter of considerable interest and is always attempted by arranging the valve to act as its own ionization gauge. Since the pressures in such valves are normally below 10^{-4} mm. of mercury the ionization method is a convenient one in that the number of positive ions formed from the residual gas is directly proportional both to the pressure of the residual gas and to the density of ionizing electron flow. Measurement

Originally published in *Brit J Appl Phys* 1950; 1:73.

BRITISH JOURNAL OF APPLIED PHYSICS

of the number of positive ions formed is achieved by attracting them to a negatively-primed electrode and determining the electron flow required to keep this electrode neutral. The magnitude of the positive ion flow per mA of ionizing electron current is thus a measure of the residual gas pressure and is known as the vacuum factor of the valve.

There appears to be little systematic work on the subject of the vacuum factor apart from a paper by Herrmann and Krieg[1] in which the factors of a number of commercial valves were measured and compared. The results showed wide variations of magnitude between valves of different types. The object of the present paper is to describe more recent work on the general characteristics of the vacuum factor and to attempt to systematize the subject in a way that has not yet been achieved.

TECHNIQUE OF MEASUREMENT

General Principle.—Determination of the vacuum factor of a triode valve consists of a measurement of the positive ion flow to the negatively-primed control grid when 1 mA of electron current is passing to the anode whose potential is of the order of 150 to 300 V. For residual gas pressures below 10^{-3} mm. of mercury the number of ionizing encounters is directly proportional both to gas concentration and density of the electron stream. Thus if

I_{RG} = Positive ion flow (usually known as reverse grid current); I_A = Anode electron current; and p = Gas pressure

then, $I_{RG} \propto p$;
$\propto I_A$;
$= CpI_A$
or, $I_{RG}/I_A = Cp$
$= k$ (1)

where k is the vacuum factor appropriate to the pressure p. C is a proportionality constant depending on the geometry of the valve structure and to the nature of the residual gas. This constant can be determined experimentally for a particular geometry by observing the vacuum factor k in a valve connected to a McLeod gauge working in the range 10^{-3} to 10^{-4} mm. of mercury. Since the bulk of the gas in a high-vacuum valve is normally in the adsorbed state on internal surfaces and only moves under thermal or electrical disturbance, the conception of a pressure in the accepted sense is probably unsound. In the present paper therefore the quantity of gas will be presented in the form of the vacuum factor measured in micro-microamperes of positive ion current per milliampere of electron current.

Precautions to be taken in Measurement.—Two major sources of error were encountered in the measurement of I_{RG}. The first and most serious was insulation leakage between anode and control or ion collector grid. This was overcome by mounting the anode on a glass insulating pillar and taking the anode connexion out

through the glass envelope of the valve at the end remote from that at which the control grid was led out. This precaution was augmented by the use of "guard rings" at the base of the anode supporting pillar and on the internal and external surfaces of the glass envelope. With these arrangements leakage from anode to control grid was less than 10 $\mu\mu$A with anode at 400 V potential. The other source of error encountered was electron emission from the control grid. This emission must be eliminated as far as possible since it is indistinguishable in the external grid circuit from a positive ion flow. The emission may be thermal or photoelectric. The thermal component can be readily measured in the grid-cathode circuit and eliminated by reducing the grid temperature. The photoelectric component is due either to visual light or to soft X-rays originating at the electron bombarded anode. The component due to visual light can be eliminated by excluding daylight from the envelope and slightly underrunning the oxide cathode. Little can be done at this stage about the X-ray photo-emission and more will be said of its effects later in the paper.

Measurement of the positive ion current presents some difficulty as it is normally in the range $10^2 \mu\mu$A to $10^5 \mu\mu$A. A two-metre reflecting galvanometer with a sensitivity of 20 $\mu\mu$A/mm. was used in the present work and found adequate. With the employment of high anode voltages (50 to 400 V) it will be apparent that the measurement of such small currents required a test set of high insulation properties. With the aid of "guard rings," however, these insulation difficulties were overcome. The test circuit employed was a conventional triode one with meters for measurement of anode current and voltage, negative grid voltage and heater voltage.

CHARACTERISTICS OF VACUUM FACTOR k

Preliminary Measurements.—Preliminary measurements were made on a series of commercial pentode receiving valves of different types and selected in the laboratory for absence of anode-grid mica insulation leakage. The valves were arranged as triodes with anode, suppressor and screen grids connected as a common "anode" maintained at 200 V. The values of k obtained varied from 2000 $\mu\mu$A/mA to 500,000 $\mu\mu$A/mA. Two factors of significance were, however, noted during the measurement. Firstly, the valves of a similar type from a common source had values of k which were fairly closely grouped. Secondly, all valves showed k to be unstable with time; some increasing but the majority decreasing in magnitude.

The Vacuum-Factor/Time Characteristic.—The characteristic of k against time t was next investigated. A valve was set as a triode with 200 V on the anode and a suitable negative voltage on the control grid. These potentials were maintained constant and the positive ion current in the control grid measured over an appropriate period. In general, all valves tested showed the same form of k/t characteristic—a rapid and smooth fall of k

BRITISH JOURNAL OF APPLIED PHYSICS

in the early stages gradually slowing up to make an asymptotic approach to a steady low value. Differences between valves showed up as a difference in time to approach the final value of k. As might be expected, valves with a high value of k at time $t = 0$ showed a longer time to achieve the final value of k. The main variant from this generality was a tendency to irregularity in the rate of change of k in the earliest stage of the test. Some valves showed an increase in k which rose to a maximum and then fell in the normal manner. This is probably an effect of secondary importance and due to outgassing of the anode during the warming-up period.

A typical result is shown in Fig. 1. The first and most interesting observation was that all valves fell to

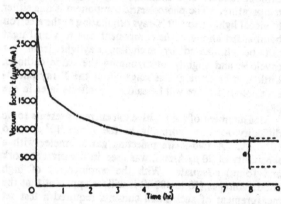

Fig. 1. Typical example of vacuum factor/time characteristic
(a) k_0 range at $E_A = 200$ V for all valves tested

a final value of k within a relatively narrow margin (300 to 800 $\mu\mu$A/mA) irrespective of the initial values which spread over the range of 2000 to 500,000 $\mu\mu$A/mA. This result was surprising as the valves tested were pentodes and triodes of different electrode geometries and from different manufacturing sources. Special triodes made in the laboratory gave similar results. The only common feature of these tests was an anode potential of 200 V. The second point of interest was the stability of k once it had reached its steady low value. Measurement over a period of 250 hr. produced no changes greater than those explainable by the experimental difficulty of measuring the small positive ion current. The third interesting feature of the k/t characteristic is that the integral $\int_{t_0}^{t} f(k)dt$ gives a direct means of comparing the number of ionic collisions that occur in the valve before k reaches its steady low value. The fall of k is due, of course, to absorption of the positive ions either on the negative grid surface or in the oxide-cathode which is maintained at earth potential and is also attractive to positive particles. Some ions reaching the negative grid will be neutralized and escape to become available for a further cycle. As k approaches its final value the bulk of the

gas will probably have found its way into the massive oxide cathode from which further escape is difficult. The integral thus gives some idea of the degree of gas poisoning to which the cathode is subjected.[2] As a first approximation it will, therefore, be concluded that the vacuum factor (or pressure) of the high-vacuum oxide-cathode valve falls to a low constant value when the valve is run under ionizing conditions. This value is in the range 300 to 800 $\mu\mu$A/mA and may indicate a common final operating pressure for such valves. The value will be described as the residual vacuum factor k_0.

CHARACTERISTICS OF THE RESIDUAL VACUUM FACTOR k_0

Variation of Efficiency or Probability of Ionization with Electron Energy. The number of ions produced per milliampere of electron flow increases from zero at the ionizing potential to a maximum in the electron energy range 90 to 120 eV and thereafter declines slowly and apparently indefinitely. All the common gases follow this general characteristic. A typical example from Compton and Van Voorhis[3] for oxygen is shown as Curve A in Fig. 2. It was pointed out by Herrmann and

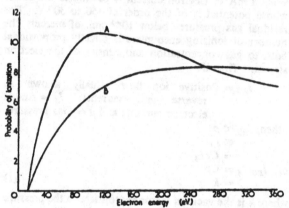

Fig. 2. Characteristic of ionization probability/electron energy. Curve A, constant velocity electron stream; Curve B, constant acceleration electron stream

Runge[4] that the normal probability characteristic is only applicable to a constant-velocity electron beam and that the accelerating stream in a triode valve requires further consideration before a probability characteristic can be derived. Assuming a constant acceleration over the whole electron path a direct integration of Curve A will lead to the required characteristic. Such an integration has been undertaken graphically and leads to a characteristic of the form Curve B. The maximum has shifted to an anode voltage of about 250 V and the characteristic is very flat between 200 to 300 V. Tests on a tungsten-filament triode confirmed the prediction at a pressure of about 10^{-3} mm. of mercury.

The k_0/E_A Characteristic. The variation of residual vacuum factor k_0 with anode voltage E_A was next

BRITISH JOURNAL OF APPLIED PHYSICS

investigated for a series of the specially prepared oxide-cathode triodes previously described. The control-grid voltage was maintained at a constant negative voltage and the anode voltage varied. Values of k_0 were obtained over the anode voltage range 70 to 200 V. During this series the anode wattage increased with each increase of anode voltage. Two results shown in Fig. 3 are typical of a total of more than twenty valves tested. In general there is an approximately linear relationship in the range 100 to 200 V and a flattening of the characteristic below 100 V. All valves show this transition within the range 100 to 130 V. Individual valves were tested up to 400 V and showed that linearity was maintained at least up to this limit. At this stage the measurements were repeated with a variant. Each increase in anode voltage was accompanied by a sufficient increase in negative grid voltage to maintain a constant anode wattage. No change in the k_0/E_A characteristic was observed. It is apparent that there is a wide difference between the k_0/E_A characteristic and its predicted form set out as Curve B in Fig. 2. An attempt to reconcile the difference will be made in the next section.

The k_0/E_H Characteristic at Constant E_A.—Variation of k_0 with the temperature of the oxide-cathode was the next point to be investigated. The cathodes were the common indirectly-heated type with mixed barium-oxide/strontium-oxide on a nickel core. The heater voltage was 6·3 V (or a nominal cathode temperature of 1000° K.). The value of k_0 was examined at $E_A = 140$ V over a range of heater voltage from 3·5 to 6·5 V, with the following results:

Heater voltage	Approximate cathode temperature	$k_0 \mu \mu A/mA$
3·50 V	870° K.	152·0
4·00 V	930° K.	151·6
4·50 V	970° K.	152·1
5·00 V	1010° K.	151·6
5·50 V	1050° K.	150·9
6·00 V	1080° K.	151·4
6·50 V	1120° K.	152·6

The residual factor k_0 is therefore invariant with the temperature of the oxide-cathode. For heater voltage higher than 7 V there is a rapid increase in reverse grid current but this was found to be due to primary electron emission from the overheated grid.

Influence of Envelope Temperature on k_0.—Increasing the envelope temperature of a valve by 25° C. showed an increase of vacuum factor but this gradually returned to the residual value k_0. Presumably this rise is due to adsorbed gas leaving the warm glass surface. Absorption of this gas in the cathode results in the vacuum factor returning to k_0. Cooling the envelope produced no measurable change in k_0. The cooling was effected by lowering a special valve into freezing alcohol and then into liquid oxygen. The tube was maintained at the lower temperature for 30 min. but no change of k_0 was noted. During these measurements the heater

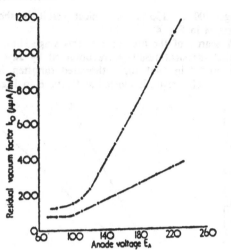

Fig. 3. Typical k_0/E_A characteristics of oxide-cathode valves

voltage was maintained at 6·3 V and it was observed that the anode current fell slightly. This was doubtless due to the radiation cooling of the cathode by the surrounding envelope at − 183° C. This invariancy with envelope temperature is informative and shows that k_0 cannot be ascribed to barium metal vapour or to a whole range of gases with boiling points higher than that of liquid oxygen.

Summary. To summarize, it appears that k_0 is:

(a) a direct function of anode voltage E_A
(b) independent of anode current
(c) independent of anode temperature providing no out-gassing occurs
(d) independent of oxide-cathode temperature
(e) independent of envelope temperature providing adsorbed gas is not evolved.

The only variable on which k_0 appears to be dependent is therefore the energy possessed by the electron when it strikes the anode.

The k_0/E_A Characteristic of a Tungsten Filament Triode.—Since the value of k_0 for the oxide-cathode valve appears unrelated to the nature of the electron emitter it follows that a similar form of k_0/E_A characteristic should be obtained from a tungsten filament triode. Furthermore the transition range (100 to 130 V) should remain unchanged. Experiments were undertaken with such a tube which was fitted with a cylindrical nickel anode, a molybdenum helical grid and a straight tungsten-filament cathode. After thorough processing, the valve was sealed at pressure of 2×10^{-6} mm. of mercury. Subsequent tests showed a close similarity in behaviour between the tungsten and oxide-cathode cases. The vacuum factor k fell with time to a final steady residual factor k_0. The form of the k_0/E_A characteristic was very similar to that of the oxide-cathode case and four experimental tubes all showed a transition effect in the

BRITISH JOURNAL OF APPLIED PHYSICS

range 100 to 130 V. A typical result is shown as Curve A in Fig. 4.

A search of the literature at this stage showed that almost identical results were obtained by Jaycox and Weinhart[5] in 1931 for a thoriated tungsten filament triode. This result is plotted as Curve B in Fig. 4 and

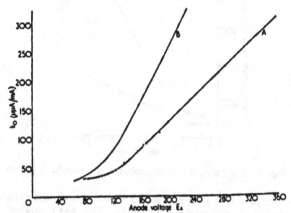

Fig. 4. k_0/E_A characteristic of tungsten-filament triodes

shows a transition effect at 120 V. The reduction of k to k_0 in the case of the tungsten filament is probably due to residual gas absorption on the negative grid and to chemical reaction with the tungsten filament.

DISCUSSION OF RESULTS

The striking difference in form of the k_0/E_A characteristic and the "probability"/E_A characteristic of Fig. 2 remains for discussion. The probability curve is another form of the k_0/E_A characteristic and should differ from it only in respect of the units employed for the ordinate axis. The difference between the two forms is explained in the following manner. The k_0/E_A characteristic obtained in the present work is regarded as the resultant of two superimposed characteristics—one being of the normal probability form and the other a linear characteristic with an intercept at 100 to 130 V. This conception is shown in Fig. 5. The component with normal probability form is regarded as due to a true positive ion flow resulting from residual gas collision processes. The linear component is probably due to a loss of electrons by the negative collector under irradiation by soft X-rays. Such X-rays are generated at the surface of the bombarded anode and their ability to release electrons from the negative collector may be regarded as a function of E_A.

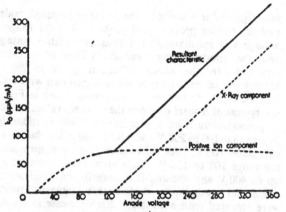

Fig. 5. Compounding of characteristics

A possible alternative to the X-ray theory exists and cannot be ruled out at this stage. The linear characteristic might be due to a true positively ionized stream of metal ions torn from the electron bombarded anode surface. It is conceivable that the number of such sputtered atoms or ions might be a function of the energy of electron impact. A final point may be made concerning the influence of electrode geometry on the general form of the k_0/E_A characteristic. Experiments with a number of geometries in the oxide-cathode case showed no change. Some recent work by the author's colleague, Dr. S. Wagener, employed a tetrode valve with a probe wire collector fitted close to the anode and his findings were a normal k_0/E_A characteristic with a transition at 100 V. The general form of the characteristic is therefore independent of valve geometry.

ACKNOWLEDGMENTS

The author wishes to express his appreciation to the Engineer-in-Chief of the Post Office for permission to publish this work and also to Mr. H. Batey for his skilled technical assistance.

REFERENCES

(1) HERRMANN, G., and KRIEG. O. *Telefunkenröhre*, **21** and **22**, p. 219 (1941).
(2) METSON, G. H., and HOLMES, M. F. *Nature, Lond.*, **163**, p. 61 (1949).
(3) COMPTON, K. T., and VAN VOORHIS, C. C. *Phys. Rev.*, **27**, p. 724 (1926).
(4) HERRMANN, G., and RUNGE, I. *Z. Phys.*, **19**, p. 12 (1938).
(5) JAYCOX, E. K., and WEINHART, H. W. *Rev. Sci. Instrum.*, **2**, p. 401 (1931).

Ultra-High Vacuum Technology

D. Alpert

Westinghouse Electric Corporation, East Pittsburgh, Pennsylvania.

Recent developments in ultra-high vacuum technology have made possible the achievement of pressures below 10^{-9} mm. Hg. Some of the important limitations on the achievement and measurement of ultra-high vacuum have been determined and total pressures below 5×10^{-12} have been attained. The new techniques have made possible a number of researches concerning the phenomena which occur at very low pressures.

During the past year, we have published a description[1,2] of an ultra-high vacuum technology which has evolved at our laboratories over a period of several years and with which it is possible to achieve pressures of 10^{-10} mm. Hg or lower in a routine and straightforward way. This ultra-high vacuum technology is based on the important pioneering work of W. B. Nottingham, P. A. Anderson, L. Apker, and others, and on a number of specific developments which have been carried out in our laboratories and have been previously described in the literature.

These engineering developments are listed as follows:

1. *The Bayard-Alpert ionization gauge.* With the publication of the paper describing this gauge, it was shown that it was possible to reduce the low pressure limit of an ordinary ionization gauge by a factor of 1,000 or more. This was accomplished by reducing the solid angle available to x-rays formed by impinging electrons at the grid of the ionization gauge.

2. *The all-metal vacuum valve.* This valve, which can be baked out at high temperatures, made it possible to seal off one part of a vacuum system from another or to transfer gases with a contamination less than one-thousandth of that contributed by ordinary grease stopcocks.

3. *Ion pumping.* We showed that it was possible to use the ionization gauge as an ion pump to evacuate systems to very low pressures. Such techniques are now being contemplated for use in evacuating large vacuum systems.

4. *Vacuum system design.* We have adopted an entirely new approach to the design of vacuum systems, an area which has for some time been considered an art rather than a science.

I haven't included in this list such other developments as a manometer for measuring pressures above 10^{-1} mm. Hg and capable of bake-out at high temperatures, a simplified mass spectrometer for measuring partial pressures of gas constituents, etc.

Using these means we have built extremely flexible systems to achieve working pressures of 10^{-10} mm. Hg or less without chemical getters or refrigerants of any kind. The first requisite for achieving such low pressures is that the entire system must be baked out at high temperatures, and Figure 1 shows that the vacuum systems are built on a series of module units, each 16 inches square, which form the base of standard size furnaces. In this figure, the valves, ionization gauges, and gas source are shown in place before final assembly. To increase the size or com-

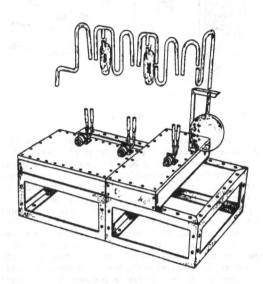

Figure 1.

plexity of a given system we simply bolt down an additional module in either direction and extend the system by making a single vertical glass seal. Figure 2 shows a system with a new design of vacuum furnace in place. With this design, it is possible to add furnaces either in the horizontal or vertical direction to give complete versatility with standard designs. There are 16 or 17 such vacuum systems in our laboratories, and many of the people who use them feel that it is easier to get 10^{-10} mm. Hg in such systems than it used to be to get pressures hundreds of times greater in conventional systems.

With these techniques, we are not only in a position to carry out experiments in gaseous electronics with gases purer than one part in a billion and in surface physics with truly clean surfaces, but also to carry out researches on the limitations on achieving even lower pressures. Although the x-ray limit of our ionization gauge is 3 or 4×10^{-11} mm. Hg, and we have demonstrated that the gauges are linear down to this region, we have thus far not been able to achieve vacuums less than roughly 4×10^{-11} mm. Hg in the sealed-off portion of a vacuum system. We have examined many possible reasons for this limitation in pressure. Information concerning the source

Originally published in *1954 Vacuum Symposium Transactions* (Committee on Vacuum Techniques), p. 69.

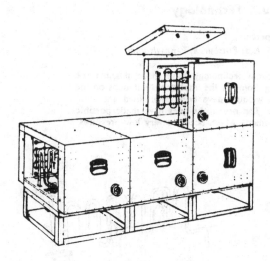

Figure 2.

of the residual gas was first obtained from measurements of the rates of pressure rise in a sealed-off system. If all pumping action (including both diffusion pumps and electrical pumping) is eliminated, we observe an easily measured residual rate of rise no matter how carefully the system has been prepared or the leaks eliminated. A typical rate of rise is shown in Figure 3 and as indicated, the rate of rise is approximately 3 or 4×10^{-11} mm. Hg/min. the value depending on the volume and surface area of the system. After an initial period, this rate of rise continues linearly for months, and though the rate of rise can be measured in the first few minutes, the actual rate corresponds to a change in pressure of one micron per century. Even after many days of seal-off no flash filament effect was observed; i.e., there was no indication

of an adsorbable gas in the system. From these observations we inferred that an inert gas, probably a noble gas, came from outside the system possibly diffusing through the glass wall of the enclosure. From the rather sketchy data available in the literature, we found that the rate of rise was not inconsistent with the hypothesis that helium, present in the atmosphere to only a few parts in a million, was diffusing through the glass walls of the system. The published data available on the permeation of helium through glass varied among different workers by a factor as high as 10 to 20, and we could not be certain that other effects might not be responsible for the measured rate of rise.

We thereupon proposed a simple experiment to determine whether the gas came from outside and, if possible, the mechanism involved. The experiment is shown schematically in Figure 4. We enclosed a carefully sealed-off ionization gauge with a pressure of 10^{-10} mm. Hg within another vacuum system. Then, if at zero time, the outer vacuum system were evacuated the rate of rise should disappear. As shown by the data in the figure, it did disappear but it took almost 60 days to do so. This strongly indicated a diffusion process and a thoroughgoing consideration of the data showed that we could get not only the permeation rate, but also the diffusion coefficient and the solubility for the gas in the glass envelope. As shown in textbooks on the subject,[3] the permeation rate is equal to the product of the diffusion coefficient and the solubility. The diffusion coefficient describes the rate at which a helium atom travels through a membrane under a given concentration gradient, while the solubility describes the concentration of helium atoms which occupy the glass membrane with a given external pressure. From the curve in Figure 4, we obtained values for the diffusion coefficient and solubility and would have liked to compare them with known values. On looking into the literature, however, we found that the diffusion coefficient had never been measured, and the solubility had been measured only once, at a temperature above 500°C.

RATE OF RISE OF PRESSURE IN A TYPICAL VACUUM SYSTEM.

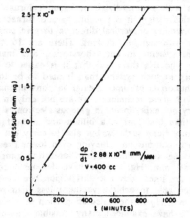

$$\frac{dp}{dt} = 2.88 \times 10^{-11} \text{ mm/MIN}$$
$$V = 400 \text{ cc}$$

Figure 3.

SEALED OFF ION GAUGE IN VACUUM

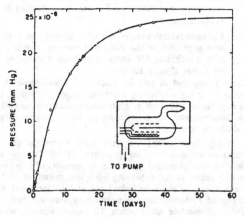

Figure 4.

We then proposed a new experiment to measure the permeation rate, k, diffusion coefficient, D, and the solubility, S, and to check the non-stationary diffusion equation for this process. The apparatus is shown in **Figure 5**.

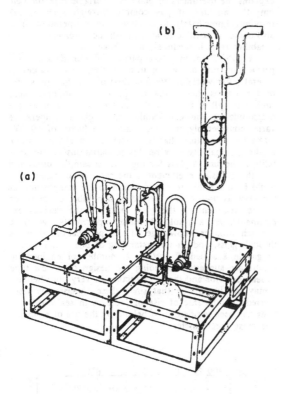

(b)

(a)

Figure 5.

It contains two individually pumped volumes separated by a diaphragm of Pyrex of thickness, d, area, A. At time = O, a pressure p_A was introduced into the first volume, and thereafter the pressure, p_R, in the receiving volume was measured. Despite our sensitive methods for measurement, it took sixteen hours to see a change in the receiving pressure due to the diffusion of helium. The differential equations which describe this situation are the usual diffusion equation, sometimes known as Fick's law, and the time dependent diffusion equation:

$$\frac{V_R}{A} \frac{dp_R}{dt} = - D \left. \frac{\partial n}{\partial x} \right|_{x = d} \tag{1}$$

$$\frac{\partial n}{\partial t} = D \frac{\partial^2 n}{\partial x^2} \tag{2}$$

An exact solution is possible, but involves an infinite series of terms which converge slowly. As described in detail in another publication, approximate solutions are possible for small times, t, giving the following solution:

$$\sqrt{t} \ \frac{dp_R}{dt} = c \, p_A \, S \, e^{-\frac{d^2}{4Dt}} \tag{3}$$

By plotting the log of $\sqrt{t} \frac{dp_R}{dt}$ vs. $1/t$, we get from the slope of the curve a value for the diffusion coefficient, D, and from the intercept we can get a value for the solubility, S. Typical data is shown in **Figure 6** for a

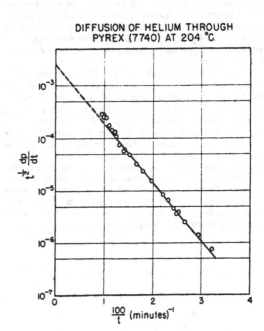

DIFFUSION OF HELIUM THROUGH PYREX (7740) AT 204 °C.

Figure 6.

temperature of 204°C. This curve demonstrates a check on the time dependent diffusion equation, and from such curves as these we get values for the diffusion coefficient as a function of temperature. In Table I, we compare our values taken from such curves with the values for the unknown gas in the walls of the sealed-off ionization gauge.

Table I
Measured Values of D and S

	Helium in Pyrex		"Unknown Gas" in Walls of Sealed-off Ion Gauge
T°C.	27	204	Room Temperature
D $\frac{cm.^2}{sec.}$	8.05×10^{-9}	4.32×10^{-7}	1×10^{-8}
S $\frac{cc\ at\ STP}{cc \times ATM}$	$.8 \times 10^{-2}$	$.8 \times 10^{-2}$	2×10^{-2} (assuming atmospheric helium)

We see, first of all, that at room temperature the value of D checked that for the unknown gas in the glass of our ionization gauge and gives indisputable evidence that the gas diffusing through the wall of the ionization gauge is helium. Using for p_A the atmospheric concentration of helium in air, we get values for S, the solubility; and again we get quite reasonable agreement between the values as measured for helium and the value for the "unknown gas." Note that the solubility does not vary with temperature. In fact, the values obtained at 27° and 204° agree with a measurement by Williams and Ferguson, the only previous measurement of this quantity, which was taken at 515°C. It should be pointed out that the values for S have an inherent error of plus or minus 25 per cent due to the great sensitivity of Equation (3) on the value for D as measured at a given temperature.

If we plot the logarithm of the diffusion coefficient of the permeability versus 1/T, we should get from the slope the energy of activation for these processes. Figure 7 shows such plots, the permeability shown plotted as circles, while the diffusion coefficient is plotted in triangles. If the solubility is a constant as a function of temperature, the slopes of the two curves should be the same, and as shown in the figure the slopes agree to within five per cent, or roughly within the experimental error. The energy of activation for these processes is approximately 6.5×10^3 cal/mole. Since the accuracy of measurement for either the diffusion coefficient or permeability is considerably better than that for solubility, these curves show the invariance of solubility of temperature with greater accuracy.

this was one of the first evidences of the temperature independence of the solubility. If the solubility varied with temperature, it is quite apparent that it would require a number of weeks before the rate of rise assumed its equilibrium value for room temperature. This characteristic explains the usefulness of glass as a variable leak for helium; the amount of gas coming through such a leak responds immediately to a change in temperature. This also makes it possible to accomplish measurements of permeability rates immediately after bake-out.

Figure 8 shows a comparison of our data for the permeation rate of helium through Pyrex with values of several other workers. We find, first of all, that the slopes of the curves (hence the energy of activation) is very similar for all of the data shown. On the other hand, the magnitude of the permeability at a given temperature varies considerably, in some cases by a factor of 20. We make the conjecture that the difference in absolute value of the permeability may be due to variations in the solubility of helium in glass for the various samples used, and that the diffusion coefficient is the same in all cases. This could be due either to the variation in the composition of the glass as used over a period of a number of years, or to the treatment of the glass prior to the various experiments. As a result of these experiments, we have a much broader knowledge of the permeation of gases through glass, and an understanding of one major source of gas in a sealed-off system. A number of engineering results are suggested including the design of a new type of leak detector and the manufacture of very pure helium. I may say in passing that since these experiments were carried out, the rate of rise of pressure in sealed-off systems has been verified to be helium by the use of a simplified mass spectrometer.

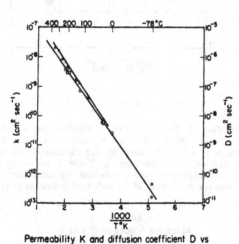

Permeability K and diffusion coefficient D vs
reciprocal temperature

Figure 7.

It is interesting to note that the magnitude of the solubility is approximately one per cent; i.e., the density of the gas within the volume of the glass is actually one per cent of the density of gas adjacent to the glass. Since the solubility is completely independent of temperature, the rate of rise of gas coming into a vacuum system assumes its equilibrium value immediately after bake-out. Indeed,

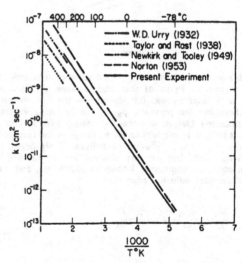

Comparison of permeability values for various
investigators

Figure 8.

We know then that helium is entering at a constant rate, C. The question which remains is: Is the value of the minimum pressure, 4×10^{-11} mm. Hg, consistent with the values for the influx of gas together with the pumping speed of the ionization gauge? From the ordinary form for the rate of reduction of pressure:

$$dp/dt = - \frac{S}{V_p} + C \qquad (4)$$

When the pumping action is exactly balanced by the incoming gas, dp/dt equals zero, and the equilibrium pressure is given by

$$p_e = C \frac{V}{S} = C\tau \qquad (5)$$

where τ is defined as V/S and is a characteristic time required to reduce the pressure in the system to $1/e$ of its original value. Since the rate of rise, C, is approximately 3×10^{-11} mm. Hg per minute, and τ has been previously measured to be of the order of one minute, the equilibrium pressure that we should expect is of the order of 3×10^{-11} mm. Hg. This is very close to the observed value for the equilibrium pressure and satisfactorily explains the ultimate pressure in a sealed-off system.

Now suppose we should like to lower this ultimate value of pressure. This audience may consider it somewhat ludicrous to go even further in lowering the pressure after having achieved total pressures of 3×10^{-11} mm. Hg with an estimated absorbable gas content of less than 10^{-13} or 10^{-14} mm. Hg. Having identified the principle gas involved and the physical reasons for this residual pressure, why should we try to achieve even lower pressures? We would do so purely out of curiosity, but as a matter of fact, we have found a number of research uses in which it is essential to have considerably lower pressures and rates of rise. For example, in our experiments on the diffusion of helium through glass, the accuracy of the measurements is limited by the diffusion of helium from the atmosphere into the receiving volume, and the experiment would be far more sensitive if it were eliminated. The effect of many parameters on the achievement of ultra-high vacua cannot be determined until some of the more important limitations are reduced or eliminated.

If, then, we propose to lower the pressure below 3×10^{-11} mm. Hg, it should be possible to do so (on the basis of Equation 5) by either of two ways: by reducing the rate of rise C, or by increasing the pumping speed S. In the first of two experiments which I should like to describe briefly, it was proposed to eliminate C entirely by enclosing one vacuum system completely within another. This experiment, carried out by Varnerin and White,[4] is shown in Figure 9. Note that by sealing the two systems to the same kovar lead, no portion of the inner glass wall was exposed to the atmosphere. The inner vacuum system, consisting of a Bayard-Alpert ionization gauge, could be sealed from the diffusion pumps and evacuated below 10^{-x} mm. Hg by ionic pumping. The outer system was highly evacuated to eliminate the external gas source. The results are shown in Figure 10.

As shown in previous papers, it is possible to distinguish between the x-ray effect and the gas ion current

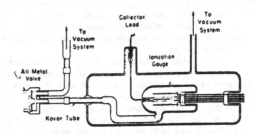

Figure 9.

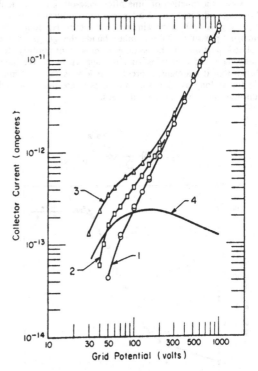

Figure 10.

in an ionization gauge by varying the grid voltage over a wide range and plotting the ion collector current as a function of this potential of a log-log scale. The resulting curve, similar to curve 3, has been shown to be the superposition of a straight line characteristic of the x-ray effect and one similar to curve 4, characteristic of the gas ionization. Under the best conditions of bake-out, Varnerin and White obtained curves like 1 and 2. Curve 1 indicates a total gas pressure so low that the curve is predominantly characteristic of the x-ray effect. Under these conditions the pressure was estimated to be less than 10^{-11} mm. Hg. Hence, the first result of the experiment was the achievement of pressures lower than any previously measured in our laboratories. For the first time, the present design of the Bayard-Alpert gauge was not adequate to estimate the total gas pressure. A second result indicated by the curves is that at extremely low pressures, greater pressure sensitivity is obtained by operating the ionization gauge at lower than normal voltages, i.e., at 70 as compared to 150 volts. And, finally, within the limits of accuracy of the experiment, kovar and copper are vacuum tight and impervious to helium.

I have previously mentioned that the lowest pressures heretofore obtained were in sealed-off vacuum systems. We showed in an earlier experiment that this was due to the fact that an oil diffusion pump acts as a source rather than a sink for contamination at pressures below 10^{-8} mm. Hg. The system, shown schematically in Figure 11, contained two continuously running ionization gauges, one in direct contact with the diffusion pump, the other separated from it by a vacuum valve. The plot of the pressure as a function of time after bake-out indicates that the gauge adjacent to the oil pump soon saturates and the reading goes up to approximately 10^{-7} mm. Hg equivalent nitrogen pressure. On the other hand, the gauge in the isolated portion of the system pumps continuously for many days and maintains an ultra-high vacuum. We have shown that it maintains a pressure as low as 2×10^{-10} mm. Hg for as long as 75 days, and we estimate that it would continue to do so for several years.

the pump speed drops to zero at about 10^{-7} or 10^{-8} mm. Hg. As many of you know, however, there is much evidence to indicate that a typical diffusion pump maintains its actual pumping speed at pressures far lower than the values of 10^{-7} or 10^{-8} mm. Hg. In any case, an experiment was carried out by J. H. Carmichael to see if one could lower the pressure by increasing the pumping speed, S, using a trap with which we have had considerable success previously. This so-called copper foil trap has previously been described,[5] but until recently we have done little in the way of quantitative experimentation. Figure 12 shows the copper foil traps in two forms. In its original form, the common re-entrant design is utilized. Corrugated copper foil is inserted as is shown in sketch b, thus forming a large number of long straws of highly conducting material. This was designed to serve two purposes: first, the high thermal conductivity provides that with the trap immersed in liquid nitrogen, the temperature of the foil remains constant as the level of liquid air varies due to evaporation. Secondly, the trap is extremely efficient with a large length-to-diameter ratio. In practice, the trap provided excellent results and showed no re-evaporation of condensed gases with a lowering of the level of liquid air. As a matter of fact, it was soon demonstrated that the

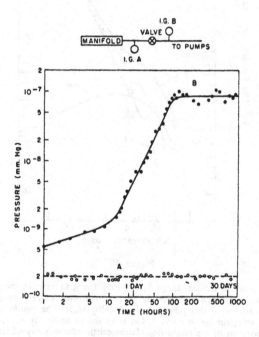

Figure 11.

Copper foil trap

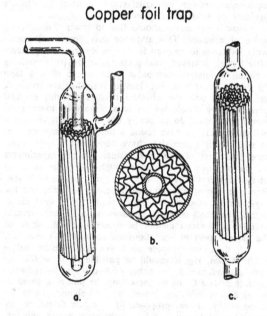

Figure 12.

We ask the question: Is it not possible to substitute a trap for the valve which would isolate the system from the by-products of the oil pumps but would utilize the larger pumping speed of the diffusion pump to decrease the ultimate pressure? By increasing the pumping speed we should obviously (see Equation 5) reduce the equilibrium pressure. You may ask, however, about the fact that all manufacturing curves for diffusion pumps show that

trap works excellently when the level of liquid air is zero, in other words, at room temperature. Thus many of the traps in our laboratory are of the second design, which are used with no refrigerants whatsoever. There are still many things we don't know about the copper trap, but to demonstrate its effectiveness without liquid nitrogen we show in Figure 13 the results of an experiment in which we have substituted a trap for the vacuum valve in the previous experiment. Again, we have two ionization gauges, one (B) in direct contact with the diffusion pumps, the other (A) on the isolated side of the trap. Both gauges are run at a low emission current so that their pumping speeds are very small. As in the previous experiment, the pressure in ionization gauge (B) rises rapidly after the first day to pressures of the order of 10^{-7} mm. Hg. The isolated portion of the system, on the other hand, is reduced in pressure significantly below 10^{-10} mm. Hg for several days before saturation effects occur and the pressure rises above 10^{-8} mm. Hg. In more recent experiments, pressures significantly below 10^{-11} mm. Hg have been obtained in ionization gauge (A), so that in this case we have achieved extremely low pressures by increasing the pumping speed, S.

This suggests a way of measuring the pumping speed, S, or the characteristic time, τ, for pressures below 10^{-7} mm. Hg. By inserting a valve between the system and the trap, we can accumulate gas in the system and then upon opening the valve can read pressure as a function of time to get a value for the pumping speed. The results of such experiments are shown in the final figure. In two successive runs, we show the pressure as a function of time and measure a characteristic time, τ, of about 6.6 secs. which corresponds to a pumping speed of one-tenth of a liter/sec. This is, in fact, the conductance of our vacuum system as accurately as we can compute it. Though this does not constitute an accurate measurement of the speed of the diffusion pump, there is absolutely no indication of a reduction in the pumping speed at pressures below 10^{-7}. This experiment was carried out with a two-stage glass diffusion pump using Octoil-S pumping fluid and indicated a front-to-back ratio of at least 10^8. While these results are not surprising, I believe that these are the first data showing this type of pumping action at pressures below 10^{-7} mm. Hg. In the particular experiments shown, the rate of rise was rather surprisingly high, approximately

10^{-11} mm. Hg per minute. It is quite possible that there was a leak in the system, but it is interesting to point out that the equilibrium pressure of 1.6×10^{-10} mm. Hg was very nearly equal to the product of the rate of rise C and the measured characteristic time τ. Thus we have a direct measure of the pumping action of an oil diffusion pump at pressures below 10^{-8} mm. Hg.

To summarize briefly,

(1) We have accounted for the limitation on the production of low pressures in sealed off systems and have demonstrated that the total pressures may be reduced either by eliminating the rate of influx of gas into a system, or by increasing the pumping speed of the evacuating mechanism.

(2) Measurements have been made of the permeability, diffusion coefficient, and solubility for helium in Pyrex glass in a single experiment. The diffusion coefficient, D, was measured for the first time, and the solubility over a range of temperatures not previously studied.

(3) We have utilized the copper foil trap to isolate a system from oil diffusion pumps to attain very low pressures.

(4) It has been demonstrated that oil diffusion pumps can be utilized directly to get pressures substantially below 10^{-11} mm. Hg, and we are now in a position in which it is essential that we develop new and better pressure gauges to measure total pressures below 10^{-11} mm. Hg.

In conclusion, I should like to acknowledge the collaboration in this work of R. S. Buritz, J. H. Carmichael, and L. J. Varnerin.

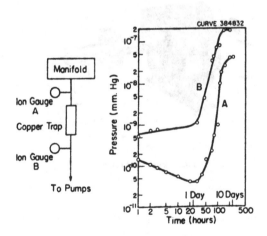

Figure 13.

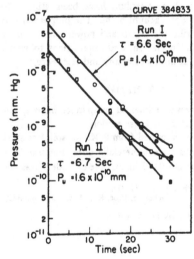

Pump down of helium by oil diffusion pumps.

Figure 14.

1. D. Alpert, J. Appl. Phys., 24, 860-876 (1953).
2. D. Alpert and R. S. Buritz, J. Appl. Phys., 25, 202-209 (1954).
3. e.g., R. M. Barrer, Diffusion In and Through Solids (Cambridge University Press, 1951).
4. L. J. Varnerin, Jr. and D. H. White, J. Appl. Phys., September, 1954.
5. D. Alpert, Rev. Sci. Instr. 24, 1004-1005 (1953).

THE REVIEW OF SCIENTIFIC INSTRUMENTS   VOLUME 29, NUMBER 5   MAY, 1958

Electronic Ultra-High Vacuum Pump

LEWIS D. HALL

Varian Associates, Palo Alto, California

(Received January 10, 1958; and in final form, February 17, 1958)

An ultra-high vacuum pump has been developed based on the combined effects of ionization, excitation and sputtering. A cold-cathode discharge in magnetic field is employed with no hot filaments nor moving parts. The ultimate vacuum attainable is not yet known but is probably below 2×10^{-10} mm Hg. A pumping speed of about 10 liters/sec for air has been recorded at 1×10^{-7} mm Hg. The pump replaces liquid nitrogen-trapped oil diffusion pumps for pumping microwave tubes.

INTRODUCTION

IT was pointed out by Penning[1] in 1937 that *pumping* of gas, i.e., removal of material from the gas phase, occurs in the Penning cold-cathode ionization gauge, which was the first electronic vacuum pump of the kind to be described here. From Penning's data one can calculate a pumping speed of about 0.02 liters/sec at pressures of the order of 10^{-4} mm Hg. Earlier workers, including Soddy and Mackenzie[2] and Vegard[3] had observed that gas clean up occurs as a result of cathode sputtering and is apparently related to adsorption of gas by the deposits of freshly sputtered material.

The combined effects of excitation in a cold-cathode discharge and gettering by atoms of a freshly sputtered reactive metal such as titanium have been utilized to provide an ultra-high vacuum pump. The purpose of this paper is to describe the pump and power supply, their characteristics and operation, and applications to which they have been put in this company and elsewhere.

DESCRIPTIVE

Figure 1 shows an experimental pump in its magnet, and Fig. 2 shows the pump in cross section.

The pump body is formed from 0.049-in. stainless steel sheet; the cathodes are commercially pure titanium. The anode wall thickness is 0.015 in. and the cathode plates are

[1] F. M. Penning, Physica 4, 71 (1937).
[2] F. Soddy and T. D. Mackenzie, Proc. Roy. Soc. (London) A80, 92 (1908).
[3] L. Vegard, Ann. Physik 50, 769 (1916).

⅛-in. thick. The insulator is a high alumina body, glazed to minimize surface electrical leakage.

Titanium was found to be the most suitable material for cathodes, from the standpoint of chemical reactivity and availability and price of the sheet. Other metals were examined, including magnesium, iron, aluminum, molybdenum, and copper, but titanium was found clearly superior in pumping speed, although magnesium and molybdenum can be used. Copper was examined with the idea of using it to differentiate between the relative importance of chemical bonding as opposed to *plastering* of gas with sputtered material, and it was found to produce no measurable pumping, on a short term basis at least. Its pumping speed is clearly many orders of magnitude inferior to those of the reactive metals. Such metals as zirconium, hafnium, vanadium, and the rare earths are also potentially promising cathode materials.

FIG. 1. Pump and magnet.

368

LEWIS D. HALL

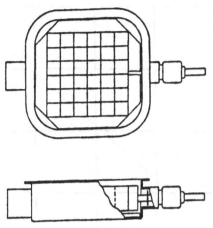

FIG. 2. Pump detail.

The power supply, which provides filtered half-wave rectified dc, consists basically of a high-voltage transformer, a condenser, a rectifier, current-limiting resistance, and a current-metering circuit. The procedure in operating the pump is to rough down with a conventional oil-sealed mechanical pump to a pressure of about 20 microns. At this point the power supply is turned on, with the valve to the roughing pump still open. The current-limiting resistance, which consists of several household lamp bulbs, limits the pump current to about 140 ma. The latter figure is subject to considerable variation depending on the number and sizes of lamp bulbs used. However, limiting the current to low values, e.g., 50 ma, increases the starting time excessively, while too high a limit, e.g., 500 ma, also increases the starting time, because of thermal effects. The initial voltage is about 400 volts. As the discharge and sputtering continue, the voltage increases, reaching a value of 3 kv, and the current decreases, because of the decrease in gas pressure caused by the pumping action. The lamp bulbs grow dim and finally extinguish, as the current drops below 20 ma. At this point, the valve to the roughing pump is closed and the system is now sealed off. The pump continues to lower the pressure in the system to some value which depends on the combined real and virtual leaks present.

If an all-metal system without organic components is used, with the roughing pump pinched off according to vacuum tube practice and the system then thoroughly baked out while pumping with the electronic pump, very low pressures can be achieved, of order 10^{-9} mm Hg or below. No other pumps are needed and no refrigeration is necessary. In a recent experiment, the pump was removed from its magnet and baked out at 520°C for 30 minutes. During bakeout, it was pumped with another electronic pump which was pinched off after bake-out and cooling were completed. In subsequent operation the

pump current was 0.003 microamp, corresponding to a pressure of about 2×10^{-10} mm Hg. A similar experiment, performed earlier, in which a Westinghouse Bayard-Alpert gauge was appended, resulted in a pressure too low to move the ion gauge amplifier needle from its zero position. The least count on the dial was 2×10^{-9} mm Hg.

CHARACTERISTICS OF PUMP AND POWER SUPPLY

As indicated previously, the bases for the pumping mechanism are ionization, excitation, and sputtering. The high degree of ionization is caused by the presence of the magnetic field. The ionization results in bombardment of the cathode plates by positive ions of the gas which is to be pumped. This bombardment causes sputtering of titanium atoms, which are ejected from the cathode plates and deposited chiefly on the anode walls. The sputtering action thus takes the place of the thermal evaporation which has been employed by Herb *et al.*[4] The gettering surfaces are the anode deposits of freshly sputtered titanium.

Earlier cold cathode discharge pumps for which data is available[1,6–7] gave pumping speeds of a few hundredths liter/sec and ultimate vacua of 5×10^{-7} mm Hg and above. Because it is becoming increasingly important to obtain ultra-high vacuum as well as to provide reasonable pumping speeds at all pressures, the present pump has been designed to provide considerably greater pumping speed and lower ultimate pressure. As shown by Wagener,[8] atoms and metastable molecules as well as ions are created in great numbers in a discharge, and the former may in some cases be almost entirely responsible for observed gettering rates, the ionic contribution being negligible; in addition it is known that ions which are driven into cathode surfaces may later be re-ejected.[9] The aim is therefore to trap excited states such as dissociated molecules (atoms), and metastable molecules, and atoms at fresh titanium surfaces on the anode walls, where they will be permanently held by chemical bonds. Some gettering occurs on cathode surfaces outside the bombarded regions, but this is considerably smaller in extent.

The subdivided anode, which is a unique feature of the present pump, is designed in accordance with these effects. It increases the effective gettering area, provides increased surface in close proximity to the discharge, intercepts sputtered atoms and prevents their being resputtered, increases the ion current at any given pressure, and permits the discharge to persist to indefinitely low pressures. The latter feature insures that the pumping

[4] Herb, Davis, Divatia, and Saxon, Phys. Rev. **89**, 897 (1953).
[5] A. M. Gurewitsch and W. F. Westendorp, Rev. Sci. Instr. **25**, 389 (1954).
[6] A. J. Gale, *C.V.T. Vacuum Symposium Transactions* (Pergamon Press, Inc., New York, 1956), p. 12.
[7] F. M. Penning, U. S. Patent No. 2,146,025 (February 7, 1939).
[8] S. Wagener, Brit. J. Appl. Phys. **2**, 132 (1951); Proc. Inst. Elec. Engrs. (London) **99**, Pt. III, 135 (1952).
[9] E. Brown and J. H. Leck, Brit. J. Appl. Phys. **6**, 161 (1955).

speed will not drop to zero until extremely low pressures are reached, since the discharge is responsible for sputtering which in turn causes pumping.

Pumping of noble gases such as argon or helium is not well understood, but it appears that it is probably accomplished by burial of ions in cathode surfaces at high velocity, or by *plastering* of atoms onto anode walls by sputtering titanium. Mass spectrometric analysis of gas evolved from cathodes and anodes on subsequent heating would undoubtedly provide a means of settling this question, and will probably be undertaken in the future.

Pumping speeds for several gases have been measured for a voltage of 3 kv and a magnetic field of 1200 gauss. The measurements are summarized in Table I. Varying the magnetic field over a wide range has little effect on pumping speed. Outside the range 10^{-7} to 10^{-8} mm Hg, measurements are difficult to make with the present apparatus. It is known, however, that somewhere above 10^{-5} mm Hg the pumping speed falls off, partly because of decrease in the output voltage of the present power supply. Pumping speeds for argon and helium apparently fall off more rapidly than for air above 10^{-5} mm Hg, as shown by the fact that it is more difficult to start a pump in an atmosphere of either of those gases. Below 10^{-7} mm Hg the pumping speed has not yet been measured; however, it does not fall to zero at 10^{-9} mm or below, and there is reason to believe that it should continue to pump at indefinitely low pressures, because the discharge should persist to those levels.[10] Future experiments in ultra-high vacuum are now being considered to explore this region further.

The pump has been calibrated as a pressure manometer by recording pump currents corresponding to the pressure readings of a Westinghouse Bayard-Alpert gauge which is attached to the pump envelope. The calibration curve, shown in Fig. 3, is quite reproducible from pump to pump. An interesting effect which has been observed at times, with a nonbaked-out ion gauge, is an increase in gauge pressure reading with a simultaneous decrease in pump ion current. This is apparently caused by absorption of gas on the electrodes of the ion gauge, resulting in a reading which is not a true reflection of the actual gas pressure in the space. The effect does not occur with a clean ion gauge.

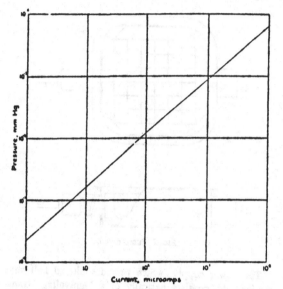

FIG. 3. Calibration curve for pressure measurement. $E = 3$ kv; $H = 1000$ gauss.

Aging and saturation effects are still under investigation, in order to clarify the factors which may limit the life of the pump. The relative roles of pump envelope and titanium need to be examined in connection with the re-evolution of gas which has been adsorbed during pumping. It has been found, however, that a pump whose base pressure has increased after long service and which has become more difficult to start can be rejuvenated by bake-out. A case in point is that of two pumps which have been in use for six months, one on each of two double window sections of a linear accelerator. After repeated gas bursts from rubber O-rings for several weeks, one pump became very difficult to start. It was removed from the system, torched briefly at rough vacuum, and replaced. Since that time it has operated for three months in the same system, with silicone rubber O-rings,[11] without incident. The pressures in these double window sections are of order 10^{-6} mm Hg and below.

The ultimate life of the pump will be limited by the supply of titanium, and in this connection, the economical use of titanium which has been observed is of interest. It has been found that in pumping such gases as air, CO_2, and hydrogen, approximately $\frac{1}{2}$ to 1 gas molecule is pumped for each atom of titanium sputtered.

An interesting aspect of the pump's behavior is its self-regulated property. Because of the nature of the pumping mechanism and the economical use of titanium, only as much material is sputtered as is necessary to pump the gas present, at a pressure, of course, which

TABLE I. Pumping speeds for various gases.

Gas	Pressure mm Hg	Speed l/sec	Time to start, min
Room air	1×10^{-7} to 1×10^{-6}	10.5	1 to 3
Hydrogen	2×10^{-7}	11.5	1 to 3
Argon	2×10^{-7}	9.0	10
Helium	3×10^{-7}	10.0	10

[10] P. A. Redhead, Bull. Radio and Elec. Engr., Div. Natl. Research Council Can. 7, No. 1, 19 (1957).

[11] Vinylloyd Company, 720 N. Broadway, Los Angeles 12, California.

370   LEWIS D. HALL

depends on the pumping speed. This means that on systems with low gas evolution, very long life can be expected. The absence of heated filaments and moving parts also contributes greatly to life. In order to obtain optimum performance and life, it is necessary to use the pump in clean systems without solder flux and preferably with a minimum of rubber present. Rubber O-rings should be avoided, and solid organic particles inside the pump itself can be disastrous, as they result in gas bursts for indefinite periods. Metal gaskets are highly recommended, and the Kel-F elastomer O-ring[11] has been found to be quite satisfactory.

The power consumption of the pump depends on the pressure at which it is operated, since this determines the ion current. As can be seen from the calibration curve of Fig. 3, it is of order milliwatts at pressures of 10^{-4} mm Hg and below. Power consumption is a maximum during the starting operation, amounting to 50 or 60 w. It is undoubtedly desirable to operate the pump under high vacuum at all times, to eliminate the necessity for recurrent starting, and to this end a high-vacuum valve of suitably high conductance is being developed. The present power supply has been designed to provide for all phases of pump operation, including starting, recovery from gas bursts which occur in the system (e.g., during processing of vacuum tube cathodes), and steady-state operation. In addition, metering is provided so that pressures in the pump can be measured over a wide range, from 10^{-4} mm Hg and above to 10^{-8} mm Hg and below. Several pumps can be operated simultaneously with one power supply, but individual current meters must be provided if pressures are to be measured.

APPLICATIONS AND REMARKS

The electronic pump is designed to replace oil and mercury diffusion pumps for a variety of applications, as well as to be used as an appendage on such systems as large microwave tubes to maintain high vacuum after the tube has been baked-out and sealed off. An advantage of the new technique, in addition to the elimination of trapping, is that it becomes economically feasible to pump small tubes individually, rather than on a manifold. Various microwave tube types have been completely processed in this way, following standard time schedules. A typical microwave tube processing schedule includes exhaust, bake-out, and cathode conversion, after which the tube is pinched off. The pump has also been used as an appendage on large microwave tubes which have been processed with electronic pumps or with oil-diffusion pumps. In pumping tubes, the limiting factor is usually the internal impedance of the tube to gas flow, rather than the pumping speed of the pump. Normal values of pumping speed may be misleading; for example, in processing a large experimental klystron, three 5 liter/sec electronic pumps in parallel successfully replaced a 4-in. oil diffusion pump, which has a nominal speed of several hundred liters/sec.

For certain applications greater pumping speed is desirable, hence designs are presently under consideration for larger pumps. There appears to be no fundamental limitation on pump size and speed, but providing the necessary magnetic field may become inconvenient for very large sizes.

ACKNOWLEDGMENTS

Grateful thanks are due Dr. R. L. Lepsen for his unfailing interest and cooperation, E. B. Hodges and W. A. Wood for expert technical assistance without which the program would not have been possible, and Dr. J. C. Helmer for devising electronic methods for solving many of the problems of measurement.

Printed in the United States
By Bookmasters